High-Speed Railway
TRAIN–TRACK–BRIDGE SYSTEMS

High-Speed Railway TRAIN–TRACK–BRIDGE SYSTEMS

A SEISMIC SAFETY TECHNOLOGY FRAMEWORK

WEI GUO
Central South University, Changsha, China

ZHIWU YU
Central South University, Changsha, China

LIZHONG JIANG
Central South University, Changsha, China

ELSEVIER
中南大學出版社
www.csupress.com.cn

Elsevier
Radarweg 29, PO Box 211, 1000 AE Amsterdam, Netherlands
125 London Wall, London EC2Y 5AS, United Kingdom
50 Hampshire Street, 5th Floor, Cambridge, MA 02139, United States

ISBN: 978-0-443-49310-2

For information on all Elsevier publications visit our website at https://www.elsevier.com/books-and-journals

Publisher: Matthew Deans
Acquisitions Editor: Glyn Jones
Editorial Project Manager: Somoshri Banerji
Production Project Manager: Maria Bernard
Cover Designer: Arumugam Kothandan

Typeset by TNQ Tech

Contents

Foreword

Bridges constitute a substantial portion of China's high-speed railway network. Consequently, the dynamic performance of the train-track-bridge system is crucial to ensuring the running safety and ride comfort of high-speed trains passing through bridges. As high-speed railway lines extend into seismically active regions, the coupled dynamic response of the train-track-bridge system under earthquakes, along with the control of train running safety, has become a core issue in system design, operation assessment, and maintenance decision-making.

This book distills the authors' long-term research on the seismic safety of high-speed railway train-track-bridge systems. Its main feature lies in the establishment of a comprehensive research framework spanning seismic disaster simulation, dynamic modeling, and seismic mitigation and control technologies, supported by laboratory experimental platforms and in-service lines. Within this framework, key findings on the seismic resistance of the coupled train-track-bridge dynamic system are transformed into quantitative indices and corresponding technical solutions.

Focusing on the coupled dynamic analysis of the train-track-bridge system, the book begins with numerical simulations and experimental methods for studying train operation on bridges under earthquakes. It then examines the seismic damage characteristics and mechanisms of key components, including bridge piers, bearings, and tracks, and analyzes damage patterns under both far-field and near-fault ground motions. On this basis, the book establishes an evaluation index system for the performance of trains running on bridges and proposes safety limits for train operation under earthquakes based on spectral intensity measures. It also discusses seismic mitigation and control technologies, including rotational friction dampers, composite energy-dissipating bearings, and tuned vibration absorbers.

This publication provides an important reference for the seismic design of high-speed railway bridges, the running safety assessment of trains on bridges, and the maintenance decision-making at the system level. It will be of great value to researchers, graduate students, and related technical professionals working in the fields of high-speed railway train-track-bridge system dynamics and bridge seismic design.

Zhai Wanming
Chair Professor, Southwest Jiaotong University, China
Academician, Chinese Academy of Sciences
International Member of the US National Academy of Engineering

Preface

High-speed railway has become one of China's most representative achievements in modern transportation engineering and large-scale system integration. Under the revised Medium- and Long-Term Railway Network Plan, the national network is being expanded through eight north–south corridors and eight east–west corridors, which will further optimize route structure and strengthen the backbone role of high-speed railway. In this context, safe operation is the essential prerequisite for the sustainable development of the network, and it is also a key determinant of the international credibility of China's high-speed railway technology.

To ensure alignment control, serviceability, and ride comfort, viaducts are widely adopted to replace embankments in high-speed railway design. As a result, bridge sections account for a high proportion of the network. Meanwhile, as lines extend into seismically active regions in central and western China, earthquake risk becomes increasingly prominent. Although China has reached a world-leading level in high-speed railway construction, the coupled dynamic behavior and operational safety of train–track–bridge systems under seismic excitation still involve unresolved scientific questions and engineering challenges. The motivation of this monograph is to address these practical demands by focusing on the seismic catastrophic behavior of high-speed railway train–track–bridge systems, and by providing systematic methods that can be verified and applied in engineering settings.

The research contributions presented can be summarized in four aspects. First, an open and reusable simulation platform for train–track–bridge systems is developed through multiplatform collaborative computing. By leveraging mature and widely used software, the platform reduces redundant development that often occurs in group-specific closed codes, and it enables verification, extension, and sharing. Second, laboratory testing methods for evaluating train running performance on bridges under earthquakes are established. Two complementary routes are developed, including a physical simulation approach for scaled models and a real-time hybrid simulation framework for full-scale investigations. Third, track structures are explicitly incorporated into the seismic analysis of the coupled system. This moves beyond conventional simplifications that treat the track as an inertial mass and therefore overlook damage and degradation of the track structure itself. Fourth, since explicit consideration of moving trains remains difficult in many existing design procedures, a dynamic mapping between derailment state and structural velocity response is established, and a velocity spectrum intensity index is proposed. In this way, operational safety can be addressed through controlling structural response indices, so that train safety performance is considered at the structural design level.

This monograph consists of eight chapters and follows a progression from numerical simulation to experimental techniques to catastrophic mechanisms, and finally to mitigation and control. Chapter 1 introduces the development of high-speed railway and reviews recent advances in seismic research on train–track–bridge systems, and outlines the scope of the work. Chapter 2 presents numerical simulation methods for train operation on bridges under earthquakes, including wheel–rail contact point detection, a cosimulation framework based on OpenSees and Simpack using a client–server architecture, and a moving element method for train–track–bridge analysis. The emphasis is placed on maximizing reuse of mature platforms to enhance efficiency and to support open sharing. Chapter 3 proposes laboratory physical simulation methods for assessing train running safety on bridges under earthquakes and for generating experimental data to support theoretical and numerical studies. It introduces a scaled testing system that combines shake tables with controlled train operation. Chapter 4 presents a real-time hybrid simulation method for train running on bridges, which provides an alternative laboratory route for full-scale train–bridge interaction studies. The framework is further extended to maglev train–bridge interaction investigations at speeds of up to 600 km/h. Chapter 5 examines seismic catastrophic mechanisms of the coupled system under far-field and near-fault earthquakes, including failure patterns of piers, bearings, and tracks, a shear strength model for bridge piers under flexural shear failure, constraint effects and damage characteristics of rail tracks, and derailment mechanisms during and after earthquakes. Chapter 6 proposes evaluation indices for train running performance under seismic excitation, investigates operational performance on simply supported and long-span bridges, and establishes corresponding speed safety limits. The spectral intensity indicator is introduced as a basis for ensuring train safety at the structural design level. Chapter 7 presents seismic mitigation and control technologies for train–track–bridge systems, including rotational friction dampers, energy-dissipating bearings, and tuned mass dampers, and proposes a performance-based energy design methodology. The integrated application of damping technologies and design methods is discussed to mitigate system-level catastrophes such as pier damage, bearing failure, track deformation, and derailment. Chapter 8 summarizes conclusions and provides an outlook on future development, key technical challenges, and potential directions.

The authors gratefully acknowledge the contributions of former and present master's and doctoral researchers in the team. This work was supported by the National Natural Science Foundation of China (NSFC), including the Young Scientists Fund (No. 51108466), the General Program (No. 51878674), and the Excellent Young Scientists Fund (No. 52022113). Additional support was provided by key and major programs of the former Ministry of Railways, China State Railway Group Co., Ltd., and China Railway Group Limited. Publication of this book was also supported by the National Fund for Academic Publication of Science and Technology.

Writing a book is an attempt to crystallize a coherent knowledge system and to offer systematic solutions to specific engineering problems. The work presented here is inevitably limited by time and experience, and imperfections may remain. We sincerely hope this volume will serve as a useful reference for researchers, designers, and practitioners, and we welcome feedback and constructive suggestions from readers.

Authors

CHAPTER 1

Introduction

Contents

1.1 Overview

1.1.1 Development History of High-Speed Railway in China

Railway development is the top priority in building a transportation powerhouse, which reflects how the development of the railway industry serves as a key indicator of a nation's comprehensive strength [1]. At present, there are two main trends in international railway development: heavy-haul freight railways and high-speed passenger railways. High-speed railways offer irreplaceable advantages in facilitating travel and driving economic growth. The world's first high-speed railway line, the Tokaido Shinkansen connecting Tokyo and Osaka, was built in Japan in 1964. It heralded the dawn of the high-speed railway era and sparked a global wave of high-speed railway research and construction. Subsequently, countries such as France, Germany, Spain, Italy, and South Korea successively developed and constructed their own high-speed railway systems [2,3].

High-Speed Railway Train–Track–Bridge Systems
ISBN 978-0-443-49310-2
https://doi.org/10.1016/B978-0-443-49310-2.00008-9

China officially entered the high-speed railway era on August 1, 2008, marked by the inauguration of the Beijing-Tianjin Intercity Railway, the nation's first domestically developed high-speed railway line with a maximum operating speed of 350 km/h. This milestone ushered China into an era of rapid high-speed railway expansion, sparking a new global wave of high-speed railway construction [4]. After over a decade of development, China's high-speed railway construction has achieved globally acclaimed accomplishments. According to the official data released by the National Railway Administration of the People's Republic of China, China's railway operating mileage reached 146,300 km by the end of 2020, including 38,000 km of high-speed railway, which accounted for 70.3% of the world's total high-speed railways. As the world's leading high-speed railway operator, China boasts the longest high-speed network, the highest transport density, and the most complex operational network.

The Medium-and Long-Term Railway Network Plan (Fa Gai Ji Chu No. 1536, 2016) [5] further proposes that China expand its railway network to approximately 175,000 km by 2025 and its high-speed railway network to 45,000 km by 2030, positioning it as the first country to establish a developed and comprehensive modern railway network. At present, China's high-speed railway system has achieved global leadership in advanced equipment and construction expertise, emerging as a hallmark of the nation's great rejuvenation. However, there is still room for improvement in the overall technological system of high-speed railways. For instance, research areas such as the dynamic performance and operational safety of train–track–bridge systems under earthquakes are still in the early stages of research, necessitating sustained and extensive efforts by relevant scientific researchers.

1.1.2 Characteristics of High-Speed Railway Construction in China

A high-speed railway network with eight lines running from north to south and eight lines running from east to west has taken shape [6], characterized by high operating speeds, long rail mileage, wide geographical coverage, and a high proportion of bridge sections [7]. In the construction of high-speed railways, bridges are extensively built to replace embankments to overcome natural obstacles, including rivers, lakes and

Table 1.1 Bridge to total length ratio of some high-speed railway lines in China.

Railway Name	Total Line Length/km	Bridge Length/km	Bridge to Total Length Ratio/%	Operation Time
Beijing-Tianjin Intercity Railway	200.0	175.4	87.7	August 2008
Guangzhou-Zhuhai Intercity Railway	142.0	133.0	94.2	January 2011
Beijing-Shanghai High-Speed Railway	1318.0	1061.0	80.5	June 2011
Beijing-Hong Kong High-Speed Railway	2193.0	1384.0	63.1	December 2012
Shanghai-Kunming High-Speed Railway	2261.0	1207.0	53.4	December 2016
Xuzhou-Lanzhou High-Speed Railway	1360.0	822.0	60.5	July 2017
Beijing-Harbin High-Speed Railway	1713.0	1269.0	74.1	December 2018

mountain valleys, to build straight and smooth tracks, reduce differential settlement in geologically challenging sections, conserve valuable land resources by minimizing farmland occupation, enhance operational safety through fully enclosed infrastructure, and resolve conflicts with existing highways and urban roads [8], resulting in a high proportion of bridge sections.

Table 1.1 presents representative high-speed railway lines in China and summarizes key parameters, including route length and bridge-to-total-length ratio. Fig. 1.1 shows the bridge to total length ratio of some high-speed railway lines in China. As shown in Table 1.1 and Fig. 1.1, the average bridge-to-total-length ratio is as high as 53.0%. Notably, the ratio of the Beijing-Shanghai High-Speed Railway is 80.5% whereas that of the Beijing-Tianjin Intercity Railway is 87.7%. The highest ratio is found on the Guangzhou-Zhuhai Intercity Railway, at 94.2% [7], where trains operate almost entirely on bridges.

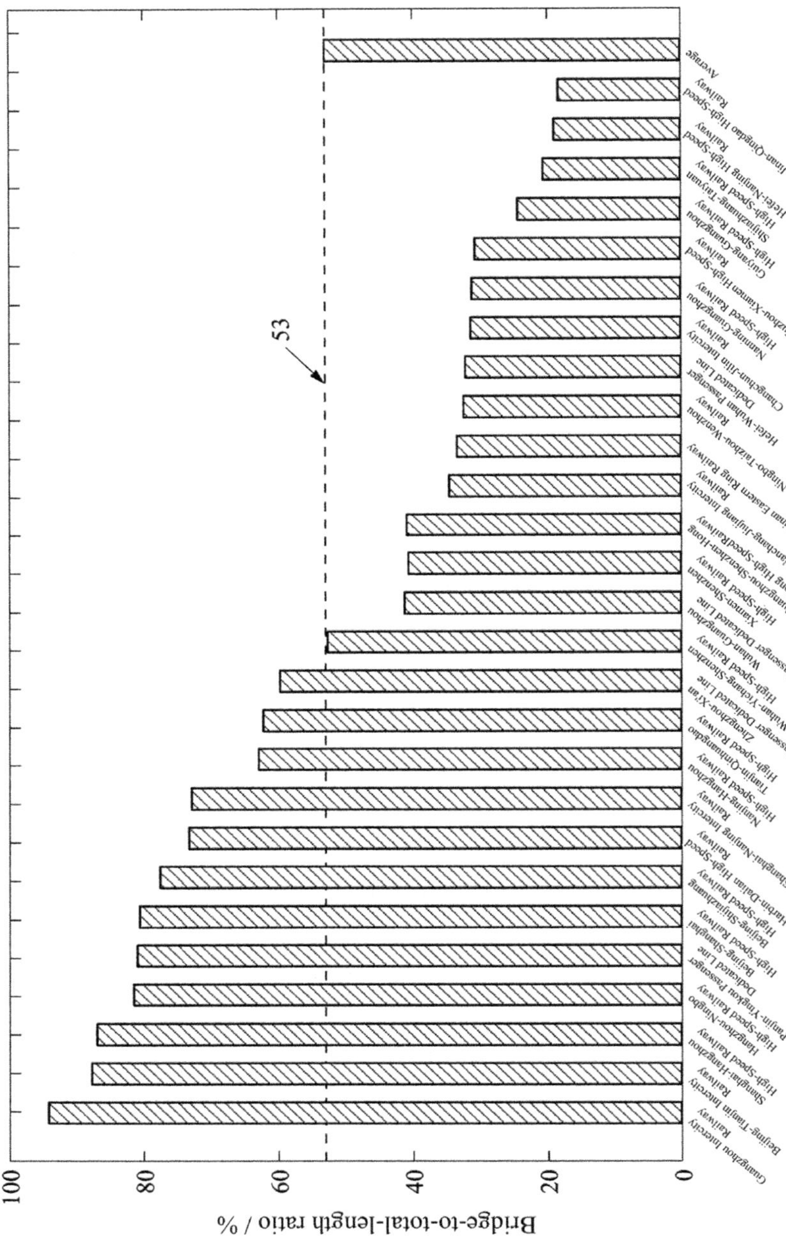

Figure 1.1 Statistics of the bridge-to-total-length ratio of some High-Speed railway lines in China.

1.2 Impact of Earthquakes on High-Speed Railway Operational Safety

China is located at the intersection of the Mediterranean-Himalayan Orogenic Belt and the Pacific Ring of Fire, with extensive seismic zone coverage and frequent seismic activity, making it one of the countries most severely affected by earthquake disasters worldwide. In recent decades, China has frequently been hit by devastating earthquakes such as the 2008 Wenchuan earthquake [9], the 2010 Qinghai Yushu earthquake [10], the 2013 Sichuan Ya'an earthquake [11], and the 2017 Sichuan Jiuzhaigou earthquake [12]. These devastating earthquakes caused significant economic losses and casualties [13,14]. As critical transportation hubs, bridges are lifeline projects under earthquakes. They must be available for safe traffic both during and after earthquakes. Otherwise, bridge damage caused by seismic forces would significantly hinder disaster relief efforts, exacerbate secondary disasters, and cause greater losses to the people and the country.

China's historical seismic records and high-speed railway network distribution show that most high-speed railway lines are either built along seismic zones or include sections crossing active faults. The most critical overlaps between high-speed railway lines and seismic zones are notably observed in the Beijing-Tianjin-Hebei region, the border areas of North China and Central and Northeast China, and the economically developed coastal areas. The high bridge-to-total-length ratio, combined with widespread seismic zones, inevitably leads to high-speed railway lines crossing or being constructed along active faults, significantly increasing the probability of high-speed trains operating on bridges during and after earthquakes. Globally, three most well-known train derailment incidents were caused by seismic activity. (1) On October 23, 2004, the Niigata Chuetsu earthquake caused a Shinkansen train to derail. The earthquake, with a magnitude of 6.8, caused the first to ninth carriages of the "Toki 325" Shinkansen train running between Urasa and Nagaoka to derail, resulting in the entire train being scrapped and the line being shut down. The incident caused catastrophic economic losses and shattered the 40-year myth of Japan's Shinkansen being free from major accidents. Fortunately, none of the 155 passengers suffered fatalities [15]. (2) On March 4, 2010, a 6.7-magnitude earthquake struck Kaohsiung, Taiwan, China. A northbound high-speed train traveling through the Xinshi section of Tainan derailed with two wheels off the tracks. The train executed an emergency brake at 280 km/h immediately after the earthquake, and all 300 passengers escaped unharmed. (3) On April 15, 2016, a 6.5-magnitude earthquake struck Kumamoto, Japan, causing a Shinkansen train to derail. Fortunately, there were no passengers aboard the train at the time [16].

As a critical lifeline infrastructure, high-speed railways play a strategic role in personnel evacuation and material transportation after an earthquake. Any safety incidents such as derailments or service suspensions would severely impede disaster relief and reconstruction efforts. Therefore, it is of great significance to study how high-speed trains can safely navigate bridges at a reasonable speed during earthquakes [17,18].

1.3 Dynamic Performance of High-Speed Trains Operating on Bridges

1.3.1 Research on Operational Performance

1.3.1.1 Development History

The problem of train–bridge coupling vibration has been a subject of research interest since the inception of the world's first railway in the United Kingdom in 1825. When trains traverse bridges, the moving vehicle loads induce structural vibrations on the bridge, which in turn interact with the entire train system, resulting in a coupling effect termed train–bridge coupling vibration. Train–bridge coupling vibration is a complex dynamic problem that can lead to train derailments and bridge collapses in severe cases [19,20]. Research on train–bridge coupling vibration has gone through three development stages in roughly chronological order.

The first stage, from 1825 to the 1960, was the early stage of research and exploration of train–bridge coupling vibration. This stage primarily focused on deriving analytical and semi-analytical solutions to train–bridge coupling vibration problems, establishing fundamental concepts such as the "impact coefficient" and "critical speed" [21].

The second stage, from 1961 to 2000, was a stage of improving research on train–bridge coupling vibration. This stage marked a transformative advancement in train–bridge coupling theory, driven by the advent of computer technology. With the improvement of computing methods and the Finite Element (FE) theory, significant progress has been made in modeling both trains and bridges within the coupling system, which directly led to the emergence of high-speed railways [22].

The third stage, from 2001 to the present, represents the integration and maturity of train–bridge coupling theory, driven primarily by the second global wave of high-speed railway construction, with China's rapid advancement as a major catalyst [23]. The train dynamics and track-bridge dynamics are theoretically combined to form a real train–bridge coupling system that enables more accurate and rational acquisition of dynamic responses. At the same time, the scope of research has expanded significantly to incorporate multiple-hazard and stochastic analysis [24,25].

1.3.1.2 Classification of Dynamic Performance Research

(1) Research on train models

The development of train models bears distinct era-specific characteristics. Prior to the 1960s, train models were generally simplified to facilitate manual derivation of analytical and semianalytical solutions. After the 1960s, with the advent and widespread adoption of transistor computers [22,26], coupled with advancements in numerical solution methods, train modeling has become higher-fidelity physical models, and various multiple-parameter models have been proposed and used in actual engineering computations. The general development course of train modeling is: constant force motion model—simple harmonic motion model—mass motion model—half-car model—single-car model—marshaled train [22]; from primary to secondary suspension systems, from single-degree-of-freedom (DOF) to multiple-DOF models; from multiple-rigid-body models to elastic body models, collectively demonstrating a consistent trend toward greater modeling complexity [27]. Figs. 1.2 and 1.3 shows a schematic diagram of the computational model for a typical train–bridge coupling system. Fig. 1.2 shows the historical development of train models for the train–bridge coupling system.

(2) Research on bridge models

The structural forms of high-speed railway bridges, with the exception of standard simply supported girders, are typically complex, and it is impractical to establish the motion differential equations of bridge structures directly to obtain analytical solutions [28]. The most common approach involves using FE software to develop bridge models and then formulating the system's equations of motion by assembling the structural mass, stiffness, and damping matrices [29–31]. In bridge FE modeling, commonly used elements include planar/spatial beam elements, plate elements, shell elements, solid elements, and truss segment elements, in which planar/spatial beam elements and their modified variants are the most frequently employed. In terms of calculation accuracy and efficiency, beam elements can satisfy the requirements of most train–bridge coupling vibration research. Plate elements, shell elements, and solid elements increase structural DOFs, particularly for complex large-span or extra-large bridges, which substantially raise computational loads, reduce efficiency,

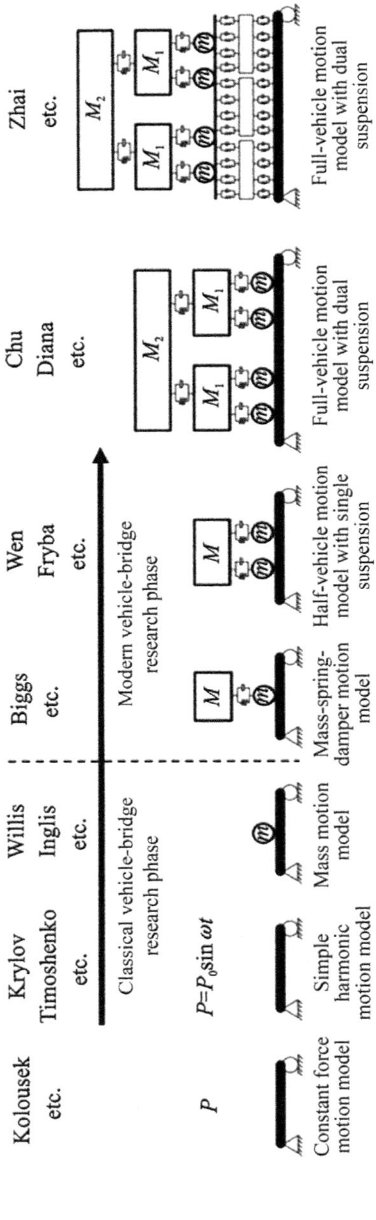

Figure 1.2 Schematic diagram of the calculation model of a typical train-bridge coupling system.

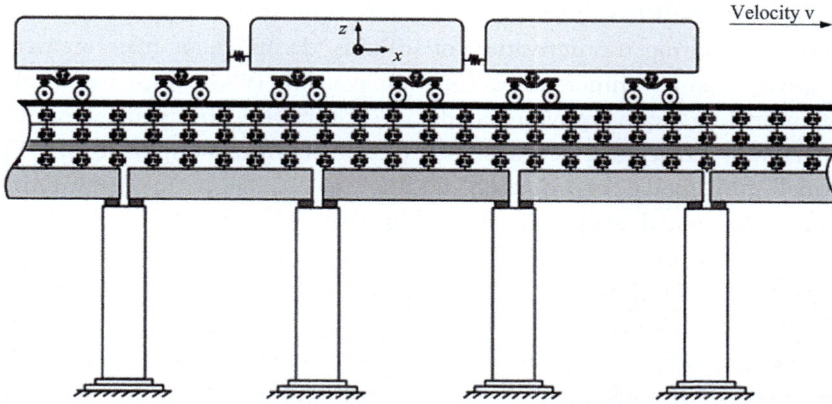

Figure 1.3 Evolution history of train-bridge coupling system train models. *M*, mass of vehicle body and bogie; *M1*, bogie mass; *M2*, vehicle body mass; *m*, wheelset mass.

and demand substantial computational resources. To address this challenge, Prof. Zeng Qingyuan proposed the concept of truss segment FEs [32,33], which dramatically reduces structural DOFs and consequently enhances computational efficiency.

(3) Track model research

High-speed railway track structures are typically complex in form, often requiring appropriate simplifications during modeling. Commonly used simplified models include the equivalent lumped-parameter model [35], continuous beam-on elastic foundation model [36], and continuous beam-on discrete elastic support model [31,37]. Among them, the continuous beam on elastic foundation model can adopt either the Euler-Bernoulli beam model or the Timoshenko beam model. The key difference is that the Timoshenko beam accounts for shear deformation and rotational inertia effects [38], although this increases computational complexity and reduces efficiency. In terms of modeling methods, they can be classified into the direct modeling method and the FE modeling method. Direct modeling typically involves establishing the differential equations of motion for each track component, or using prederived track beam elements [39] or train–track–bridge coupling elements [40]. A key drawback of this method is that when adopting new track structural models, the differential equations of motion for the track system must be rederived, which requires significant effort and results in poor adaptability.

In contrast, the FE method directly models the system through software without requiring the derivation of stiffness, damping, or mass matrices. However, most commercial FE software available lacks robust capabilities for modeling train moving loads or accurately simulating complex wheel–rail contact interactions. This often necessitates exporting the FE model into modal coordinates, leading to challenges in determining appropriate modal truncation orders [34,41].

(4) Research on wheel–rail relationship

Wheel–rail technology is a key technology for high-speed railways. The wheel–rail relationship constitutes the core of wheel–rail technology and has always been a critical research subject in train dynamics. Among its aspects, the wheel–rail contact geometry relationship serves as the foundation for subsequent train system solutions. In 1926, Carter first proposed the rolling contact theory [42], establishing the foundation for analyzing wheel–rail contact relationships using creepage and creep forces. By adopting the assumption of an elastic half-space in contact with a cylinder, he successfully resolved the two-dimensional elastic contact problem of wheel–rail systems, delineated the adhesion-slip zones within the contact patch, and derived calculation formulas for longitudinal and lateral creepages. In the 1960s, building upon Carter's research, Johnson and Vermeulen [43–45] established the relationship between longitudinal/lateral creepages and tangential creep forces in three-dimensional rolling contact under spin-free conditions, and experimentally validated the correctness of wheel–rail contact theory. The scholar who later synthesized wheel–rail rolling contact theory was Dutch Prof. Kalker [46,47]. From the 1960s to the early 21st century, Kalker successively proposed the linear creep theory [48], simplified theory [49], and three-dimensional nonlinear exact rolling contact theory [50], and developed computational software for wheel–rail rolling contact analysis such as FASTSIM, DUVOROL, and CONTACT, with CONTACT in use to this day. Kalker made outstanding contributions to the theoretical research of wheel–rail rolling contact. Prof. Shen Zhiyun et al. incorporated wheel–rail spin effects and nonlinear corrections to propose the internationally recognized "Shen's Theory" (Shen-Hedrick-Elkins Theoretical Model) [51]. Prof. Jin Xuesong has conducted in-depth research on wheel–rail rolling contact theory and experimental studies, and compiled research findings in his monograph *Creep Theory and Experimental Studies of Wheel-Rail Systems* and relevant papers [52–54].

(5) Research on track irregularity excitation

Track irregularities serve as a primary excitation source in train–bridge coupling systems, directly influencing wheel–rail interaction forces, train vibrations, ride comfort, and operational safety [55]. They are typically characterized using one-sided power spectral density (PSD) functions [56–58]. The PSD of track irregularities is typically obtained through field measurements on existing railway lines. Typical track irregularities are illustrated in Fig. 1.4. Commonly used track irregularity spectra include Germany's high-interference spectrum for conventional railways and low-interference spectrum for high-speed railways, the US Class 6 track irregularity spectrum, Japan's Shinkansen spectrum, and China's spectra for both conventional and high-speed railway lines [59]. Because train–bridge coupling vibration analysis is generally conducted in the time domain, it is necessary to convert the track irregularity power spectrum into the time domain. The conversion from stochastic track irregularity power spectra to time-domain track mileage samples can be achieved through the secondary filtering method [31], the trigonometric series method [60], the white noise filtering method, and the Auto Regressive Moving Average (ARMA) model method.

(6) Research on train–bridge coupling vibration under multiple disasters

In actual high-speed railway operations, both railway lines and trains may be subjected to various loads and environmental effects, such as an

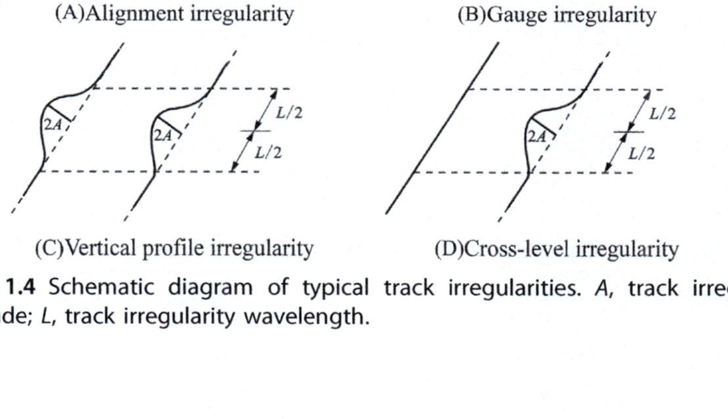

Figure 1.4 Schematic diagram of typical track irregularities. *A*, track irregularity amplitude; *L*, track irregularity wavelength.

earthquake load [61,62], wind load [63,64], snow load, temperature load [65,66], collision load, and other factors. Under these loads, structures such as tracks and bridges, along with running trains, experience varying degrees of influence. This, in turn, alters the vibration characteristics of the train–bridge coupling system and undermines operational safety. In the past, owing to limitations in theoretical understanding and computational capabilities, research progress in this field was slow. With the resurgence of high-speed railway development and the rapid advancement of computer technology, the study of train–bridge coupling vibrations under multiple disasters has regained significant attention among scholars [67,68].

(7) Research on stochastic train–bridge coupling vibration

Studies have shown that the train–bridge coupling system exhibits stochastic vibration characteristics during actual operation. Research by Tan Sui and Zhu Yan also supports this conclusion through extensive on-site measurement data [69,70]. This is primarily attributable to the stochasticity of the structural parameters and input excitations in this system. For instance, the damping ratios of track–bridge substructures, material elastic moduli, and fastener support stiffness exhibit probabilistic variations. Meanwhile, input excitations of the coupling system, such as track irregularities and seismic loads, demonstrate strong stochasticity, which is further compounded by the probabilistic nature of wheel–rail interactions. The superposition of these multiple stochastic factors leads to stochastic dynamic responses in the coupling system, a phenomenon that has attracted significant scholarly attention and research. Early research on stochastic vibrations of the train–bridge coupling system primarily employed the stochastic moving load method [28]. However, constrained by the theoretical and computational capabilities of the time, this approach failed to account for the impact of the locomotive structure, resulting in discrepancies between research findings and actual conditions. Particularly with the rapid development of modern high-speed railways, tremendous changes in structural forms and materials of bridges, tracks, and trains have rendered relevant early studies obsolete for practical applications, although they retain theoretical reference value. Current methods for investigating stochastic vibrations of the train–bridge coupling system include the Monte Carlo method, stochastic perturbation method, pseudoexcitation method [67], and probability density evolution method.

1.3.2 Dynamic Performance Evaluation Standards

1.3.2.1 Evaluation Standard for Bridge Dynamic Performance

In accordance with domestic and international railway bridge design and inspection standards, the primary criteria for evaluating railway bridges generally include natural frequency, deflection, lateral amplitude, bridge end rotation angle, and vibration acceleration. However, because of variations in dynamic performance evaluation standards across different national codes and space constraints, this section provides only a brief overview of China's railway bridge evaluation standards. For detailed evaluation indicators, please refer to Zhai Wanming's *Vehicle-Track Coupling Dynamics* [27].

(1) Bridge natural frequency

International and domestic regulations on natural frequency of bridges typically account for train operating speeds. When considering the design speed of a bridge, the natural frequency of the girder must satisfy this equation:

$$\frac{v_{max}}{f_0} \leqslant \left[\frac{v}{f_0}\right]_{\lim} \tag{1.1}$$

where $v_{\max}$ is the maximum train operating speed; f_0 is the natural frequency of the bridge; and v is the operating speed of the train.

Code for Design of High-Speed Railway (TB 10621—2014) stipulates that the vertical natural frequency of a simply supported girder should not be lower than these limits (in Hz):

$$f = \begin{cases} \dfrac{80}{L} & (L \leqslant 20\text{m}) \\ 23.58L^{-0.592} & (20m < L \leqslant 96\text{m}) \end{cases} \tag{1.2}$$

where L represents the wavelength of track irregularities.

China's Railway Bridge Inspection Code (Tie Yun Han No. 120, 2024) stipulates the typical measured values of the minimum lateral natural frequency for various types of simply supported bridge spans during normal passenger and freight train operations, as specified in Table 1.2. Additionally, it requires the minimum lateral natural frequency of bridge spans to meet safety criteria to prevent wheel derailment of empty or mixed freight trains, along with the typical lateral natural frequency values of bridge piers.

Table 1.2 Typical measured values of minimum lateral natural frequency for bridge spans.

Bridge category	**Structural type**				
	Plate beams and trusses without a deck system		**Plate beams and trusses with a deck system**		
	Ordinary bridge steel	**Low-alloy steel**	**Ordinary bridge steel**	**Low-alloy steel**	**Prestressed concrete girder**
Minimum lateral natural frequency f/Hz	$f \geqslant 100/L$	$f \geqslant 90/L$	$f \geqslant 100/L$	$f \geqslant 90/L$	$f \geqslant 90/L$

(2) Bridge deflection

China's Code for Design of Railway Bridge and Culvert (TB 10002–2017) stipulates that for girder bridge spans, the lateral overturning stability coefficient under the most unfavorable combination of calculated loads should not be less than 1.3. For reinforced concrete cantilever girder bridge spans, the longitudinal overturning stability coefficient under vertical live loads corresponding to 30% exceeding the allowable stress also should not be less than 1.3.

The vertical deformation limits of the girder should comply with these provisions.

1. Under the static live load of trains, the vertical deflection of the girder should not exceed the limits specified in Table 1.3.
2. For statically indeterminate structures such as arch bridges, rigid frames, and continuous girder bridges, the influence of temperature on vertical deflection must be considered. The vertical deflection should be determined based on the most unfavorable condition listed subsequently and must comply with the limits specified in Table 1.3.
   - **a.** The sum of the deflection caused by the vertical static live load of the train and the deflection value caused by 0.5 times the temperature.
   - **b.** The sum of the deflection caused by 0.63 times the vertical static live load of the train and the deflection value caused by all temperatures.

Table 1.3 Allowable value of vertical deflection for bridge girder.

Design Standard		$L \leqslant 40$ m	$40\text{m} < L \leqslant 80$ m	$L > 80$ m
Intercity railway	200 km/h	$L/1750$	$L/1600$	$L/1200$
	160 km/h	$L/1600$	$L/1350$	$L/1100$
	120 km/h	$L/1350$	$L/1100$	$L/1100$
Mixed passenger-freight railway	200 km/h	$L/1200$	$L/1000$	$L/900$
	160 km/h	$L/1000$	$L/900$	$L/800$
Heavy-haul railway	120 km/h and below	$L/900$	$L/800$	$L/700$

The deflection limits in the table apply to simply supported double-track girders with three or more spans. For continuous girders with three or more spans in one unit, the allowable vertical deflection should be 1.1 times the tabulated value. For continuous girders with two spans in one unit or double-track simply supported girders with two or fewer spans, the deflection limit should be 1.4 times the tabulated value. For single-track simply supported or continuous girders, the deflection limit should be 0.6 times the corresponding value for double-track bridges. The symbol L in the table denotes the span length of the simply supported girders or the checked span of a continuous girder.

3. For prestressed concrete beams on ballastless track bridges with a design speed of 200 km/h or above, the vertical residual creep deflection after track installation should not exceed 10 mm when the span is less than or equal to 50 m; when the span is greater than 50 m, the vertical residual creep deflection should not be greater than L/5000 and should not be greater than 20 mm.

4. For prestressed concrete girders on ballasted track bridges of high-speed railways, the vertical residual creep deflection after track installation should not exceed 20 mm.

Code for Design of High-Speed Railway (TB 10621–2014) specifies the vertical deflection limits for concrete structures with spans less than 96 m, as detailed in Table 1.4. For the purpose of calculating the vertical deflection of arch bridges and rigid frame bridges, in addition to considering the static effect of ZK live load, the influence of temperature deformation should also be taken into account.

In addition, China's Railway Bridge Inspection Code (*Tie Yun Han No. 120, 2004*) stipulates that under the static live load of trains (converted to ZH live load), the normal values of the measured vertical deflection-span ratio are shown in Table 1.5.

(3) Lateral bridge amplitude

The lateral amplitude of a bridge reflects the geometric, physical, and dynamic properties of the bridge structure under train loading related to transverse stiffness. To ensure operational safety and passenger comfort, it is necessary to impose limits on the lateral vibration amplitudes of bridges. China's Railway Bridge Inspection Code (*Tie Yun Han No. 120, 2004*) specifies the safety limits $[A_{\max}]5\%$ and its normal value $[A_{\max}]5\%$ for midspan lateral amplitudes (measured at the load plane) of simply supported bridge spans, where 5% denotes a 5% exceedance probability. This section focuses solely on safety limits, as shown in Table 1.6. In the research and design of bridges for the Qin-Shen Passenger Dedicated Line high-speed

Table 1.4 Vertical deflection limits of China's high-speed railway bridges.

Design Speed/ ($km \cdot h^{-1}$)	Bridge Span		
	$L \leq 40$ m	40m < $L \leqslant 80$ m	$L > 80$ m
250	L/1400	L/1400	L/1000
300	L/1500	L/1600	L/1100
350	L/1600	L/1900	L/1500

Table 1.5 Common values of vertical deflection of bridges in China's railway bridge inspection code.

Bridge category	Structural type		Allowable deflection
Steel girder	Plate girder	Ordinary bridge steel	$L/1200$
		Low-alloy steel	$L/950$
	Truss girder	Ordinary bridge steel	$L/1500$
		Low-alloy steel	$L/1250$
Reinforced concrete girder	Normal height	$(h/L = 1/9–1/7)$	$L/4000$
	Low height	$(h/L = 1/15–1/13)$	$L/1900$
Prestressed concrete girder	Normal height	$(h/L = 1/13–1/11)$	$L/1800$
	Low height	$(h/L = 1/16–1/14)$	$L/1300$

test section, international practices were referenced to propose that the maximum lateral amplitude (half-peak value) at the bridge midspan should satisfy the condition: $A_{max}/26$ mm.

Code for Design of High-Speed Railway (TB 10621-2014) stipulates that under the influence of lateral swaying force of trains, centrifugal force, wind load, and temperature, the horizontal deflection of bridge girders should not exceed 1/4000 of the calculated span of the girders; for ballastless track bridges, the transverse relative displacement between adjacent rail support points at girder ends should not be greater than 1 mm.

(4) Vibration acceleration of bridges

According to *Eurocode 1: Actions on structures–Part 2: Traffic loads on bridges*, when calculating bridge's vibration acceleration, at least three vibration modes should be considered, with the frequency range extending up to $f_{lim} = max[30, \lambda f_0]$, and the first-order natural frequency of the bridge takes a λ value of 1–2. The limit values for vertical vibration acceleration of bridges are:

$$[a_v]_{max} = \begin{cases} 0.35\,g \text{ (ballasted track bridge)} \\ 0.50\,g \text{ (ballastless track bridges and open} - \text{deck bridges)} \end{cases} \tag{1.3}$$

where av is vertical vibration acceleration; g is gravitational acceleration.

Code for Design of High-Speed Railway (TB 10621-2014) stipulates that the vertical vibration acceleration limits for bridge decks under a strong vibration frequency of 20 Hz are 0.35 g for ballasted bridge decks and 0.50 g for ballastless bridge decks.

Table 1.6 Operational safety limit of lateral amplitude at bridge midspan $[A_{max}]5\%$.

Bridge category	Structural type			Amplitude limit $[A_{max}]5\%$
Steel girder	Plate girder or truss girder without a deck system			$L/5500$
	Plate girder or truss girder with a deck system	Plate girder		$L/6000$
		Truss girder	$L \leq 40$ m	$L/6500$
			$40 \text{ m} < L \leq 96$ m	$L/(75L + 3500)$
Reinforced concrete girder, prestressed concrete girder				$L/9000$

According to experimental research by Japan's Railway Technical Research Institute, for lateral vibration, the risk of train derailment increases significantly when bridge's lateral vibration acceleration reaches 0.1-0.2 g, making this range the critical threshold. In addition, China's railway bridge inspection code stipulates a maximum lateral vibration acceleration limit of 0.14 g in the load plane when a train passes through a bridge. The *Technical Specifications for Bridge Maintenance and Repair of Comprehensive Test Section of the Qinhuangdao-Shenyang Passenger Dedicated Line* imposes a limit of 0.1 g for bridges with a design speed of 300 km/h.

1.3.2.2 Evaluation Standard for the Operational Safety of Trains Crossing Bridges

(1) Derailment coefficient

The evaluation index for preventing wheel derailment is the derailment coefficient, defined as the ratio of the lateral force Q_1 (or Q_2) at the wheel–rail interface on one side to the dynamic wheel load P_1 (or P_2). The force relationship is illustrated in Fig. 1.5, from which the critical derailment state can be derived as:

$$\frac{Q}{P}=\frac{\tan\alpha-\mu}{1+\mu\tan\alpha} \tag{1.4}$$

where Q represents lateral wheel–rail force; P denotes static wheel load; α is the flange angle; and μ indicates friction coefficient between the wheel flange and rail.

The safety criterion for the vehicle derailment coefficient is:

$$\begin{cases} \dfrac{Q}{P}=1.2\ \text{(danger limit)} \\ \dfrac{Q}{P}=1.0\ \text{(allowable limit)} \end{cases} \tag{1.5}$$

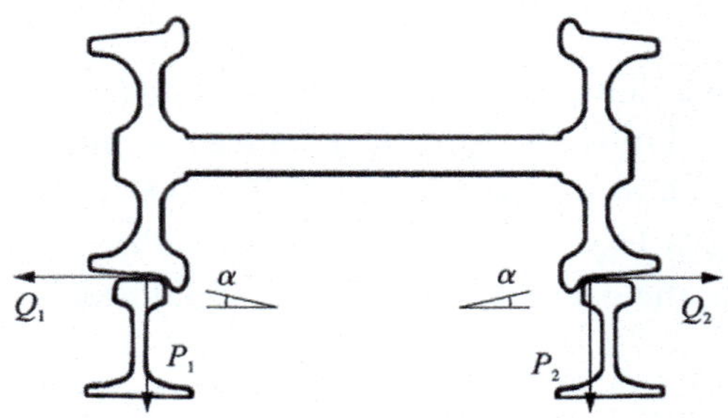

Figure 1.5 Force diagram of wheelset derailment coefficient.

Table 1.7 Vehicle derailment coefficient safety limit.

Vehicle type	Derailment coefficient Q/P	
	Curve radius $250\ \text{m} \leq R \leq 400\ \text{m}$	Other lines (radius $R > 400\ \text{m}$)
Passenger trains, EMUs	≤ 1.0	≤ 0.8
Locomotives	≤ 0.9	≤ 0.8
Freight trains	≤ 1.2	≤ 1.0

The danger limit and allowable limit generally apply to low-speed derailment scenarios. When the derailment coefficient does not exceed the danger limit, it is considered safe; staying within the allowable limit is desirable. The safety criteria for vehicle derailment coefficients specified in China's *Specification for Dynamic Performance Assessment and Testing Verification of Rolling Stock* (GB/T 5599-2019) are detailed in Table 1.7.

International model tests have been conducted on derailment phenomena caused by unstable self-excited hunting motion during high-speed train operations. According to these tests, lateral forces act only for brief durations. Moreover, when the duration of the lateral force exceeds 0.05 s, it is classified as rail climbing; below 0.05 s, it constitutes rail jumping. The relationship between wheelset derailment and duration is illustrated in Fig. 1.6, with the corresponding safety criteria specified as follows.

When the duration of lateral force exceeds 0.05 s:

$$\begin{cases} \dfrac{Q}{P} \leqslant 1.0 \text{ (Danger limit)} \\ \dfrac{Q}{P} \leqslant 0.8 \text{ (Allowable limit)} \end{cases} \tag{1.6}$$

When the duration of lateral force is less than 0.05 s:

$$\frac{Q}{P} = 0.04/\mathrm{t} \tag{1.7}$$

In high-speed railway design, Japan's *Design Standards for Railway Structures* and the UNIFE's *Eurocode 1: Actions on structures-Part 2: Traffic loads* on bridges adopt the criterion of $Q/P \leqslant 0.8$. China's *Specification for Testing of Complete High-speed EMUs* stipulates that for EMUs operating above 200 km/h, the derailment coefficient should adopt $Q/P \leqslant 0.8$. *Code*

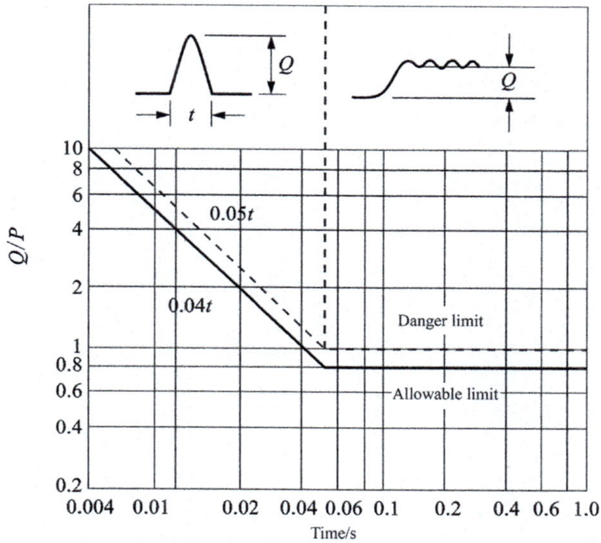

Figure 1.6 Relationship between wheelset derailment and duration.

for Design of High-Speed Railway (TB 10621-2014) restricts this derailment coefficient limit to passenger trains with design speeds of 250-350 km/h.

(2) Wheel-rail lateral force

China's *Specification for Dynamic Performance Assessment and Testing Verification of Rolling Stock* (GB/T 5599-2019) specifies the limits for wheel–rail lateral force Q as shown in Table 1.8.

Table 1.8 applies to the safety assessment of track gauge widening caused by lateral forces. In fact, concrete sleepers are now widely adopted in railway tracks, though China currently lacks corresponding standards for such infrastructure. For concrete sleeper tracks, the maximum permissible unilateral Q-force should be determined based on the lateral resistance characteristics of fasteners and the anti-overturning limit of rails [27]. International practices from European, American, and Japanese Shinkansen railways generally set the allowable lateral force at 0.4 times the axle load ($Q \leqslant 0.4P_w$), where P_w denotes the static axle load (in kN).

(3) Wheelset lateral force

In addition to wheel–rail lateral force standards, some countries have also established limits for wheelset lateral force H (the vector sum of left and right wheel–rail lateral forces Q_1 and Q_2 on a wheelset). The equation for wheelset lateral force limit, proposed by the Société Nationale des

Table 1.8 Regulations on wheel–rail lateral force *Q*.

	High-speed railway and intercity railway	Mixed traffic railway	Heavy-haul railway	
			Locomotives	Freight trains
Wheel–rail lateral force Q/kN	$\leqslant 10 + P_0/3$	$\leqslant 80$	$\leqslant 0.90[15 + (P_{ST1} + P_{ST2})/2]$	$\leqslant 0.85[15 + (P_{ST1} + P_{ST2})/2]$

Notes: P_0, P_{ST1}, and P_{ST2} denote static wheel loads.

Chemins de fer Français (SNCF) and adopted by European/American railways and the Union Internationale des Chemins de fer, is expressed as:

$$\begin{cases} H \leqslant 10 + \dfrac{P_{\mathrm{w}}}{3\ (\text{Limit value})} \\ H \leqslant 0.85 \times \left(1.0 + \dfrac{P_{\mathrm{w}}}{3}\right) (\text{Recommended value}) \end{cases} \tag{1.8}$$

China's *Specification for Dynamic Performance Assessment and Testing Verification of Rolling Stock* (GB/T 5599-2019) exhibits notable problems and deficiencies in its provisions regarding wheelset lateral force *H*. After analyzing these problems, Prof. Zhai Wanming proposed the following equation to determine the allowable limit for wheelset lateral force:

$$H \leqslant 5\mathrm{Pwt} \tag{1.9}$$

where P_{wt} denotes the static wheel load (in kN).

China's *Interim Provisions for the Design of New 200km/h Mixed Passenger and Freight* Traffics Railways specifies that the wheelset lateral horizontal force must satisfy $H \leq 80$ kN.

1.3.2.3 Evaluation Standard for Operational Stability of Trains Crossing Bridges

The operational stability of trains is typically assessed using two indicators: vehicle body vibration acceleration and operational stability. The vehicle body vibration acceleration indicator primarily reflects the amplitude of train vibration. The train operational stability indicator evaluates the vibration level from a statistical perspective by considering multiple factors, including vibration amplitude, frequency and duration. For passenger trains, this indicator can reflect passengers' fatigue level during travel and it is also referred to as the passenger ride comfort indicator.

(1) Vehicle body vibration acceleration

Regarding the evaluation criteria for vehicle body vibration acceleration of passenger trains, China's *Specification for Dynamic Performance Assessment and Testing Verification of Rolling Stock* (GB/T 5599-2019) only includes recommended provisions for the average maximum vibration acceleration of vehicle bodies under operating speeds below 140 km/h. This clearly does not meet the operational requirements of China's high-speed railways. *Eurocode 1: Actions on structures-Part 2: Traffic loads on bridges* only specifies evaluation criteria for vertical vibration acceleration of vehicle bodies, but does not account for lateral vibration acceleration, as detailed in Table 1.9.

Table 1.9 Evaluation criteria for vertical vibration acceleration of passenger train vehicle bodies.

Grade	Excellent	Good	Acceptable
Vertical vibration acceleration of vehicle body/(cm·s^{-2})	100	130	200

In China's early dynamic analysis of train–bridge systems for the *Qinhuangdao-Shenyang Passenger Dedicated Line, and subsequently in the Interim Provisions for the Design of New 200 km/h Mixed Passenger and Freight Railways* and *Code for Design of High-Speed Railway* (TB10621-2014), the evaluation criteria for vehicle body vibration acceleration of trains on bridges were established as follows:

$$\begin{cases} a \leqslant 0.13\,g\ (\text{Half} - \text{peak value, vertical}) \\ a \leqslant 0.10\,g\ (\text{Half} - \text{peak value, lateral}) \end{cases} \tag{1.10}$$

where a is the vibration acceleration of the vehicle body; g represents gravitational acceleration.

Regarding the evaluation criteria for vehicle body vibration acceleration of trains, China's railway industry standard-*Appraisal Method and Evaluation Standard for Dynamic Performance Test of Railway Vehicles* (TB/T 2360-1993)-specifies the maximum vibration acceleration limits for train vehicle body dynamics, as detailed in Table 1.10.

(2) Ride index

There are many methods to evaluate train vibration stability and ride comfort, such as ISO 2631 evaluation method, equal comfort level method, and ride index method. Among them, the ride index method has been widely applied since its introduction in the 1940s. China has long used the ride index method to evaluate the ride comfort of trains, with this approach being particularly applied to train–bridge dynamic analyses to

Table 1.10 Evaluation grades of maximum vibration acceleration of China's railway vehicles (Eurocode 1).

Grade	Excellent	Good	Acceptable
Vertical vibration acceleration of vehicle body/(cm·s^{-2})	2.45	2.95	3.63
Lateral vibration acceleration of vehicle body/(cm·s^{-2})	1.47	1.98	2.45

assess stability during bridge crossings. The Sperling index is adopted for Chinese locomotives, and the calculation equation is:

$$W = 3.57\left[\frac{A^3}{f}F(f)\right]^{0.1} \tag{1.11}$$

where W is the ride index; A is the vehicle body vibration acceleration, in g; f is the vibration frequency, in Hz; and $F(f)$ is the frequency correction coefficient (Table 1.11).

As train vibrations exhibit stochastic characteristics with continuously varying acceleration and frequency, the evaluation process requires dividing the target acceleration waveform into frequency bands. The ride comfort index (W) is calculated separately for each band based on its specific acceleration and frequency, with the composite index for the entire waveform determined by the following equation:

$$W = \sqrt[10]{W_1^{10} + W_2^{10} + W_3^{10} + \cdots + W_N^{10}} \tag{1.12}$$

where N is the total number of groups in the entire band. The grade standard of the Sperling ride index is shown in Table 1.12.

Table 1.11 Frequency correction coefficient in the ride index calculation equation.

Vertical Vibration (Hz)		Lateral Vibration (Hz)	
0.5–5.9Hz	$F(f) = 0.325 f^2$	0.5–5.4Hz	$F(f) = 0.8 f^2$
5.9–20Hz	$F(f) = 400 f^2$	5.4–26Hz	$F(f) = 650 f^2$
>20Hz	$F(f) = 1$	>26Hz	$F(f) = 1$

Table 1.12 Grade standard of the Sperling ride index.

Ride Index	W	Ride Quality
1	1	Just noticeable
2	2	Clearly noticeable
3	2.5	More pronounced but not unpleasant
4	3	Strong, irregular but still tolerable
5	3.25	Very irregular
6	3.5	Extremely irregular, unpleasant, annoying prolonged exposure intolerable
7	4	Extremely unpleasant; prolonged exposure harmful

Table 1.13 China's train ride comfort evaluation grades.

Ride comfort grade	Evaluation	Locomotives	Passenger vehicles	Freight trains
Grade 1	Excellent	<2.75	<2.5	<3.55
Grade 2	Good	2.75–3.10	2.5–2.75	3.5–4.0
Grade 3	Acceptable	3.10–3.45	2.75–3.0	4.0–4.25

The ride comfort evaluation grades for locomotives, passenger trains and freight trains prescribed by China's *Specification for Dynamic Performance Assessment and Testing Verification of Rolling Stock* (GB/T 5599-2019) is listed in Table 1.13, where identical evaluation grades are applied to both vertical and lateral assessments.

1.4 Summary

This chapter provides a systematic overview of the development history and characteristics of China's high-speed railway, highlighting the impact of earthquakes on railway operational safety. It summarizes the types of high-speed trains, tracks, and bridges, examines the computational models for train-track-bridge systems and dynamic coupling solution methods, and investigates the current research and evaluation criteria for the dynamic performance of high-speed trains operating on bridges.

At present, China's high-speed railway network has taken shape, characterized by a high proportion of bridge sections. Meanwhile, as China is situated at the intersection of the Mediterranean-Himalayan Orogenic Belt and the Pacific Ring of Fire, a region characterized by frequent seismic activity, research on the operational safety of high-speed trains on bridges under earthquakes holds significant practical importance. China's EMUs are mainly divided into CRH series EMUs and CR series EMUs. CR series EMUs represent a new generation of high-speed trains independently developed by China with full intellectual property rights. CRTS I, CRTS II, and CRTS III ballastless track slabs are mainly used, each featuring distinct structural characteristics. Additionally, four main types of bridges are predominantly employed in high-speed railway lines: standard-span bridges, long-span concrete girder bridges, composite structure bridges, long-span deck arch bridges, and long-span cable-supported bridges. The high-speed train-track-bridge system is a highly coupled, closely connected large system, and the computational model for a sophisticated high-speed train-track-bridge system can more accurately characterize the dynamic responses of the system under load. Train calculation model, track calculation model, bridge calculation model and wheel–rail contact model together form a complete train–track–bridge coupling model. The weak coupling method and the strong coupling method are used as dynamic coupling solution methods. Research on the dynamic performance of high-speed trains operating on bridges has gone through three stages, with key focuses encompassing wheel–rail contact relation, track irregularity excitation, numerical analysis method and other aspects. The evaluation standards for operational dynamic performance are established with reference to specifications, systematically analyzing bridge natural frequencies, deflection, lateral vibration amplitudes, and vibration acceleration.

References

[1] Lu DF. Railway development is the top priority in building a transportation powerhouse: contributing more to sustainable economic and social development: Excerpt from the report at the Work Conference of China Railway Corporation. China Railway 2018;(1):1–6 [In Chinese].
[2] Qian LX. The world's high-speed railway technology. Beijing: China Railway Publishing House; 2003 [In Chinese].
[3] Qian LX. A review of the technological development of high-speed trains with a speed of 350 $km.h^{-1}$ in the world. China Railw Sci 2007;(4):66–72 [In Chinese].
[4] Lu CF. High-speed railway in China. Beijing: China Railway Publishing House; 2013 [In Chinese].
[5] Medium- and long-term railway plan [EB/OL]. (2016-07-13) [2025-10-15]. https://zfxxgk.ndrc.gov.cn/web/iteminfo.jsp?id=366.
[6] Kang X, Jiang LZ, Bai Y, et al. Seismic damage evaluation of high-speed railway bridge components under different intensities of earthquake excitations. Eng Struct 2017;152:116–28.
[7] He X, Wu T, Zou Y, et al. Recent developments of high-speed railway bridges in China. Struct Infrastr Eng 2017;13(12):1584–95.
[8] Wei YX, Zuo DY. Research on the conditions and decisions of using a bridge instead of an embankment for high-speed railway. Railway Investig Surv 2004;(3):9–12 [In Chinese].
[9] Wu ZB, Zhou XX, Xie W, et al. Statistics of Wenchuan Earthquake investigation and some thoughts on seismic design of buildings. Build Struct 2010;S2(40):144–8 [In Chinese].
[10] Lin FX, Xie CD, Zhao SF, et al. Study on the spatiotemporal characteristics of the 2010 Yushu Ms7.1 earthquake sequence. Plateau Earthq Res 2017;29(4):12–7 [In Chinese].
[11] Zeng ZX, Wang J. Lushan earthquake: a case for the successful medium-term prediction [J]. Geosci Front 2013;20(3):21–4 [In Chinese].
[12] Liang HL, Shen Y, Mao L, et al. Comparison of building damage characteristics of the Lushan MS7.0 and Jiuzhaigou MS7.0 earthquakes. South China J Seismol 2019;39 (3):14–22 [In Chinese].
[13] Lai M, Fan ZZ, Liu XH, et al. Experiment on a population distribution statistical method for human loss assessment in earthquake disasters. Earthq Res Sichuan 2017; (4):21–3 [In Chinese].
[14] Huang XK. Study on the economic loss assessment methods of earthquake disasters in the Chinese. Harbin: Institute of Engineering Mechanics, China Earthquake Administration; 2015 [In Chinese].
[15] Brabie D, Andersson E. An overview of some high-speed train derailments: means of minimizing consequences based on empirical observations. Proc Inst Mech Eng F J Rail Rapid Transit 2008;222(4):441–63.

[16] International Online: 6.5-magnitude earthquake hits Kumamoto, Japan: Shinkansen train derails [EB/OL]. (2016-04-15) [2025-10-15]. http://www.xinhuanet.com//world/2016-04/15/c_128898878.htm.
[17] Nie TY, Guo W, Jiang ZL, et al. Influence of pier height on the safety of trains running on high-speed railway bridges during earthquakes. J Cent South Univ 2024;31(6):2102–15.
[18] Guo W, He X, Jiang LZ, et al. The effect of high-speed railway train on the seismic response of bridge under earthquakes. Eng Struct 2024:303:117446.
[19] Willis R. Appendix of report of the commissioners appointed to application of iron to railway structures [M]. London: William Clowes and Sons; 1849.
[20] Willis R. Preliminary essay to the appendix B: experiments for determining the effects produced by causing weights to travel over bars with different velocities. [C]// Report of the commissions appointed to inquire into the application of iron to railway structures. London: William Clowes and Sons; 1849.
[21] Wang Fu T. Vehicle system dynamics. Beijing: China Railway Publishing House; 1981 [In Chinese].
[22] Sun JY. A brief history of computer development. Mech Weekly 2014;13:234 [In Chinese].
[23] Wang Y, Guo W, Wu W, et al. Study on the vehicle-continuous bridge coupled vibration characteristics and influencing factors of 600 km/h high-speed maglev trains [J]. Journal of Southeast University (natural science Edition). 2026. p. 1–16.
[24] Zhai WM, Han ZL, Chen ZW, et al. Train-track-bridge dynamic interaction: a state-of-the-art review. Veh Syst Dyn 2019;57(7):984–1027.
[25] Yang GW, Wei YJ, Zhao GL, et al. Research progress on the mechanics of high-speed rails. Adv Mech 2015;45:217–460 [In Chinese].
[26] Tao JH, Liu RT, Xu K, et al. A brief history of computer development in China [J]. Sci Technol Rev 2016;34(14):12–21 [In Chinese].
[27] Zhai WM. Vehicle-track coupling dynamics. 4th ed. Beijing: Science Press; 2015 [In Chinese].
[28] Fryba L. Vibration of solids and structures under moving loads, vol 1. Groningen: Noordhoff International Pub; 1973.
[29] He X, Zhang N, Guido DR. Dynamic analysis of high-speed railway bridge under articulated trains. Comput Struct 2003;81(26–27):2467–78.
[30] Mao JF, Yu ZW, Xiao YJ, et al. Random dynamic analysis of a train-bridge coupled system involving random system parameters based on the probability density evolution method. Probab Eng Mech 2016;46:48–61.
[31] Yu ZW, Liu HY, Guo W, et al. A general spectral difference method for calculating the minimum safety distance to avoid the pounding of adjacent structures during earthquakes. Eng Struct 2017;150:646–55.
[32] Zeng QY, Yang P. The "set-in-right-position" rule for forming structural matrices and the finite truss-element method for space analysis of truss Bridge. J China Railw Soc 1986;8(2):48–59 [In Chinese].
[33] Zeng QY, Tian ZQ, Yang Y, et al. Truss segment finite element method for space truss driving vibration calculation. Bridge Constr 1985;(4):1–17 [In Chinese].
[34] Bowe CJ, Mullarkey TP. Unsprung wheel-beam interactions using modal and finite element models. Adv Eng Software 2008;39(11):911–22.
[35] Sato Y. Abnormal wheel load of test train. Permanent Way (Tokyo), 1973;14(53):1–8.
[36] Ahlbeck DR, Meacham HC, Prause RH. The development of analytical models for railroad track dynamics. Ker RAD. Railroad Track Mechanics & Technology. Oxford: Pergamon Press; 1978. p. 239–63.
[37] Zhai WM. Vehicle-track coupled dynamics: theory and applications. Beijing: Science Press; 2020.

[38] Clark RA, Dean PA, Elkins JA, et al. An investigation into the dynamic effects of railway vehicles running on corrugated rails. J Mech Eng Sci 1982;24(2):65–76.
[39] Lei X, Mao L. Dynamic response analyses of vehicle and track coupled system on track transition of conventional high-speed railway. J Sound Vib 2004;271(3.5):1133–46.
[40] Cheng YS, Au FTK, Cheung YK. Vibration of railway bridges under a moving train by using bridge-track- vehicle element. Eng Struct 2001;23(12):1597–606.
[41] He X, Zhang N. Dynamic analysis of railway bridge under high-speed trains. Comput Struct 2005;83(23–24):1891–901.
[42] Carter FW. On the action of a locomotive driving wheel. Proceedings of the Royal Society of London. Series A, Containing Papers of a Mathematical and Physical Character 1926;112(760):151–7.
[43] Vermeulen PJ, Johnson KL. Contact of nonspherical elastic bodies transmitting tangential forces. J Appl Mech 1964;31(2):338.
[44] Johnson KL. The effect of spin upon the rolling motion of an elastic sphere on a plane. J Appl Mech 1958;25(3):332–8.
[45] Johnson KL. A note on the influence of elastic compliance on sliding friction in ball bearings. J Basic Eng 1960;82(4):899.
[46] Schwab AL, Yan F, Ding JJ, et al Kalker JJ. 1933–2006: a life in rolling contact. Foreign Rolling Stock 2011;48(5):44–6 [In Chinese].
[47] Kalker JJ. Three dimensional elastic bodies in rolling contact. Delft; Dordrecht: Kluwer Academic Publishers; 1990.
[48] Kalker JJ. On the rolling contact of two elastic bodies in the presence of dry friction. Delft: Delft University of Technology; 1967.
[49] Kalker JJ. A fast algorithm for the simplified theory of rolling contact. Veh Syst Dyn 2007;11(1):1–13.
[50] Kalker JJ. The computation of three-dimensional rolling contact with dry friction. Int J Numer Methods Eng 1979;14(9):1293–307.
[51] Shen ZY, Hedrick JK, Elkins JA. A comparison of alternative creep force models for rail vehicle dynamic analysis. Veh Syst Dyn 2007;12(1-3):79–83.
[52] Jin XS. Wheel-rail creep theory and its test research. Chengdu: Southwest Jiaotong University; 1999 [In Chinese].
[53] Zhong S, Xiao X, Wen Z, et al. The effect of first-order bending resonance of wheelset at high-speed on wheel-rail contact behavior. Adv Mech Eng 2013;5:296106.
[54] Jin XS, Shen ZY. An analytical solution for adhesion/slip borderline in rolling contact patch under pure spin. Proceedings of the 16th IAVSD Symposium on Vehicle System Dynamics. 1999. Southampton.
[55] Ning YZ. Comparative analysis and reflections on track irregularity spectra of high-speed railways in China and abroad. Railway Qual Contr 2016;44(4):11–5 [In Chinese].
[56] Garg VK, Dukkipati RV. Dynamics of railway vehicle systems. Amsterdam: Elsevier; 1984.
[57] National Railway Administration. Irregularity spectrum for high-speed railway ballastless track (TBT 3352-2014). Beijing: China Railway Publishing House; 2014 [In Chinese].
[58] Luo X. Research on the management value of emergency repair of track irregularities. Beijing: China Railway Publishing House; 1988 [In Chinese].
[59] Kang X, Liu XB, Li HY, et al. PSD of ballastless track irregularities of high-speed railway. Sci Sin 2014;44(7):687 [In Chinese].
[60] Xing GS. Random vibration analysis. Beijing: Seismological Press; 1977 [In Chinese].
[61] Wu XW. Running safety assessment of railway vehicles under the earthquake excitations. Chengdu: Southwest Jiaotong University; 2016 [In Chinese].

[62] Qiao H. Research on several key issues in dynamic response analysis of the vehicle-bridge system under earthquake. Beijing: Beijing Jiaotong University; 2018 [In Chinese].
[63] Xu L, Zhai W. Cross wind effects on vehicle-track interactions: a methodology for dynamic model construction. J Comput Nonlinear Dynam 2019;14(3):031003.
[64] Li YL, Xiang HY, Wang B, et al. Dynamic analysis of wind-vehicle-bridge coupling system during the meeting of two trains. Adv Struct Eng 2013;16(10):1663–70.
[65] Lou P, Zhu J, Dai G, et al. Experimental study on bridge-track system temperature actions for Chinese high-speed railway. Arch Civ Mech Eng 2018;18(2):451–64.
[66] Gou HY, Yang LC, Mo ZX, et al. Effect of long-term bridge deformations on safe operation of high-speed railway and vibration of vehicle-bridge coupled system. Int J Struct Stabil Dynam 2019;19(9):1–25.
[67] Wang Y, Guo W, Jiang L, et al. Vibration characteristics and operational safety assessment of high-speed maglev vehicle-bridge coupling systems under near-field earthquakes. J China Railw Soc 2025;47(7):97–107.
[68] Zeng C, Guo W, Jiang L, et al. Validation of a scaled dynamic test system for simulating a high-speed train passing bridges under seismic excitation. Earthq Eng Struct Dynam 2024;53(12):3729–53.
[69] Tan S. Spatial random vibration analysis and performance evaluation of train-track-bridge coupling systems. Changsha: Central South University; 2019 [In Chinese].
[70] Zhu Y. Theory and application of random vibration of vehicle-bridge system. Chengdu: Southwest Jiaotong University; 2011 [In Chinese].

CHAPTER 2

Numerical Simulation Methods for Train Operation on High-Speed Railway Bridges under Earthquake Conditions

Contents

High-Speed Railway Train–Track–Bridge Systems
ISBN 978-0-443-49310-2
https://doi.org/10.1016/B978-0-443-49310-2.00002-8

2.1 Overview

Based on the current development of high-speed railways and considering the distinctive characteristics of China's high-speed railway system, this chapter highlights the necessity and significance of studying train–track–bridge systems under earthquake conditions.

Section 2.2 summarizes the strengths and limitations of the strong-coupling and weak-coupling methods currently employed in train–bridge coupling vibration analysis. Building upon this, a novel normal contact-point detecting algorithm for flange contact scenarios was proposed. A MATLAB-based visual computational platform for space wheel–rail contact relationships was developed, and the accuracy and universality of the normal point detection theory are verified through numerical comparison with the commonly used wheel–rail vertical point detection theory.

Train–bridge coupling vibration analysis, which is crucial for the design of high-speed railway bridges, can ensure the operational safety and riding comfort of trains. To address the inefficiency in modeling train–track–bridge systems, Section 2.3 presents a novel simulation strategy for train–bridge coupling vibration analysis by introducing the Client–Server (C/S) technology. This simulation strategy treats the superstructure (train) and substructure (track–bridge supporting structure) as independent servers, while treating wheel–rail interactions between the superstructure and substructure as clients. The clients and servers exchange data in real time through network communication technology. This section also elaborates on the CS technical architecture and the coupling computation process of key components.

Section 2.4 implements the train–bridge coupling vibration analysis based on the OpenSees platform and using the CS technology. Numerical cases demonstrate the ability of this simulation method to quickly apply the robust nonlinear and seismic analysis functions of the OpenSees software to solve train–track–bridge system problems, establishing it as a convenient, practical, and straightforward approach for coupling vibration analysis of such systems.

Section 2.5 develops the Simpack and OpenSees co-simulating high-speed train–track–bridge spatial coupling vibration system (SOTTB) based on the CS architecture using the multibody dynamics software Simpack and FE software OpenSees. Two distinct simulation approaches—co-simulation and integrated FE modeling—are employed to conduct time–history analysis on a single-span simply supported beam bridge. This section further applies the SOTTB program to establish a refined computational model for the train–track–bridge spatial coupling vibration system. This model can be used to evaluate train comfort and safety, as well as the dynamic performance of bridge structures. Finally, the SOTTB program is utilized to investigate train operational safety under lateral earthquakes and track irregularity excitations, and the influence of train running speed and earthquake intensity on train operational safety is discussed.

Section 2.6 presents a novel moving wheel–rail element model. Furthermore, based on the nonlinear vertical wheel–rail contact relationship, this section develops an innovative two-dimensional wheel–rail coupling element model, which is implemented on the OpenSees FE software platform. Building upon the aforementioned moving element model, it elaborates the construction methodology of a train–track–bridge system simulation platform, and builds a simulation platform suitable for train–track–bridge coupling vibration analysis under earthquake conditions with the open-source FE analysis software OpenSees as the integrated platform. The integration methods for moving elements are subsequently described in both vertical and spatial dimensions. The accuracy of vertical moving elements is validated through comparative analysis with classical theoretical solutions. The developed simulation platform provides an analytical tool for investigating train–track–bridge coupling vibrations under seismic actions.

2.2 Wheel–Rail Contact Point Detection Method

In the train-bridge spatial coupling vibration analysis, the detection of wheel–rail contact points is crucial. The accuracy of the contact point detection method directly affects the determination of wheel–rail contact parameters and the calculation of wheel–rail interaction forces, which consequently influences subsequent dynamic analyses. Therefore, it is essential to first establish an appropriate contact point detection approach. Building upon previous research, this section proposes the contact point detection method based on maximum normal embedding depth, hereafter referred to as the normal-based point detection method [1].

2.2.1 Principle of Normal-Based Point Detection Method

During the strongly constrained motion of trains on rails, the train wheelset is subjected to normal constraint forces and tangential creep forces from the rail. The first step in determining these forces is to identify the contact point position between the wheel and rail. In the solution process of the train–bridge coupling system, determining the wheel–rail contact points in an accurate and efficient way remains a persistent challenge in the study of wheel–rail contact geometry. In previous studies, the vertical point detection method based on wheel–rail contact traces was widely recognized as a simple yet effective approach, based on the widely accepted assumption that the contact point can be identified by the minimum vertical distance. However, in the vertical point detection method for wheel–rail contact calculation, it is implicitly assumed that the contact patch follows a Hertzian elliptical distribution, and the identified contact point is conventionally adopted to represent the entire contact patch.

In actual wheel–rail contact scenarios, the most representative contact point of the contact patch should be the center point of the wheel–rail contact patch. However, the vertical point detection method exhibits significant deviation from this center point during large wheelset lateral displacements or flange contact conditions, as illustrated in Fig. 2.1. Under large lateral displacement conditions, abnormal single-point flange contact may occur between the wheel tread profile and rail section. When vertically displacing the wheelset by the interbedding depth (i.e., vertical clearance), the maximum normal interbedding depth within the

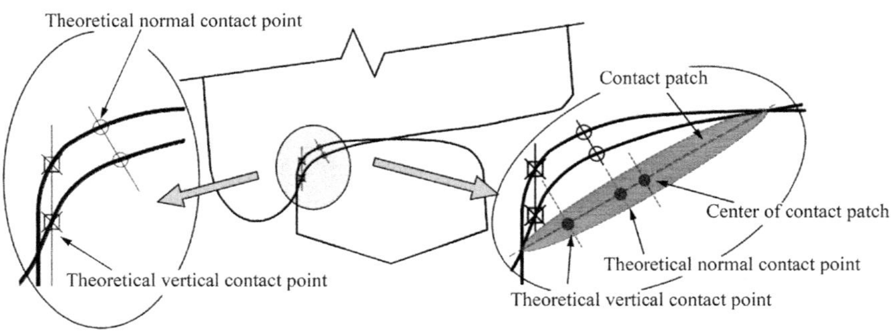

Figure 2.1 Schematic comparison of normal-based point detection method versus vertical projection method.

overlapping area is significantly smaller than the minimum vertical distance identified by the vertical point detection method.

In addition, *Code for Design of High-Speed Railway* (TB 10621—2014) [2] adopts a derailment coefficient limit of 0.8. According to Nadal's formula, to reach or exceed the derailment limit of 0.8, wheel flange contact or rail climbing inevitably occurs at this threshold. While the vertical point detection method is suitable for railhead contact scenarios, its accuracy significantly deteriorates under conditions of large wheelset lateral sway or yaw motion. Based on the above considerations, this section proposes the normal-based point detection method, which provides a more rational characterization of the wheel–rail contact patch center compared to the vertical point detection method.

2.2.2 Wheel–Rail Contact Point Calculation

Through coordinate transformation, the equations of the wheel tread trace and railhead profile in the global coordinate system can be obtained, enabling various methods to identify potential contact points between the rail cross-section and wheel tread profile. The minimum vertical distance method [3] is a commonly used approach. Based on Hertzian assumptions, this method treats the wheelset and rail as rigid bodies with a single-point contact, where the contact point is regarded as the center of the actual contact patch. The procedure involves artificially elevating the wheelset to a certain height and gradually lowering it until the first contact point is identified as the wheel–rail contact point, which is simultaneously regarded as the center of the actual contact patch. Mathematically, this entails finding the minimum vertical distance between discrete points on the left/right wheelset tread profile (trace) and left/right railhead profile. Analysis indicates that during this process, the point of minimum vertical distance during wheelset descent corresponds to the initial contact point. However, this initial contact point does not necessarily represent the actual wheel–rail contact location, nor can it be regarded as the true center of the contact patch.

Assuming constant wheelset and rail profiles with mutual embedding allowed, the wheel–rail contact point within the mutually embedded area between the wheel tread profile and railhead can be determined using the contact point detection method based on maximum normal embedding depth. When mutual embedding is permitted between the wheelset and rail, the normal vectors at their contact point no longer coincide. In other words, during the contact point detection process, the detection should be conducted separately along the normal direction of the wheelset tread and the normal direction of the railhead curve. Subsequently, the maximum normal embedding depth values should be determined based on the respective normal vectors of both the wheelset and rail. The final step involves selecting the greater of these two values as the definitive wheel–rail contact point. For the current analysis limited to straight track conditions, the longitudinal coordinate can be considered noninfluential to wheel–rail contact characteristics.

2.2.3 Calculation Steps of Normal-Based Point Detection Method

The computation of the normal-based point detection method can be implemented through MATLAB programming, which can be summarized into the following procedural steps.

(1) Determine the horizontal coordinate range of the embedded area. In the global coordinate system, the embedded area's transverse coordinate range is determined by computing the difference between the wheelset tread profile curve and the railhead contour curve, followed by a sign analysis to identify the embedded area. This provides conditions for subsequent traversal and point detection. It is worth noting that in the above process of determining the mutually embedded area, there is no requirement that the embedded area must be continuous, so this method is also applicable to the case of multipoint contact, however, multipoint contact is not considered temporarily in this section.

(2) Obtain the vertical coordinates through discrete interpolation of the transverse coordinates. Specifically, the transverse coordinates within the embedded area are discretized, followed by interpolation at these discrete points to derive corresponding vertical coordinate values, which will be subsequently utilized for identifying the contact point.

(3) Obtain the coordinates of contact points. Formulate the contact point equations based on the derived formulas and solve them to obtain the corresponding coordinate values.

(4) Calculate the embedding depth. Based on Step (3), compute the mutual embedding depth values between wheel and rail, and store the results in matrix form.

(5) Iterate through coordinates. Traverse the transverse coordinates of the embedded area and repeat Steps (2)–(4) to obtain all embedding depth values.

(6) Determine the wheel–rail contact point. From the results in Step (5), identify the point with the maximum embedding depth and its corresponding index, thereby determining the wheel–rail contact point. Store these data properly for subsequent calculation of wheel–rail contact parameters.

2.2.4 Visualization of Spatial Wheel–Rail Contact Relationship

To compare the differences between the normal-based point detection method and the vertical point detection method and verify their accuracy and universality, Liu et al. developed a wheel–rail contact geometry calculation and analysis platform (WRCG) [4] using MATLAB graphical user interface, and applied for corresponding software copyright. The computational platform can account for all six degrees of freedom (DOFs) of both train wheelsets and rails. It supports various wheel tread profiles (such as LM, LMA, S1002, and tapered tread profiles) and different rail specifications (such as CHN75, CHN60, CHN50, and UIC60 rails). It incorporates two distinct contact point detection algorithms: the normal-based point detection method and the vertical point detection method. It also accommodates variations in some initial parameters.

2.2.5 Numerical Examples and Verification

The computational analysis employs the normal-based point detection method with comparative validation against the vertical point detection method. It employs LM tread profiles and CHN60 rails as the benchmark case, with a 1435 mm track gauge, a 1353 mm wheelset back-to-back distance, and a 1:40 standard rail cant angle. Other parameters are established according to relevant standards and specifications. This method considers extreme operating conditions, with a 10.5 mm sway displacement and a 0.5° yaw angle. A 1.5 mm heave displacement is introduced to enhance the visual interpretability of the normal-based point detection method; the simulation permits wheel–rail embedding while disregarding the movements of both left and right rails; all computations and visualization are performed on the WRCG platform, yielding the results documented in Figs. 2.2 and 2.3.

Fig. 2.2 shows the three-view diagrams of contact traces for both left and right wheelsets under this working condition. As illustrated in Fig. 2.2, when yaw motion occurs, the contact traces of the wheelset are no longer confined to the *Y-Z* plane but also distribute along the *X*-axis direction,

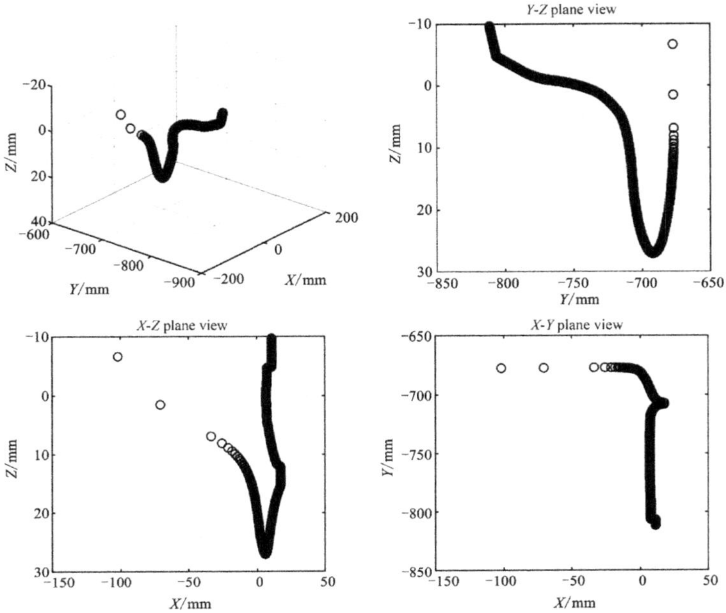

(a) Left wheel-rail contact trace in three views

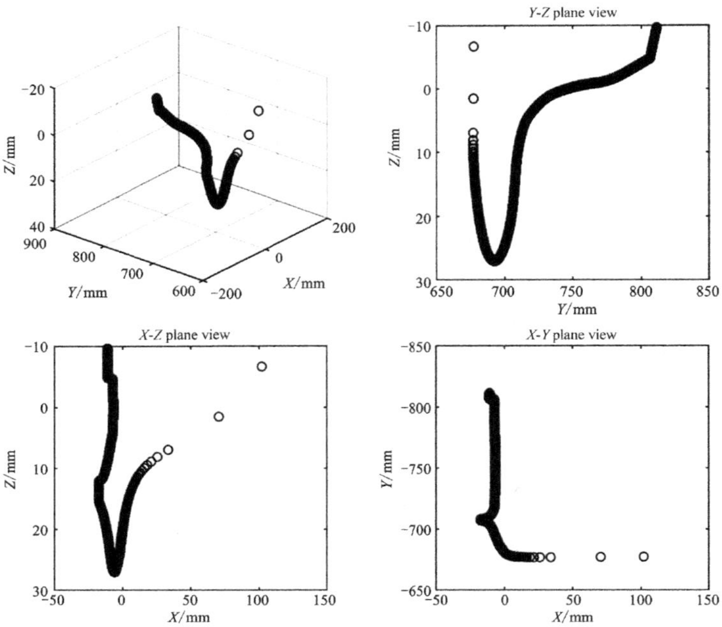

(b) Right wheel-rail contact trace in three views

Figure 2.2 Three-view drawings of left/right wheelset contact traces.

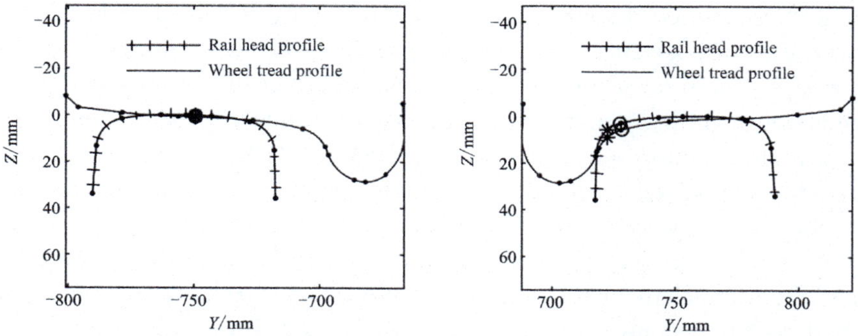

(a) Left-side contact point comparison diagram (b) Right-side contact point comparison diagram

Figure 2.3 Enlarged view of left/right contact point pairs (asterisks indicate vertical point detection method, circles denote normal-based point detection method).

indicating that the contact points on the wheelset are no longer limited to the generatrix of the wheelset. Gan and Zeng's research demonstrates that when yaw motion occurs, the wheel–rail contact trace may exhibit shadowing effects [5,6]. Under large sway motion (10.5 mm) combined with a 0.5° yaw angle, for the left side, both methods yield nearly identical contact points, while for the right side, the normal-based point detection method locates contact points closer to the center of the contact area, with a distance of 2.93 mm between the two contact points, which is less than the 3.18 mm distance obtained by the trace method. This confirms the rationality of the normal-based point detection method.

Through comparative analysis, the normal-based point detection method identifies contact points closer to the center of the wheel–rail contact area, making them more representative. Under large sway displacement and roll conditions, the contact points do not converge entirely toward the wheel flange root, demonstrating that the identified contact point is reasonable (Fig. 2.3). In Section 2.6, all geometric contact points of wheel–rail interactions are identified using the normal-based point detection method.

2.3 Co-simulation Technology for Train–Track–Bridge System Simulation under Earthquake Conditions

From the early simplified constant force motion model on simply supported girders to the current multi-rigid-body and nonlinear track–bridge coupling system models, train–bridge coupling vibration research has developed into a well-established discipline over a century and a half [7–10]. However, with the rapid advancement of high-speed railways, it remains a challenge for researchers to accurately, efficiently, and concisely simulate the interactions within train–track–bridge systems, and to ensure operational safety and comfort under complex conditions (e.g., earthquakes, crosswinds, and pile-soil coupling). Solutions for train–bridge coupling vibration problems can be categorized into four types: the analytical method, the decoupled iterative method, the coupling system holistic solution method, and the finite element (FE) method. Among these, the decoupled iterative method has become mainstream due to its widespread adoption by researchers, leading to the development of specific train–bridge interaction elements. For instance, Yang et al. [11,12] treated the train as a sprung mass, assuming continuous contact between wheelsets and the bridge, and derived train–bridge interaction elements. Later, Lou et al. [13,14] extended this approach to develop train–bridge interaction elements incorporating two-stage suspension vehicle models, as well as nonuniformly partitioned bridge–track interaction elements. However, the aforementioned train–bridge interaction elements neither incorporated the nonlinear Hertzian spring contact at the wheel–rail interface nor accounted for the wheel–rail separation. To address these limitations, this section introduces a new simulation strategy based on CS (client-server) technology. The following section demonstrates the implementation procedure using OpenSees as the representative simulation platform.

2.3.1 Principles of CS Technology

The fundamental principle of CS technology involves establishing two servers using existing commercial analysis software (e.g., OpenSees): one for the vehicle substructure and another for the track–bridge substructure, both responsible for dynamic analysis of their respective subsystems; a C++based wheel–rail coupling client program was developed to process wheel–rail interactions and ensure the kinematic coordination of forces, displacements, velocities, and accelerations between the train system and the track–bridge system.

As the train–bridge coupling represents a typical interaction between two substructures, the upper vehicle structure and the lower support structure (track–bridge coupling system) are exclusively linked through wheel–rail interactions. As shown in Fig. 2.4, the entire analysis framework consists of three parts: train server, track–bridge server, and coupling client.

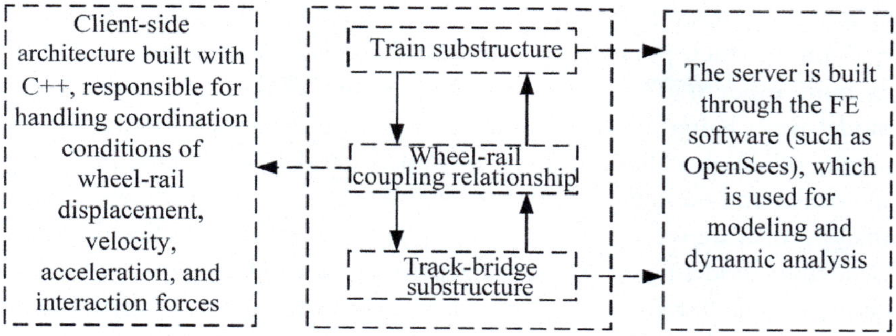

Figure 2.4 Client-server technical architecture diagram.

The train server and track–bridge server are equivalent, both established using analysis software (e.g., OpenSees); the coupling client is programmed in C++. These servers remain memory-resident and execute dynamic analysis of both substructures, receive and execute commands from the client, and perform single-step execution only upon client instruction. The client handles the coordination of displacements, velocities, accelerations, and forces between the upper and lower structures.

This simulation strategy offers many advantages, including the following. (1) Decoupled iteration. It eliminates the need for reassembling and decomposing global stiffness/damping matrices at every timestep. (2) Reduced data transfer. The servers remain resident in memory, unlike ANSYS-Simpack co-simulation [7] which requires importing/exporting substructure modal data and boundary conditions at each time step. This minimizes data transmission while enabling flexible and efficient monitoring of stress–strain level responses, surpassing traditional train–bridge coupling analyses limited to element-level responses. (3) Implementation simplicity. It requires no modification of any internal code of the analysis software (e.g., OpenSees). Instead, it only needs to establish an interface channel using the TCL (Tool Command Language). In the entire analysis process, this channel is set up once and reused multiple times, eliminating the need for repeatedly loading and releasing model data, thus ensuring highly efficient data transfer. (4) More comprehensive considerations. It can account for the nonlinear Hertz contact between wheels and rails, incorporate wheel–rail separation phenomena, and include displacement corrections for track beam elements, thereby making the simulation results more accurate. (5) Significant modeling advantages. When OpenSees is adopted as the server for modeling, it fully leverages OpenSees' capabilities in nonlinear seismic analysis and its extensive material/element libraries, enabling the incorporation of nonlinear damping and spring characteristics in train models. This approach liberates the track–bridge substructure model from being confined to linear structures or simple beam-slab elements, while also allowing users to update model parameters, load characteristics, and boundary conditions at any time step. Alternatively, if other analysis software is employed as the server, their respective strengths—such as flexible modeling—can be utilized to accommodate the structural diversity along extensive high-speed railway lines (e.g., bridges, tracks, subgrades, and soil layers), making the system adaptable to various train–bridge scenario analyses.

2.3.2 Coupling Calculation Process

This section details the fundamental computational procedure of CS technology in train–bridge coupling vibration analysis. Within the current time step, the predicted wheelset displacements are applied to the train model, and the train server performs dynamic analysis to obtain the contact forces at the wheel–rail contact forces. These forces are then reversely applied to the track–bridge structure, where the track–bridge server conducts dynamic analysis to derive the nodal displacements of the tracks and bridges. Subsequently, the displacements at the wheel–rail contact points on the track beam elements are interpolated using Hermite cubic shape functions. Accounting for track irregularities, the client program evaluates whether the wheelset displacements, track beam contact point displacements, wheel–rail embedding depths, and track irregularity values satisfy the coordination condition. If not, the Newton–Raphson algorithm is employed to update the initially predicted wheelset displacements, and the calculations are repeated until the convergence criterion for coordination is met. Upon satisfaction, the current time-step data are submitted, and the computation proceeds to the next time step.

The core of the entire wheel–rail system lies in the coordinated calculation of forces and displacements between the two subsystems. In the aforementioned computational process, forces always satisfy the action–reaction relationship, maintaining coordination. However, the displacements are contrary—the predicted wheelset displacements of the train are not necessarily true values and may not satisfy the geometric relationship with the wheelset contact point displacements, wheel–rail embedment depths, and track irregularity values. For instance, in vertical train–bridge coupling vibration analysis, the geometric coordination condition for wheel–rail interaction is expressed as:

$$\boldsymbol{\psi} = \boldsymbol{U}_{\mathrm{bc}} - \boldsymbol{U}_{\mathrm{w}} - \boldsymbol{R}_{\mathrm{n}} + \Delta \boldsymbol{U} \tag{2.1}$$

where $\boldsymbol{\psi}$ is the equivalent nonequilibrium displacement vector between the two subsystems in the current iteration; $\boldsymbol{U}_{\mathrm{bc}}$ is the displacement vector of the track beam at the wheel contact point; $\boldsymbol{U}_{\mathrm{w}}$ is the predicted wheel-–rail displacement vector (input value); $\Delta \boldsymbol{U}$ is the wheel–rail embedding depth vector; $\boldsymbol{R}_{\mathrm{n}}$ is the track irregularity vector.

During the analysis time step, the displacement of the track beam at the wheel contact point U_{bc} depends on the nodal displacements of the elements, which in turn are determined by the wheel–rail interaction forces acting on the track and bridge systems. Since the wheel–rail force at the train's wheel contact node is a function of the input excitation $\boldsymbol{U}_{\mathrm{w}}$, the displacement of the track beam element at the wheel contact node can be expressed as a function $\boldsymbol{U}_{\mathrm{bc}} = g(\boldsymbol{U}_{\mathrm{w}})$. Similarly, the wheel–rail embedding depth $\Delta \boldsymbol{U}$ is also a function of $\boldsymbol{U}_{\mathrm{w}}$, i.e., $\Delta \boldsymbol{U} = \mathrm{h}\ (\boldsymbol{U}_{\mathrm{w}})$. Therefore, Eq. (2.1) can be rewritten as:

$$\boldsymbol{\psi} = g(\boldsymbol{U}_{\mathrm{w}}) + h(\boldsymbol{U}_{\mathrm{w}}) - \boldsymbol{U}_{\mathrm{w}} - R_{\mathrm{n}} \tag{2.2}$$

The condition for wheel–rail displacement compatibility is satisfied when $\boldsymbol{\psi}$ in Eq. (2.2) equals zero or falls below a sufficiently small predefined convergence threshold. Since $\boldsymbol{U}_{\mathrm{w}}$ (the predicted wheelset displacement vector) is an unknown variable, it may not necessarily make the result in Eq. (2.2) exactly zero. Therefore, multiple iterations are typically required to determine the $\boldsymbol{U}_{\mathrm{w}}$ that meets the criteria. Mathematically, various methods can be employed to locate the root of Eq. (2.2). This section adopts the Newton–Raphson algorithm for root-finding. First, by taking the partial derivatives of both sides of Eq. (2.2) with respect to each component of the vector $\boldsymbol{U}_{\mathrm{w}}$, the equivalent tangent stiffness matrix can be derived:

$$\boldsymbol{\psi}' = \frac{\partial g(\boldsymbol{U}_{\mathrm{w}})}{\partial u_i} + \frac{\partial h(\boldsymbol{U}_{\mathrm{w}})}{\partial u_i} - \boldsymbol{I} \tag{2.3}$$

where $\boldsymbol{\psi}'$ represents the tangent slope of equivalent unbalanced displacement; $\boldsymbol{I}$ denotes the identity matrix with all diagonal elements equal to 1; u_i refers to the displacement of the i-th wheelset. Eq. (2.3) can employ the perturbation method during computation to obtain the equivalent tangent stiffness ψ'. During the differentiation process, the track irregularity vector $\boldsymbol{R}_{\mathrm{n}}$ remains constant within the time step, resulting in zero partial derivatives with respect to each component of $\boldsymbol{U}_{\mathrm{W}}$. After determining the equivalent tangent stiffness ψ', the Newton update method yields the new displacement excitation for the next iteration as:

$$\boldsymbol{U}_{\mathrm{W}}^{k+1} = \boldsymbol{U}_{\mathrm{W}}^{k} - [\psi']^{-1} \cdot \boldsymbol{\psi}^{k} \tag{2.4}$$

In the aforementioned procedure, the convergence criterion for judging wheel–rail geometric relationships is:

$$|\boldsymbol{\psi}^{n,k}| = |\boldsymbol{U}_{\mathrm{bc}}^{n,k} - \boldsymbol{U}_{\mathrm{W}}^{n,k} - \boldsymbol{R}_{\mathrm{n}} + \Delta \boldsymbol{U}^{n,k}| \leqslant e \tag{2.5}$$

where e represents the wheel–rail geometric compatibility tolerance, which can be set to a sufficiently small value as required; in this book, $e = 10^{-8}$ m is adopted; the superscript n (where $n \geqslant 1$, integer) denotes the n-th computational time step, while the superscript k (where $k \geqslant 1$, integer) represents the iteration count.

The detailed computational procedure is illustrated in Fig. 2.5, with the $(n + 1)$-th time step serving as an exemplar for explanation.

(1) Extract the track irregularity vector $\boldsymbol{R}_{\mathrm{n}}$ and wheelset displacement vector $\boldsymbol{U}_{\mathrm{W}}^{n}$ at the n-th time step ($n \geqslant 1$, integer). For $n = 1$ (initial condition), the initial wheelset displacement $\boldsymbol{U}_{\mathrm{W}}^{1}$ may be assigned a small downward nonzero value as the starting condition. Assign $\boldsymbol{U}_{\mathrm{W}}^{n}$ to the wheelset displacement vector $\boldsymbol{U}_{\mathrm{W}}^{n,k}$ at the k-th iteration, where the superscript k retains its previously defined meaning.

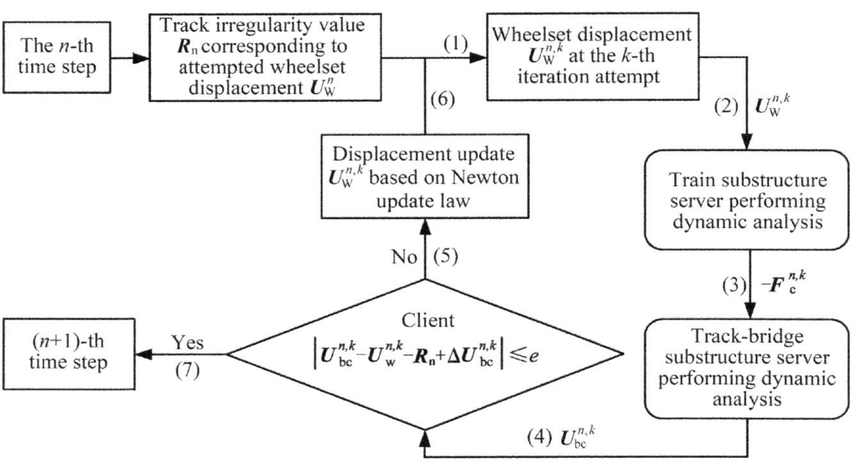

Figure 2.5 Calculation flowchart.

(2) Apply the displacement vector $\boldsymbol{U}_{\mathrm{w}}^{n,k}$ to the train wheelset. The train system server then performs dynamic analysis to obtain the contact force vector $\boldsymbol{F}_{\mathrm{c}}^{n,k}$ at the wheelset–rail interface.

(3) Apply the contact force vector $\boldsymbol{F}_{\mathrm{c}}^{n,k}$ in the reverse direction to the track–bridge substructure system. The substructure system server then performs a dynamic analysis to obtain the displacement vector $\boldsymbol{U}_{\mathrm{bc}}^{n,k}$ at the wheel–rail contact point on the track girder.

(4) Transmit the displacement vector $\boldsymbol{U}_{\mathrm{bc}}^{n,k}$ to the client program. The program then evaluates whether $\boldsymbol{U}_{\mathrm{w}}^{n,k}$ and $\boldsymbol{U}_{\mathrm{bc}}^{n,k}$ satisfy the geometric compatibility condition according to Eq. (2.5). If "Yes": directly proceed to Step (7); if "No": execute Step (5).

(5) Derive the new wheelset displacement vector $\boldsymbol{U}_{\mathrm{w}}^{n,k+1}$ for the next iteration according to Eq. (2.4).

(6) Assign the updated wheelset displacement vector $\boldsymbol{U}_{\mathrm{w}}^{n,k+1}$ to $\boldsymbol{U}_{\mathrm{w}}^{n,k}$, and reexecute Step (2) – Step (5).

(7) Submit the result data of the current step if the geometric coordination condition is met. Complete the calculation of the (n+1)-th time step, and calculate the next time step.

2.4 Train–Track–Bridge System Simulation under Earthquake Conditions Based on OpenSees Platform

To address low efficiency in modeling train–track–bridge systems, this section leverages the CS communication principle described previously to efficiently and rapidly perform simulation analysis within the open-source FE software OpenSees [15,16]. As illustrated in Fig. 2.6, compared to co-simulation, it adds the "attempt one analysis step" command in the static and dynamic analysis steps, which can perform shutdown iteration [17,18]. This enables pause-and-iterate functionality without requiring data rollback—if convergence is not achieved, the analysis can conveniently revert to the state of the previous time step. Consequently, there is no need to transmit the entire model information, and only boundary information needs to be exchanged. The client coordinator handles convergence checking, submitting the current state only after overall computational convergence is achieved. This method divides the train–track–bridge coupling system into three subsystems using the wheel–rail contact interface as the boundary: the train subsystem, the track–bridge subsystem, and

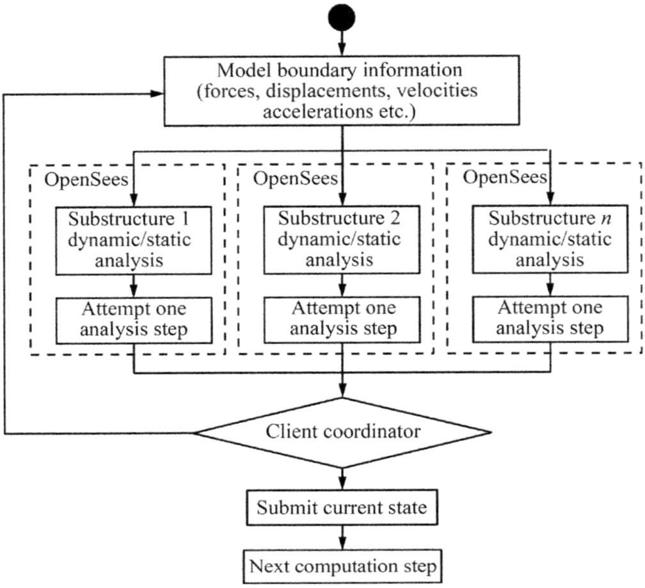

Figure 2.6 Improved OpenSees operating mode.

the wheel–rail contact relationship subsystem. Both the train and track–bridge subsystems are simulated and encapsulated as servers within the OpenSees software, with equal priority and no hierarchical distinction between them; the wheel–rail contact relationship subsystem functions as the client coordinator. Through network communication technology, the client coordinator and servers perform real-time, step-by-step data interactions, which significantly reduces the data exchange between subsystems. This approach uses OpenSees, which offers various advantages, which offers various advantages, including its extensive library of elements and materials, numerical algorithms, and powerful capabilities for seismic and nonlinear analysis. Consequently, train–bridge coupling can be rapidly performed on the unified OpenSees platform, and OpenSees alone suffices for train–bridge coupling vibration simulation, thereby lowering the threshold for simulation. Finally, through numerical case analysis, the accuracy and reliability of the method proposed in this section as well as modeling efficiency and flexibility are demonstrated.

As illustrated in Fig. 2.7, the entire analytical framework comprises three components: the train server, track–bridge server, and coupling client. Each server primarily performs dynamic and static analyses of its respective

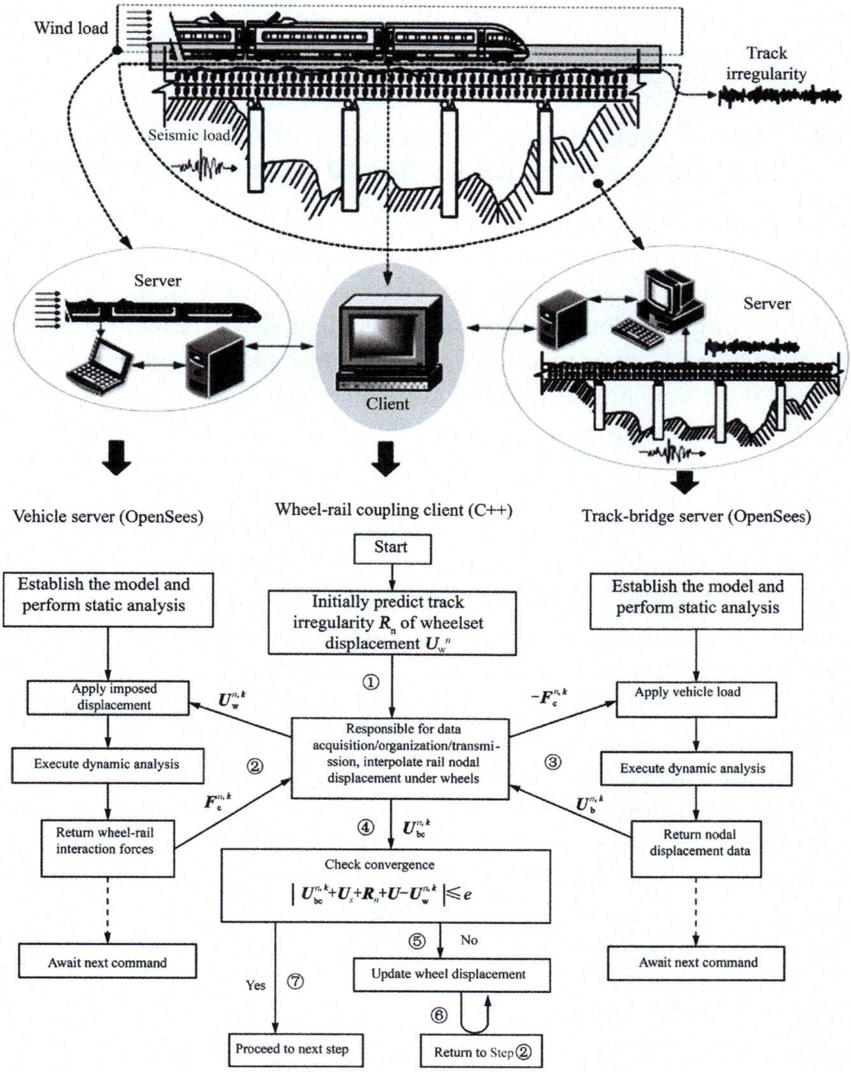

Figure 2.7 Technology principle and calculation flowchart of the OpenSees simulation for a train–bridge coupling system.

substructure, executing single-step computations only upon receiving a command from the client coordinator [17]. The OpenSees rapid simulation technique enhances efficiency in modeling train–track–bridge coupling vibration systems, effectively accounting for nonlinear contact springs and wheel–rail separation scenarios. It serves as a practical, easy-to-learn analysis method for train–bridge coupling vibrations.

2.5 Train–Track–Bridge System Simulation under Earthquake Conditions Based on Simpack Platform

In traditional numerical simulation research methods, any modification to the research object necessitates reprogramming, leading to time-consuming development, debugging difficulties, and a lack of visualization. Moreover, independently developed train–bridge coupling system analysis programs by different researchers tend to be closed-source and lack universality. To address these limitations, this section presents the Simpack and OpenSees train–track–bridge spatial coupled vibration system (SOTTB) program [19] for high-speed railways based on the CS architecture. This program effectively combines the strengths of both software platforms, overcoming the shortcomings of conventional programming approaches for train–bridge coupling vibration analysis. The SOTTB program enables rapid modeling of sophisticated train–track–bridge coupling vibration systems. Meanwhile, since it employs the well-established commercial software Simpack for multibody dynamics and the open-source OpenSees for FE analysis, the SOTTB program exhibits excellent generality and broad applicability.

This section first introduces the framework and implementation process of the SOTTB program, then elaborates on the modeling approaches and parameters for each component within the framework, and finally indirectly validates the correctness and feasibility of the SOTTB program through a numerical example of time–history analysis on a single-span simply supported girder bridge.

2.5.1 SOTTB Program Framework

The high-speed train–track–bridge system dynamic analysis model consists of a train subsystem and a track–bridge subsystem. The high-speed train and the track are coupled through wheel–rail interaction, with the wheel–rail contact surface serving as the interface. In this section, the train subsystem adopts a double-axle four-wheelset model, while the track–-bridge subsystem consists of a double-block ballastless track and a single-track multispan simply supported girder bridge. The system excitation inputs consider uniform seismic excitation and self-excitation from track irregularities. A schematic diagram of the spatial coupling vibration system for the high-speed train–track–bridge interaction, based on co-simulation between Simpack and OpenSees, is provided in Fig. 2.8.

The SOTTB program consists of a superstructure subsystem and a substructure subsystem. The superstructure subsystem, as illustrated in Fig. 2.9, is composed of a train model established with Simpack, a biblock ballastless track model, and a linear elastic girder model. The substructure subsystem, as shown in Fig. 2.10, comprises a fiber-based bridge pier model and a bearing model constructed using the FE software OpenSees. Under earthquake conditions, bridge girders typically remain elastic, whereas piers and bearings are prone to entering elastoplastic states. Therefore, the superstructure subsystem employs a linear elastic girder model, while the substructure subsystem utilizes fiber-based pier and bearing models. To enhance computational efficiency while preserving spatial characteristics, the girder dynamic analysis model adopts rigid-arm spatial girder elements, and the piers are simulated using spatial girder elements. The superstructure subsystem and substructure subsystem are coupled using the upper nodes at the bearing locations as boundaries, achieving their interaction through force equilibrium conditions and displacement compatibility relationships at these bearing nodes.

This section describes the real-time data interaction between the superstructure subsystem and substructure subsystem using the co-simulation (CS) technology [17]. First, the superstructure subsystem model is established in Simpack, with a parameter transfer interface created to serve as the superstructure subsystem server, while the substructure subsystem model is built in OpenSees as the substructure subsystem server. Second, the Simpack co-simulation module SIMAT, representing the superstructure

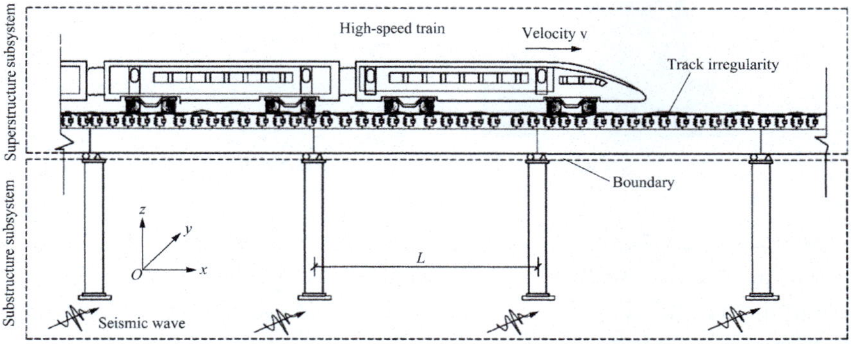

Figure 2.8 Schematic diagram of a spatial coupling vibration system for the high-speed train–track–bridge interaction based on the SOTTB program.

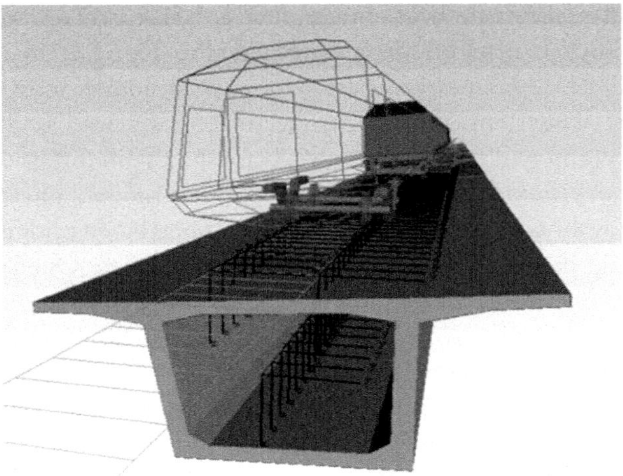

Figure 2.9 Superstructure subsystem.

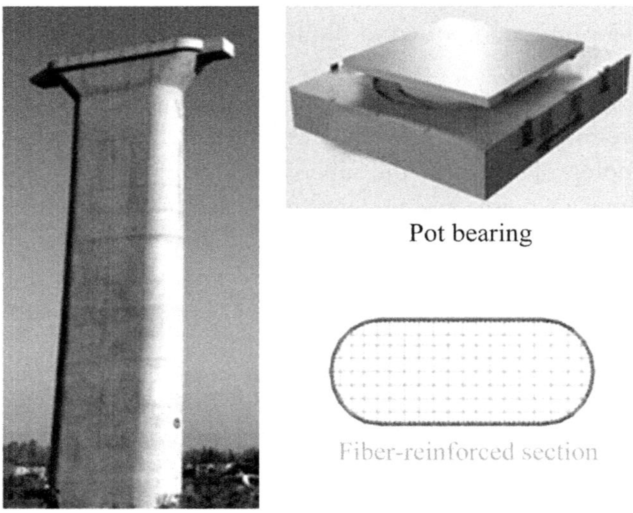

Figure 2.10 Substructure subsystem.

subsystem, is invoked in Simulink. By writing S-functions or M-functions based on the CS technology in Simulink as the client, real-time data communication between the superstructure and substructure subsystems is performed.

The composition and simulation process of the SOTTB program are illustrated in Fig. 2.11. The superstructure and substructure are bounded at the bearing locations. The dynamic response of the system in the time domain can be determined through real-time data interaction between the superstructure subsystem and substructure subsystem, combined with the force equilibrium conditions and displacement compatibility relationships at the boundaries.

In the superstructure subsystem, this section treats the high-speed train, track, and bridge girder as an integrated dynamic system, where the train and track are interconnected at the wheel–rail contact interface through wheel–rail relationships and interactions, as illustrated in Fig. 2.12. First, a multibody rigid train model is established. Next, the multibody dynamics software and FE software interface modules FlexTrack and Flexible are combined to develop elastic rail and girder models through the FE substructure technique (DOF reduction). The fasteners are simulated using force elements. The wheel–rail contact geometry relationship and force calculation method are established by configuring the corresponding

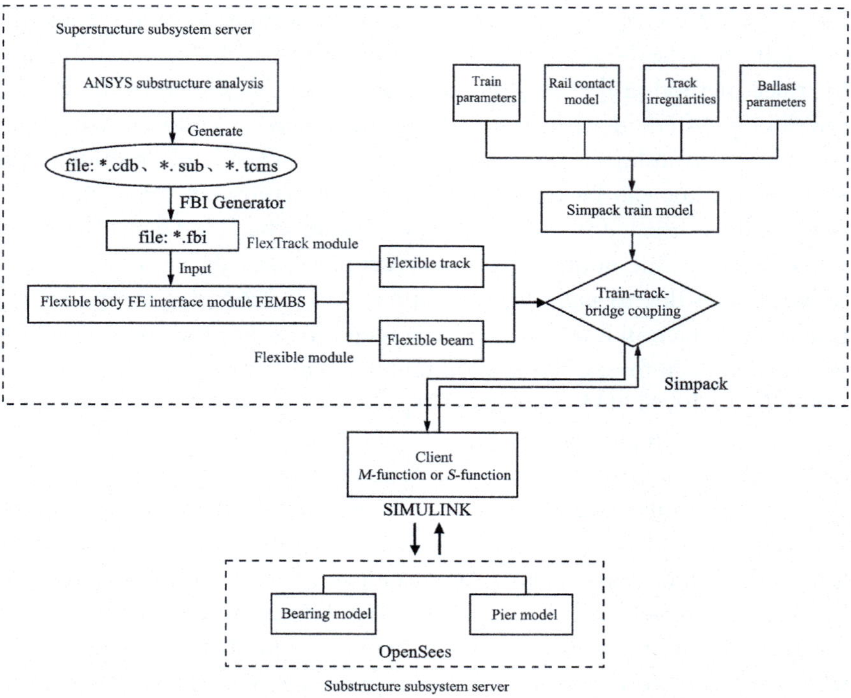

Figure 2.11 Composition and co-simulation flowchart of the SOTTB system.

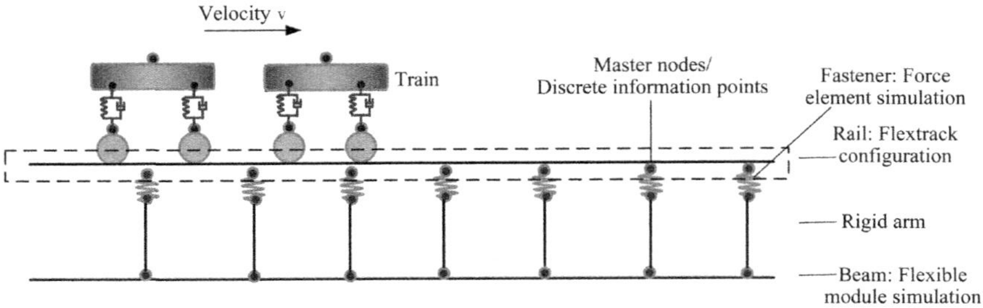

Figure 2.12 Schematic diagram of vehicle–track coupling in the superstructure subsystem.

wheel–rail module in multibody system dynamics software. Finally, the train and track interact at the wheel–rail contact interface, where the master nodes of the rail are treated as discrete information points. By enforcing force equilibrium and displacement compatibility conditions at the interface, force and displacement data exchange is achieved between the train and rail, thereby realizing dynamic coupling between the two subsystems.

The wheel–rail contact geometry relationship serves as the foundation for analyzing wheel–rail interaction forces. According to Wang Kaiwen's research, the wheel–rail geometric relationship is established typically using the trace method [20]. The basic functionality of the wheel–rail module in multibody system dynamics software Simpack is to establish wheel–rail contact relationships. By configuring the wheel–rail module, the wheel–rail force can be established during simulation. In this section, the wheel–rail normal force is calculated using the Hertzian theory of nonadhesive elastic contact. The microscopic elastic sliding between two rolling bodies due to slight velocity differences is defined as creep, and the resulting interaction forces are termed creep forces. For the computation of wheel–rail creep forces, this section employs the simplified Kalker's theory through the FASTSIM algorithm [21].

The track irregularity is a stochastic excitation characterized by uncertain factors such as damage and deformation in rails, sleepers, and ballast beds, which collectively constitute its random nature. The common geometric irregularities of tracks primarily manifest in four fundamental forms: alignment irregularity, gauge irregularity, vertical profile irregularity, and cross-level irregularity. These irregularities are derived from track spectra calculations, where track spectra represent statistical functions of track irregularities as stationary random processes, with variations existing among different countries. Since China has yet to establish a comprehensive standard system for track spectra across various railway types, this section employs Germany's low-disturbance spectrum to generate track irregularities.

2.5.2 Analytical Model of Train System

This section establishes a train model based on multibody system dynamics. The model adopts a double-axle four-wheelset configuration, consisting of multiple rigid bodies with 6 DOFs including vehicle bodies, bogies, and wheelsets. The vehicle body is connected to the bogies through secondary suspensions, while the bogies are linked to the wheelsets via primary suspensions. The suspension systems are simulated using linear elastic springs and viscous dampers. The train dynamic model comprises a total of 42 dynamic DOFs, as detailed in Table 2.1.

The train is modeled using the commercial multibody dynamics software Simpack, where the vehicle body, bogies, and wheelsets are treated as rigid bodies with their mass and inertia properties defined through body elements. The kinematic connections between components are defined using joints and constraints, while the DOFs for each part are explicitly defined. The primary and secondary suspension systems are modeled as massless spring-damper systems, which Simpack accurately simulates through various force element types in its multibody dynamics framework.

In the multibody dynamics software Simpack, preload analysis on a train model under self-weight conditions is performed to maintain the initial geometric configuration while achieving equilibrium. This technique involves applying nominal forces in the opposite direction of gravity within force elements to achieve train equilibrium without altering the

Table 2.1 DOFs of the train model.

Rigid Body Name	Surge	Sway	Heave	Roll	Pitch	Yaw
Vehicle body	x_c	y_c	z_c	θ_c	φ_c	ψ_c
Bogie ($i = 1, 2$)	x_{bi}	y_{bi}	z_{bi}	θ_{bi}	φ_{bi}	ψ_{bi}
Wheelset ($j = 1, 2, 3, 4$)	x_{wj}	y_{wj}	z_{wj}	θ_{wi}	φ_{wi}	ψ_{wi}

model's original geometry. The schematic diagram in Fig. 2.13 illustrates the principle of nominal force analysis in Simpack multibody dynamics. The model's validity is verified through nominal force analysis by evaluating whether the system's maximum residual acceleration is small enough [22]. A multirigid body system model with residual acceleration small enough, usually no more than 0.01 m/s^2, confirms correct component positioning and kinematic connections. When all model parameters are verified, this validates the train model's integrity. Otherwise, the main elements of the model need to be checked and modified.

2.5.3 Bridge and Track Models

The CS technical architecture primarily consists of three components: two terminal servers and a middleware client that facilitates data exchange between them. The terminal servers are composed of a superstructure subsystem and a substructure subsystem, respectively, while the intermediate section employs S-functions or M-functions as the client to enable data communication between the superstructure subsystem and the substructure subsystem.

The superstructure subsystem is modeled in the multibody dynamics software Simpack, where both the rail and girder are constructed as elastic bodies (flexible body models). The substructure system is composed of bridge piers and bearings. During earthquakes, bridge piers and bearings are particularly vulnerable to damage, whereas the girder typically remains in

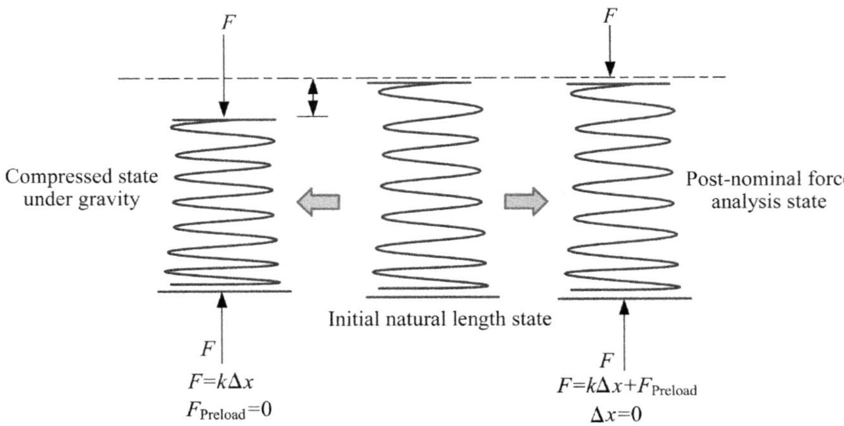

Figure 2.13 Schematic diagram of nominal force analysis in multibody systems. *F*, vehicle body gravitational force; $F_{Preload}$, nominal preload force; *k*, spring stiffness; Δx, static deflection under self-weight.

the elastic state. This section therefore, adopts a co-simulation approach combining Simpack for multibody dynamics and ANSYS for FE analysis, developing the elastic girder model in Simpack and establishing pier and bearing models in the open-source FE software OpenSees. This hybrid simulation technique leverages the respective advantages of multibody dynamics software Simpack and FE software ANSYS, reducing DOFs for enhanced computational efficiency and considering the nonlinear factors of the bridge structure for improved fidelity to real-world bridge behavior under seismic conditions.

In multibody dynamics, the position of the elastic body is determined by the gross motion of the floating coordinate and the elastic deformation of the elastic body relative to the floating coordinate. The generalized coordinates (typically Cartesian coordinates) define the gross motion of the floating reference frame, while elastic deformations are characterized by FE nodal coordinates or modal coordinates [7]. Within the multibody dynamics software Simpack, elastic deformations are resolved through the modal superposition technique. This technique introduces modal coordinates to transform the physical coordinate system into modal space, decoupling the original vibration differential equations of the system. Consequently, the decoupled vibration differential equations in modal coordinates can be solved independently.

Complex FE structural models typically require extensive node definitions, resulting in high computational DOFs. Consequently, the global stiffness and mass matrices formed after structural discretization become extremely large, thus consuming significant computational resources during structural analysis. In the multibody dynamics software Simpack, the implementation of flexible bodies involves reducing the DOFs of complex structures. By utilizing relatively few modal coordinates to describe structural motion while maintaining computational accuracy, this approach enhances computational efficiency. In this section, the classic Guyan [23] DOF reduction method is adopted to reduce the DOFs of the rail or girder structure.

The FE models of elastic rails and simply supported box girders are established using spatial girder elements in FE software ANSYS. Based on the Guyan DOF reduction method, substructure analysis is performed in ANSYS to generate a file containing the structural mass matrix, stiffness matrix, and damping matrix (.sub), a file containing node information and element information (.cdb), and a mode shape file (.tcms). Through Simpack's preprocessing program, standardized files (.fbi) representing the rail super-element and the girder super-element are generated, respectively. The workflow for generating Simpack standardized files is illustrated in Fig. 2.14.

The track model adopts a twin-block ballastless track configuration. The rails of the twin-block ballastless tracks are directly connected to the bridge deck, with sleepers fully bonded to concrete track slabs, resulting in a significant mass of the underrail foundation. Consequently, the vibration of the twin-block ballastless tracks primarily manifests in rail vibration. This section models the rail as a Euler–Bernoulli beam supported at discrete points, while incorporating other track components as participating masses in the bridge dynamic model. The rail possesses three principal vibrational DOFs: vertical, lateral, and torsional.

Under earthquake conditions, bridge girders typically remain elastic, whereas piers and bearings are prone to entering elastoplastic states. Therefore, an elastic girder model is developed in the multibody dynamics software Simpack. The cross-sectional dimensions of the prestressed box girder are shown in Fig. 2.15. To enhance computational efficiency while accounting for spatial characteristics, the girder is modeled using spatial beam elements with rigid arms. Additionally, the component mode synthesis method is employed to reduce the DOFs of the girders.

The performance of bridge piers depends on the properties of both the reinforcing steel and concrete, although these two materials exhibit significantly different characteristics. When piers enter the nonlinear phase, the accuracy of numerical simulations for reinforced concrete components is determined by the appropriate constitutive parameter settings for both materials. In this section, the seismic simulation FE software OpenSees is employed to establish a fiber-based pier model. OpenSees provides robust nonlinear analysis capabilities and an extensive library of nonlinear

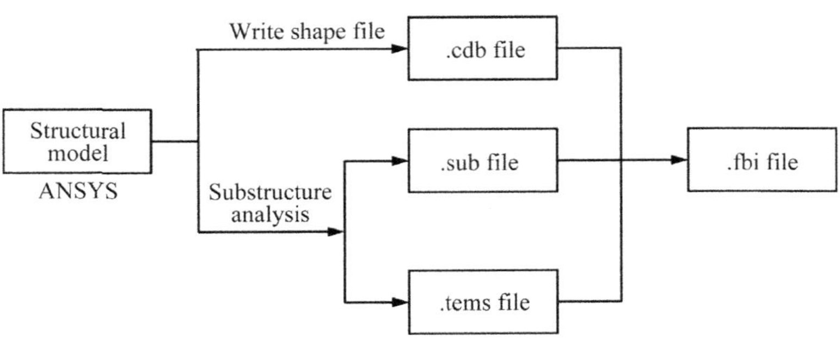

Figure 2.14 Workflow for generating Simpack standardized files.

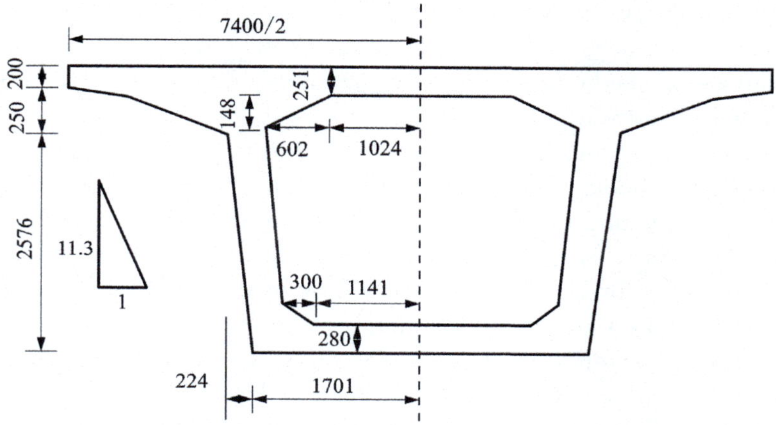

Figure 2.15 Box girder cross-sectional dimensions.

elements and material models. OpenSees enables the development of refined pier models while ensuring both accuracy and stability in structural nonlinear analysis. The fiber pier model first discretizes the pier cross-section into multiple fibers according to specific partitioning principles, and then assigns to them constitutive relationships that accurately describe the mechanical behavior of steel and concrete to establish pier models that closely reflect real-world pier conditions. The fiber pier model enables precise simulation of reinforced concrete component performance.

2.5.4 Numerical Case Studies

2.5.4.1 SOTTB Program Verification Cases

The verification of the co-simulation program for high-speed train–bridge–track coupling systems, developed through the integration of multibody dynamics software Simpack and FE software OpenSees, primarily encompasses two critical components: (1) Validation of high-speed train–track coupling computations, namely wheel–rail force calculations; (2) verification of the co-simulation methodology between the multibody dynamics software Simpack and the FE software OpenSees.

The high-speed train model in the superstructure subsystem interacts with the track model through wheel–rail forces, which are calculated by configuring the corresponding wheel–rail module in Simpack. The commercial multibody dynamics software Simpack provides accurate and reliable wheel–rail contact simulation and force computation methods. Therefore, this section confirms the correctness and reliability of the high-speed train–track coupling calculations. Consequently, the verification of the correctness of the SOTTB program primarily hinges on validating the co-simulation methodology between multibody dynamics software Simpack and FE software OpenSees.

Taking a 32 m single-span simply supported girder bridge as an example, this book validates the correctness of the co-simulation methodology integrating multibody dynamics software Simpack and FE software OpenSees by comparatively analyzing two distinct simulation methodologies for evaluating the bridge's dynamic response under both seismic excitation and vertical loading conditions. The study first establishes a simply supported girder bridge model based on a co-simulation approach; then, a complete bridge model is developed with OpenSees, where both the girder and piers are modeled with linear elastic spatial beam-column elements. The same external loads are applied to this simply supported bridge to solve for its dynamic response. The correctness of the co-simulation methodology integrating multibody dynamics software Simpack with FE analysis software OpenSees is verified by comparing the consistency of dynamic responses obtained from the simply supported bridge model.

The comparative results of acceleration responses are presented in Fig. 2.16 for the bridge midspan and Fig. 2.17 for the pier top. As shown in Figs. 2.16 and 2.17, the lateral acceleration responses at the bridge midspan and the pier top obtained from the co-simulation approach exhibit consistency with those from the OpenSees global modeling solution. The vertical acceleration responses at both bridge midspan and pier top demonstrate identical steady-state characteristics between the two

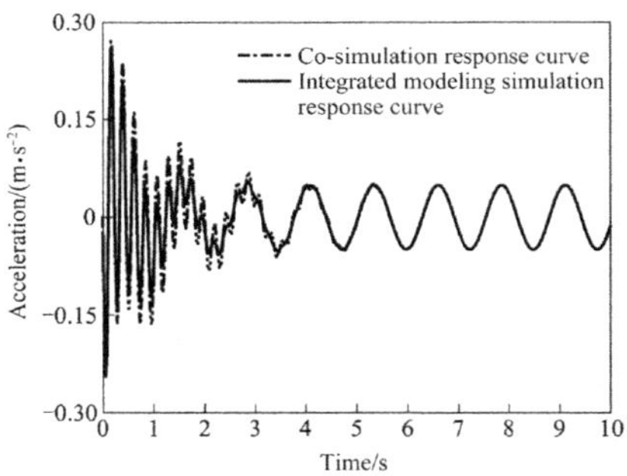

(a) Comparison of vertical acceleration time-history curves at mid-span

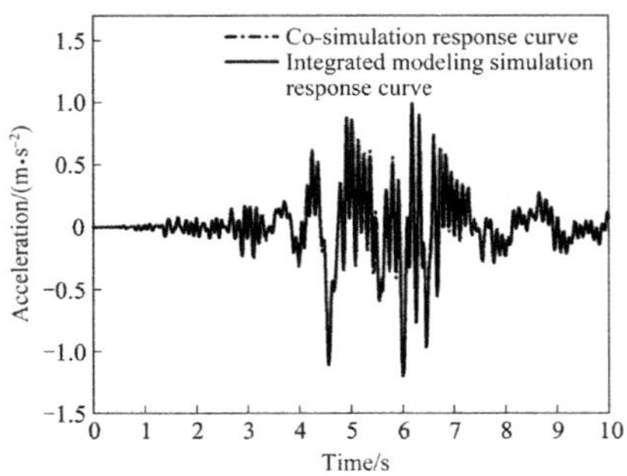

(b) Comparison of lateral acceleration time-history curves at mid-span

Figure 2.16 Midspan acceleration response comparison.

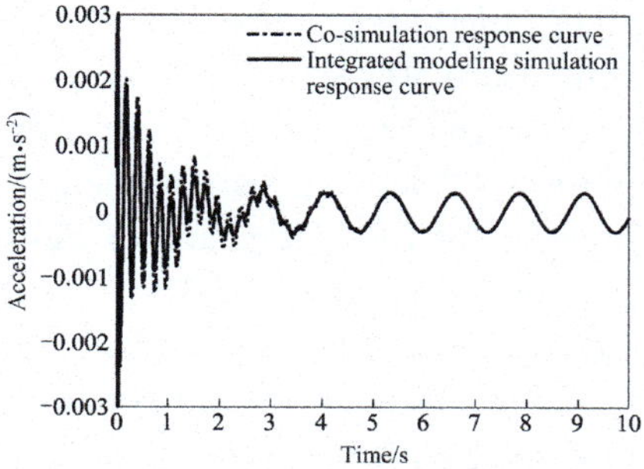

(a) Comparison of vertical acceleration time-history curves at pier top

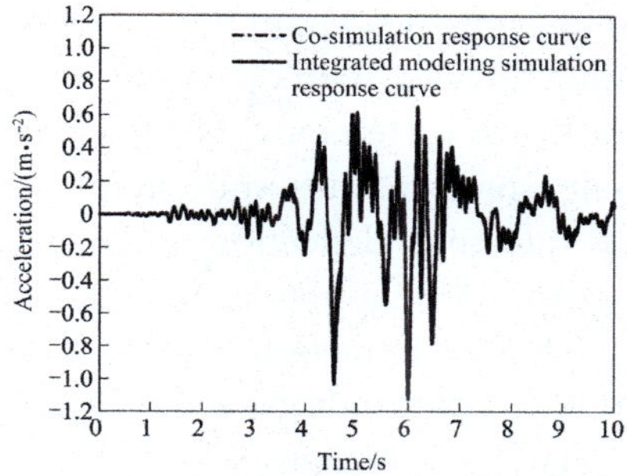

(b) Comparison of lateral acceleration time-history curves at pier top

Figure 2.17 Pier-top acceleration response comparison.

methods. These findings confirm the correctness and applicability of the co-simulation methodology integrating Simpack with OpenSees. This provides indirect verification for the correctness and applicability of the SOTTB program.

2.5.4.2 Study on Simulation of High-Speed Train–Track–Bridge Coupling Vibration for Multi-Span Simply Supported Bridge

The evaluation of dynamic performance of train–track–bridge coupling systems serves as a crucial component in related research endeavors. The dynamic performance evaluation system for train–track–bridge coupling systems primarily comprises the train operational safety evaluation, ride comfort evaluation, and evaluation of train–track dynamic interaction performance. To further validate the feasibility and correctness of the SOTTB program, this section develops a high-speed train–track–bridge coupling vibration model for a six-span simply supported girder bridge based on the SOTTB program. The vibration performance of the bridge is evaluated according to relevant specifications, while the dynamic characteristics of the train–bridge system at different speeds are investigated by analyzing train operation dynamic performance indicators, including ride comfort and operational safety.

Taking a single-track high-speed railway line as the research object, this study establishes a refined computational model of the high-speed train–bi-block ballastless track–bridge coupling system using the SOTTB program. The model systematically incorporates track vertical profile and alignment irregularities to investigate the operational stability and operational safety of high-speed trains, and bridge response characteristics under varying speeds.

The high-speed train model adopts a CRH2 EMU formation consisting of four vehicles, with operating speeds ranging from 50 km/h to 300 km/h in 50 km/h increments. The general layout diagram of the high-speed railway line (superstructure subsystem) is shown in Fig. 2.18. The variation curves of mid-span deflection and acceleration peak of the bridge under high-speed trains passing at different velocities are shown in Fig. 2.19.

The computational results indicate that the response of the bridge, train ride comfort indices, and safety indexes all increase with rising operational speeds, which aligns with the general principle; when the train operates at 200 km/h, the responses of all components within the high-speed train–track–bridge coupling system demonstrate characteristics consistent with their respective structural properties; similar response patterns are observed when the train operates at other velocities. Therefore, this study further validates the correctness and feasibility of the co-simulation approach (SOTTB program) that integrates multibody system dynamics Simpack with FE OpenSees.

2.6 Train–Track–Bridge System Simulation under Earthquake Conditions Based on the Moving Element Model

2.6.1 Moving Element Model

The moving wheel–rail element model, as illustrated in Fig. 2.20, consists of a wheelset node and a series of underlying beam element nodes that may come into contact with it, addressing time-varying contact problems through element activation and deactivation. When the wheelset moves onto the *i*-th beam element, that particular element becomes activated (referred to as the active zone), while all other beam elements remain inactive, forming the corresponding inactive zone. Within the active zone, wheel–rail interactions exist between the activated beam element and the

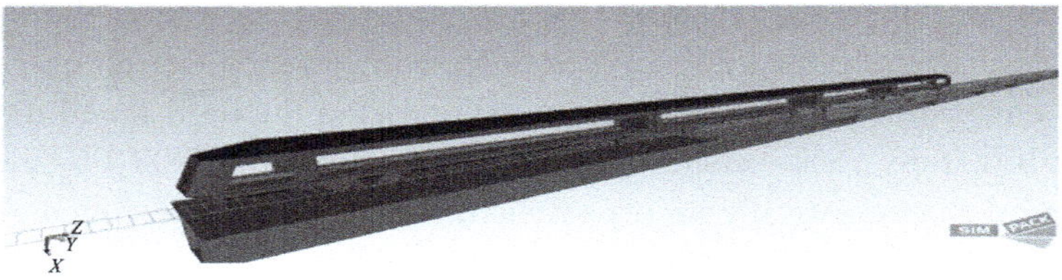

Figure 2.18 General layout diagram of the high-speed railway line (superstructure subsystem).

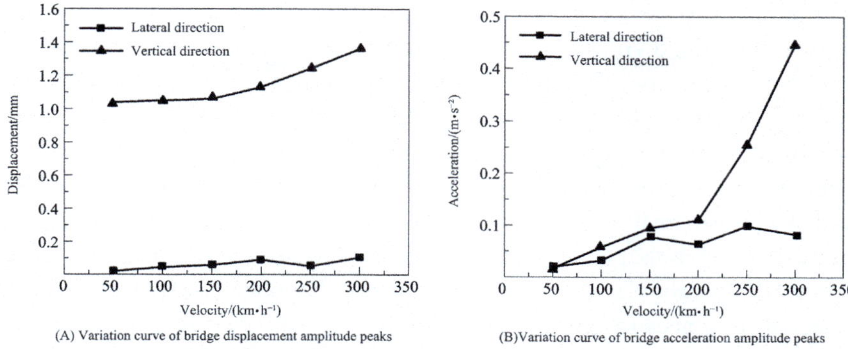

Figure 2.19 Variation curves of mid-span response peaks under different train velocities.

wheelset node. Through relevant theoretical derivations, key element characteristics such as displacement vectors, force vectors, and stiffness matrices of the moving wheel–rail element model can be obtained. This approach creates and packages the wheel–rail interaction at the element level, enabling direct application in train–bridge coupling vibration modeling and analysis just like conventional FEs.

2.6.2 Integration of the Moving Wheel–Rail Element Model into the OpenSees Platform

To overcome the shortcomings of the strong coupling and weak coupling methods, our research team proposed a moving element model by combining their advantages. Based on train–bridge dynamic theory and FE theory, we derived calculation equations for the stiffness matrices of vertical and spatial moving elements. After obtaining the stiffness matrices of the vertical and spatial moving elements, they were integrated into existing FE analysis software such as ANSYS, Simpack, and UM through programming, enabling dynamic analysis of the train–track–bridge system.

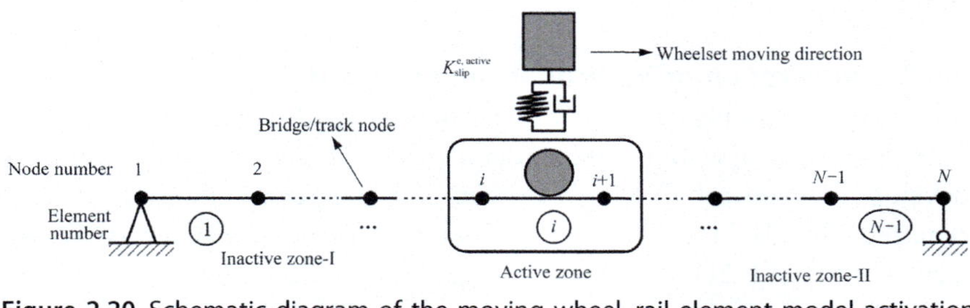

Figure 2.20 Schematic diagram of the moving wheel–rail element model activation state.

Although these commercial FE software packages provide user programming interfaces, their proprietary and non-open-source nature makes them inconvenient to use. To study the dynamic response of train–track–bridge systems under seismic excitation, where bridge and track structures may exhibit nonlinear behavior, the study in this section selects OpenSees, an open-source seismic analysis FE software, with powerful seismic analysis and nonlinear analysis functions, as the integration platform to build a simulation analysis platform suitable for the train–track–bridge coupling system. The modular program structure of OpenSees provides excellent external interfaces for element development. Without altering OpenSees' internal data structure, the moving elements can be quickly integrated into its element library, allowing direct modeling and calculation of the train–track–bridge system on the OpenSees platform.

2.6.2.1 Introduction to OpenSees Software

OpenSees is an object-oriented FE analysis software framework primarily written in C++. Object-oriented programming (OOP) features strong encapsulation, user-friendly interfaces, and inheritance, decomposing programs into objects and data organized and encapsulated through classes. The main architecture of OpenSees consists of four classes: Domain, ModelBuilder, Analysis, and Recorder, each containing numerous subclasses. These classes perform distinct functions: ModelBuilder is responsible for model construction, Analysis handles complex analyses, Recorder manages postanalysis data recording and processing, and all three are associated with Domain, which aggregates and collects all model data, as shown in Fig. 2.21.

The Element class is directly managed by the Domain class and has no direct association with other classes. During element development, data are only retrieved and exchanged from the Domain, significantly simplifying the development process. The vertical moving element (WheelRail) and spatial moving element (WheelRail3D) developed in this section are both stored under the Element class.

2.6.2.2 Integration of Vertical Moving Elements

The vertical moving element can be derived with analytical expressions [24], making its development and debugging relatively straightforward. To develop the vertical moving element in OpenSees, only two files are required: the class file WR2D.h and its implementation file WR2D.cpp. The class file WR2D.h is critical, as it defines the data used, corresponding processing functions, and external interfaces. The collaboration diagram

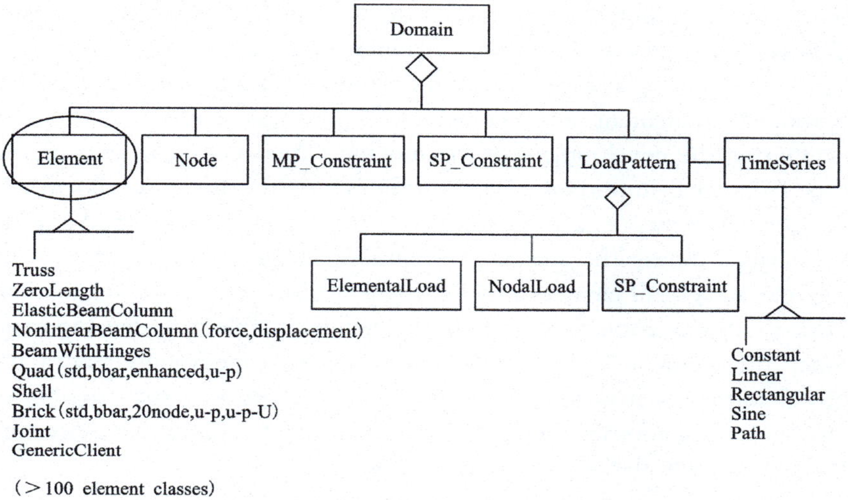

Figure 2.21 Domain architecture diagram.

between the developed vertical wheel–rail element and the main program is very complex. However, due to the encapsulation characteristics of OOP, the actual programming process does not require attention to be given to the relationships between every function, as in procedural programming, but only to the relationships between classes and their data interfaces. The inheritance hierarchy of the vertical moving wheel–rail element model WR2D is shown in Fig. 2.22. It is relatively simple and intuitive, with only three inheritance levels. The key focus is the single

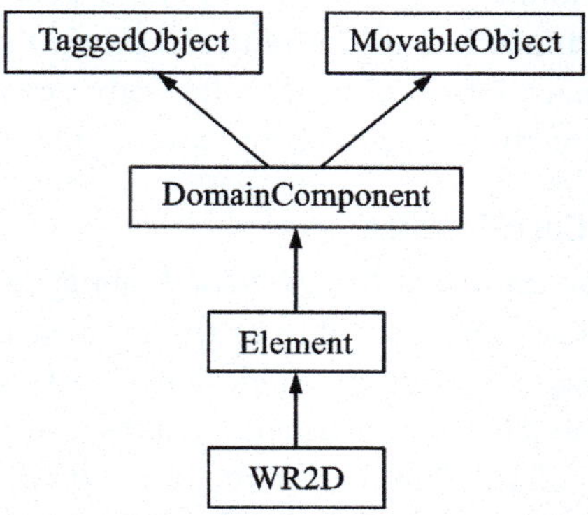

Figure 2.22 Inheritance hierarchy of the vertical moving wheel–rail element model WR2D.

inheritance relationship from the general element class to the WR2D class. This is the reason why OpenSees is employed as the integration platform, as its open-source nature and OOP framework significantly reduce the workload for element development.

During the development of the moving wheel–rail element model in OpenSees, one important consideration is that the node coordinates of ordinary elements (e.g., beam elements, truss elements) are fixed and do not change during the analysis. In contrast, the node coordinates of the moving wheel–rail element model change as the wheelset moves forward during the simulation, increasing the complexity of element development. Since the element stiffness matrix is ultimately superimposed onto the global stiffness matrix for solving the dynamic equation, the problem of changing node coordinates can be addressed by activating and deactivating the moving wheel–rail element model nodes.

2.6.2.3 Integration of Spatial Moving Elements

The integrating process and approach for the spatial moving element are similar to those for the vertical moving element. However, since the spatial moving element stiffness matrix lacks analytical expressions, it must be solved with the numerical perturbation method. Thus, the integration of the spatial moving element must account for the iteration and convergence of wheel–rail displacements and forces. The calculation of wheel–rail contact forces in spatial elements is inherently complex, with computational difficulty increasing exponentially when different theoretical approaches are employed. To enhance the computational efficiency, a lookup table method is similarly adopted for determining creep forces in the elements. It should be noted that in the spatial moving wheel–rail element model formulation, the nonlinear Hertzian contact spring with nonconformal contact relationships is inherently considered, while track irregularities are directly imported from external files.

2.6.3 Numerical Case Studies

2.6.3.1 Verification Example for Vertical Moving Elements

To verify the accuracy and reliability of the proposed moving wheel–rail element model analysis method, a single-wheel vehicle crossing a simply supported girder model was established, as shown in Fig. 2.23. All simulation parameters are labeled in the figure. The vehicle body is simulated as a sprung mass, with no consideration of damping in the train or the simply supported girder. The element length is set as 1 m, with modeling

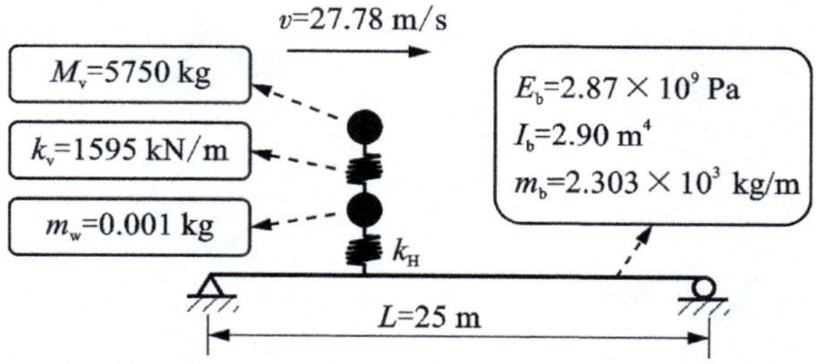

Figure 2.23 Single-wheel vehicle crossing a simply supported girder model. M_v, vehicle mass; k_v, vehicle body stiffness; m_w, wheelset mass; E_b, elastic modulus of bridge beam; I_b, moment of inertia of bridge beam; m_b, mass per unit length of bridge beam; k_H, nonlinear Hertzian spring stiffness at the wheel–rail interface.

performed in OpenSees using elastic beam-girder elements. The analysis time step is 0.005 s. The results of the moving element analysis (moving element solution) are compared with those of the CS method [25,26], Biggs' analytical solution [8], Yang's solution [12], and Bowe's solution [27], as shown in Figs. 2.24 and 2.25. For the train response, the results of the five methods are largely consistent, with only minor differences in the analytical solution. For the bridge response, the mid-span node displacement responses calculated by all methods are largely consistent. In terms of acceleration responses, significant differences are observed among computational results obtained through different methods. While all solutions exhibit consistent overall variation trends, their fluctuation intensities demonstrate notable variations. The analytical solution exhibits the least fluctuation, followed by Bowe's results. The moving element method's results are comparable in fluctuation amplitude to those of the CS and Yang methods but differ in phase and exhibit higher-frequency

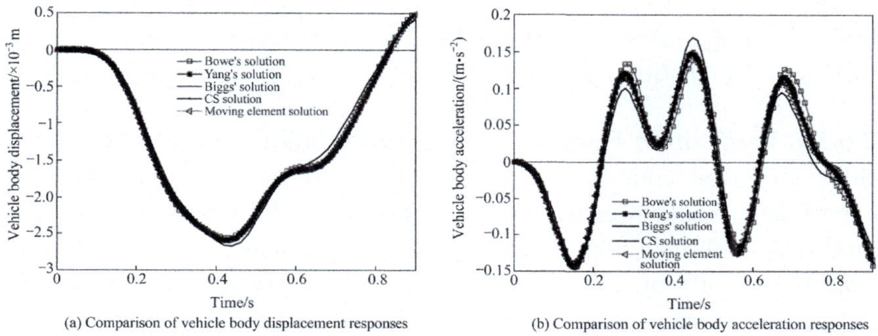

Figure 2.24 Vehicle body response comparison.

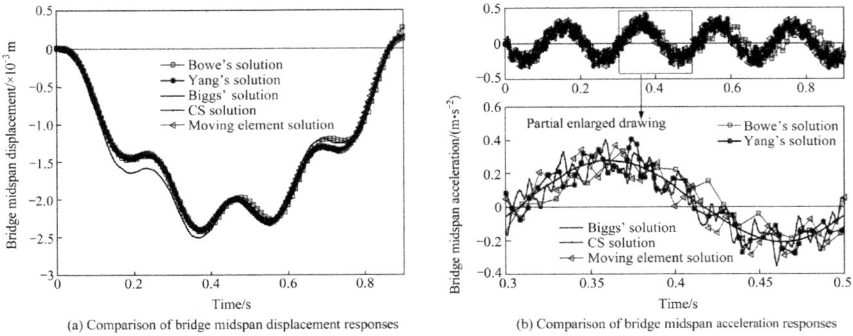

Figure 2.25 Bridge response comparison.

fluctuations. This is because Biggs' analytical solution considers only the first-order mode, neglecting higher-order modes, leading to discrepancies in the train and bridge responses compared to other simulation results. The bridge mid-span node acceleration response follows a sinusoidal trend. Both the moving element method and the CS method account for the nonlinear Hertz spring between the wheel and rail, resulting in higher fluctuation frequencies compared to Yang's method. Bowe's method uses a linear Hertz spring and a longer time step, leading to great differences. Although the moving wheel–rail element model method and the CS technical analysis method offer good computational accuracy, the CS method requires significantly more computation time than the moving element analysis method. For the same 10 span, 32 m simply supported girder model, the CS technical analysis method takes 30–60 min, while the moving wheel–rail element model method requires only about 5 min. Thus, the moving element method offers advantages in both efficiency and precision. The above analysis demonstrates that the vertical moving wheel–rail element model method effectively simulates vertical train-–bridge coupling vibration problems, accurately reflecting nonlinear contact springs and high-frequency responses.

2.6.3.2 Verification Example for the Spatial Moving Element

Since analytical solutions are unavailable for the spatial moving element, experimental validation was adopted. To verify the accuracy of the previously established OpenSees-based spatial train–track–bridge system simulation platform, the research team first constructed an experimental platform for a train–bridge interaction under earthquake conditions using the available equipment. Subsequently, the world's first scaled experimental study of train operation on bridges under earthquakes was conducted. By comparing the experimental data with the simulation results, the reliability of the OpenSees train–track–bridge system simulation platform was validated. This provides a robust tool for subsequent research on the dynamic behavior and operational safety of extra-long continuous steel truss bridges.

Based on the physical scale model for train operations on bridges and using the OpenSees train–track–bridge system simulation analysis platform built by the research team, this study established an OpenSees FE model for indoor scaled tests of high-speed train operational safety on bridges under seismic excitation. By comparing the FE model with the experimental data, the accuracy of the OpenSees train–track–bridge system simulation analysis platform was validated, demonstrating that the OpenSees simulation platform based on the moving element model can effectively simulate the train–track–bridge coupling vibration problem under earthquake conditions. This is a practical dynamic analysis approach for train–bridge coupling systems. Relevant verification details are presented in Chapter 3.

2.7 Summary

In recent years, China's high-speed railway network has developed rapidly, extending into seismically active regions such as central and western China. With a growing number of high-speed railway bridges and their growing proportion in total track length, the probability of high-speed trains operating on bridges under earthquakes has risen significantly. Consequently, it is of great practical significance to study the operational safety of high-speed trains running on bridges under earthquake conditions.

In this chapter, a series of innovative studies were carried out from the perspective of numerical simulation to address this problem, and useful conclusions were drawn as follows.

(1) Based on the development trends and typical characteristics of China's high-speed railways, this chapter highlights the necessity and significance of investigating train–track–bridge coupling vibration under earthquake conditions.

(2) It proposes the maximum normal embedding depth contact point detection theory that balances efficiency and applicability. Research results demonstrate that this method identifies more representative contact points under large lateral displacement conditions. A MATLAB-based visualization platform for spatial wheel–rail contact relationship calculation was developed, confirming the applicability and accuracy of the normal contact point detection method.

(3) For the first time, CS technology has been introduced into train–bridge coupling vibration analysis, establishing a novel numerical simulation strategy. The client and server communicate via sockets. Compared with conventional train–bridge analysis approaches, it enables direct acquisition of structural responses at the stress–strain level.

(4) To address the inefficiency in modeling train–track–bridge systems, this study proposes an OpenSees rapid simulation technique that enables coupling vibration analysis on a unified OpenSees simulation platform. The research results demonstrate that the fast simulation technique provides a practical, convenient, and easy-to-learn train–bridge vibration analysis method.

(5) This chapter demonstrates the correctness and feasibility of the co-simulation approach integrating the multibody dynamics software Simpack with OpenSees. Furthermore, it compares the seismic response and operational safety indices of the high-speed train–ballastless track–bridge coupling system under seismic and nonseismic conditions. In addition, the chapter analyzes the influence of the train speed and seismic load on the dynamic response of the coupling system and the operational safety of high-speed trains.

(6) It also presents a novel moving wheel–rail element model that is suitable for train–bridge coupling vibration analysis. Additionally, a new 2D wheel–rail coupling element based on nonlinear contact relationships has been developed, and the source code has been made publicly available.

(7) The integration methods for moving elements are described in both vertical and spatial dimensions based on the OpenSees integration platform. This chapter validates the accuracy of the vertical moving wheel–rail element model, demonstrating the flexibility, nonlinear analysis capability, and seismic performance of the vertical moving element.

References

[1] Liu H, Yu Z, Guo W, et al. A novel method to search for the wheel–rail contact point. Int J Struct Stabil Dynam 2019;19(11). 1950142.

[2] TB 10621-2014. Code for design of high-speed railway. Beijing: China Railway Publishing House; 2014 [In Chinese].

[3] Zhai WM. Vehicle-track coupled dynamics. 4th ed. Beijing: Science Press; 2015 [In Chinese].

[4] Liu HY, Yu ZW, Guo W, et al. 3D wheel-rail contact visualization & computing platform V1.0 China: 2019SR0087958. Beijing: National Copyright Administration; 2019-01-24 [In Chinese].

[5] Gao H, Dai HY, Ni PT. Algorithm of wheel-rail contact point for flexible wheelset. J China Railw Soc 2012;34(5):26–31 [In Chinese].
[6] Zeng YQ, Wen B, Yu WD, et al. Calculation and research on wheel-rail 3D contact geometry based on projection contour. China Railw Sci 2012;(6):51–9 [In Chinese].
[7] Cui SA. Refined simulation research on train–bridge coupling vibration based on multibody system dynamics and finite element method. Chengdu: Southwest Jiaotong University; 2009 [In Chinese].
[8] Biggs JM. Introduction to structural dynamics. New York: McGraw-Hill College; 1964.
[9] Fryba L. Vibration of solids and structures under moving loads. Netherlands: Springer; 1972.
[10] Zhai WM. Vehicle-track coupled dynamics. 3rd ed. Beijing: Science Press; 2007 [In Chinese].
[11] Yang YB, Yau JD. Vehicle-bridge interaction element for dynamic analysis. J Struct Eng 1997;123(11):1512–8.
[12] Yang YB, Wu YS. A versatile element for analyzing vehicle-bridge interaction response. Eng Struct 2001;23(5):452–69.
[13] Lou P, Yu ZW, Au FT. Rail-bridge coupling element of unequal lengths for analyzing train-track-bridge interaction systems. Appl Math Model 2012;36(4):1395–414.
[14] Lou P, Zeng QY. Formulation of equations of vertical motion for vehicle-track-bridge system. J China Railw Soc 2004;26(5):71–80.
[15] McKenna FT. Object-oriented finite element programming frameworks for analysis, algorithms and parallel computing. Berkeley: University of California; 1997.
[16] OpenSEES official website. 2020. http://OpenSEES.berkeley.edu/wiki/index.php/Modeling_Commands.
[17] Gu Q, Ozcelik O. Integrating OpenSees with other software-with application to coupling problems in civil engineering. Struct Eng Mech 2011;40(1):85–103.
[18] Peng P, Tomofuji H, Wang T, et al. Development of peer-to-peer (P2P) internet online hybrid test system. Earthq Eng Struct Dynam 2006;35(7):867–90.
[19] Guo W, Li JL, Liu HY. The analysis of the running safety of high-speed trains on a bridge by using refined simulation considering a strong earthquake. Eng Mech 2018;35(S1):259–77 [In Chinese].
[20] Wang KW. The track of wheel contact points and the calculation of wheel/rail geometric contact parameters. J Southwest Jiaot Univ 1984;(1):89–99 [In Chinese].
[21] Kalker JJ. A fast algorithm for the simplified theory of rolling contact. Veh Syst Dyn 1982;11:1–13.
[22] Miao BR, Xiao SN, Jin DC. Research on modeling and analysis of a complex multibody system by using Simpack. Mech Sci Technol 2006;25(7):813–6 [In Chinese].
[23] Guyan RJ. Reduction of stiffness and matrices. AIAA 1965;3:380.
[24] Liu HY. Dynamic response and operational safety evaluation of high-speed railway long-span continuous steel truss bridges under seismic loads based on the moving element model. Changsha: Central South University; 2019 [In Chinese].
[25] Liu HY, Yu ZW, Guo W. Vertical vibration analysis of vehicle-bridge coupling system based on OpenSees. In: Proceedings of the 26th National Conference on structural Engineering (Volume II), Engineering Mechanics; 2017. p. 470–4 [In Chinese].
[26] Liu HY, Yu ZW, Guo W. A fast modeling technique for the vertical train-track-bridge interactions. Shock Vib 2019:1–14.
[27] Bowe CJ, Mullarkey TP. wheel-rail contact elements incorporating irregularities. Adv Eng Software 2005;36(11–12):827–37.

CHAPTER 3

Physical Experiment Simulation Methods for High-Speed Railway Train-Track-Bridge Systems Under Earthquake Conditions

Contents

High-Speed Railway Train–Track–Bridge Systems
ISBN 978-0-443-49310-2
https://doi.org/10.1016/B978-0-443-49310-2.00007-7

3.1 Overview

Research on theoretical analysis and numerical simulation of high-speed railway operational safety under earthquake conditions remains relatively scarce compared to experimental and on-site seismic damage investigations of high-speed train rolling stock, bridges, tunnels, and other infrastructure. The lack of field measurement data has resulted in insufficient validation for theoretical and numerical simulation studies. Meanwhile, due to the unpredictability of seismic events and the challenges associated with field testing, it is difficult to plan when and how to obtain operational data under earthquake conditions through actual measurements [1]. Therefore, the experimental system for high-speed train operations on bridges in the laboratory is of great value.

Seismic simulation tests are constrained by the limitations of shake table facilities, typically requiring scale model tests. However, contemporary large-scale train operational simulations predominantly employ full-scale testing, creating a methodological contradiction. For simulation tests for high-speed train operational safety on bridges under earthquake conditions, scaled model testing is imperative, which imposes stringent requirements on train operation simulation. This necessitates research on similitude theory and experimental systems to evaluate and implement such testing. The critical challenge of such tests lies in leveraging existing shake table facilities to simultaneously achieve structural damage simulation for civil engineering components and operational safety simulation for trains within a newly constructed experimental system [2]. Our research team has established an integrated similitude design framework for concurrent simulations of seismic damage and train operations, with its reliability theoretically demonstrated and its feasibility systematically evaluated through the experimental system. Utilizing existing shake table facilities, we have completed such large-scale tests, with numerical simulations validating the reliability of the similitude design. This work culminates in the development of comprehensive methodologies and large-scale experimental systems capable of executing such tests, thereby laying the technical foundation for establishing model test systems to simulate high-speed train operational safety on bridges under earthquake conditions.

In fact, shake table tests that can simulate earthquakes have been carried out for decades, with their similitude design principles and theories now relatively mature and widely accepted in the industry. In contrast, train operation tests remain comparatively scarce, mostly through prototype testing and field operational trials [3]. The model test system for simulating high-speed train operational safety on bridges under earthquake conditions faces significant challenges in both similitude theory and experimental equipment. To address these challenges, our research team has developed a dynamic simulation test system for simulating high-speed train operational safety on bridges under earthquake conditions. The system's key aspects are detailed as follows.

(1) Similitude theory and design

The core of this research lies in the feasibility and rational design of scale model testing for train operations, with particular emphasis on identifying which experimental data can reflect prototype characteristics. Furthermore, achieving similarity in wheel–rail force simulation constitutes a critical technical challenge. This chapter will address these aspects through theoretical derivation, establishing the similitude theory and design methodology for bridge–train system model testing under earthquake conditions.

(2) Shake table equipment performance

The model test system for simulating high-speed train operational safety on bridges under earthquake conditions is constructed based on existing shake table equipment. To ensure that the shake table meets the specific experimental requirements, testing and evaluation must be conducted. In particular, the impact effects on the shake table under extreme conditions, such as train derailment, must be thoroughly evaluated for feasibility to ensure test viability and to prevent accidental equipment damage.

(3) Building of the test system

After completing the similitude design of the bridge train operation test system under earthquake conditions and verifying that the shake table equipment meets performance requirements, the next step involves designing and constructing an experimental system capable of conducting train operation tests under earthquake conditions. Some technical details in the setup construction, such as train speed control and safety measures, were resolved. Subsequently, a laboratory implementation method has been proposed to provide practical solutions for system construction.

The research team designed the test vehicle model and bridge model using the scaled-down theory, and built a high-speed train bridge operation test platform using the shake table array system at Central South University (Fig. 3.1). For the first time, the team conducted seismic-induced train operation tests on bridges under various working conditions. The rail irregularity was measured using the deflection excitation

Figure 3.1 Shake table array system at Central South University.

method, and multiple testing devices were employed for data acquisition, yielding reliable experimental results. Finally, the simulation platform was compared and verified based on the experimental data, and the comparison results fully demonstrated the correctness, applicability, and accuracy of the constructed simulation platform. This shake table test represents the world's first indoor safety experiment of high-speed trains operating on bridges under earthquake conditions, providing a valuable reference for similar studies.

3.2 Similitude Design of a High-Speed Railway Train-Track-Bridge System for Scale Model Testing

3.2.1 Similitude Design Theory

Model tests are relative to prototype tests. For large-scale structural prototypes that cannot be tested at full scale due to cost or equipment limitations in engineering, model tests are typically adopted. A well-designed model test can comprehensively reflect the key parametric characteristics of the prototype structure. For physical-mechanical models, different model design control parameters are selected based on the test objectives. Similitude condition design methods include the equation method and the dimensional analysis method. The equation method requires a clear understanding of the entire physical process to establish similarity coefficient relationships. The dimensional analysis method has broader applicability, as it does not require explicit control equations of the physical process; similarity relationships and coefficients can be derived through dimensional transformation to establish a parameter system adhering to similitude laws. Given the complexity of the seismic train-bridge interaction model tests, this book adopts the dimensional analysis method for its simplicity and ability to capture primary response laws and patterns.

In a force–time–length dimensional system, similarity relationships are typically determined by three parameters, from which other parameters can be derived. Thus, model design essentially revolves around selecting and designing these three key parameters. For example, for buildings and bridge structures under earthquake conditions, three parameters (mass, acceleration, and length) are usually selected. For soil and soil liquefaction tests, the time scale ratio is prioritized, with time listed as a design parameter. For train-bridge interaction tests under earthquake conditions, the critical parameters are geometric length, elastic modulus, and acceleration.

This book proposes a fractional similitude design method for a high-speed railway train operational safety model test under earthquake conditions and the ABC Theory of the operation system suitable for train-bridge coupling vibration tests, as shown in Fig. 3.2.

A critical aspect of scale model tests for train operational safety on bridges under earthquake conditions is whether the equivalence between the physical scale model and the full-scale physical models holds. Due to the complexity of train-bridge coupling systems, it is generally recognized that the dynamic performance of full-scale prototypes cannot be directly derived from physical scale models (experimental models) through similarity transformation. Unlike conventional seismic testing of structures, shake table model tests cannot be directly applied to evaluate the seismic performance of prototype structures. However, from an alternative perspective, if the high-speed train-bridge system under seismic excitation is treated as the prototype itself in experimental tests, the test directly captures the prototype's dynamic performance. If the dynamic performance of the test prototype can be well simulated through a numerical simulation techniques, this demonstrates the reliability and accuracy of the simulation techniques. Consequently, applying the same simulation technique to simulate real-world train-bridge coupling systems in actual railway operations would also exhibit comparable reliability and accuracy. In other

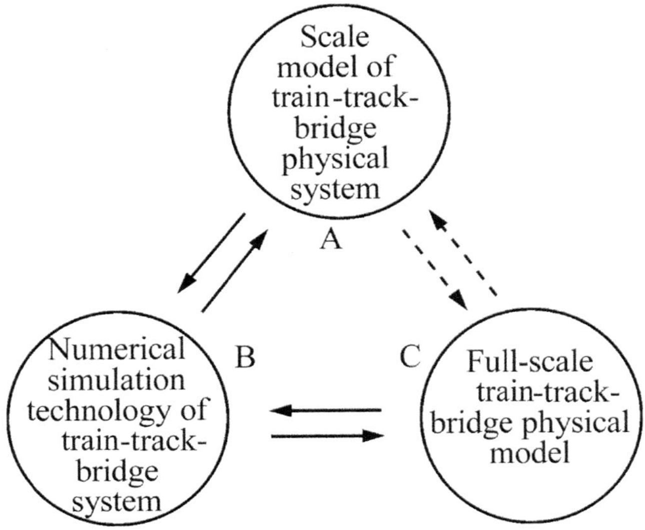

Figure 3.2 ABC Theory of a train operating system. A, scale train-track-bridge physical model; B, train-track-bridge numerical model (simulation technology); C, full-scale train-track-bridge physical model.

words, this technique bypasses direct equivalence between the scale model and the full-scale physical model. Instead, it utilizes a scale train-track-bridge physical model to validate the train-track-bridge numerical model (simulation technology), and then employs the validated numerical model to further validate the full-scale train-track-bridge physical model. This establishes an ABC logical relationship from A to B and B to C between the physical scale model, numerical model (simulation technology), and full-scale physical model in the train-bridge coupling system under earthquake conditions, termed the "ABC theory" for train operational systems. By applying the ABC theory, the dynamic performance of full-scale train-track-bridge systems can be indirectly investigated through physical scale models, thereby achieving indirect transformation between the two systems.

3.2.2 Similitude Design Parameters

Based on the analysis provided by scholars such as Wang Yanhua et al., shake table tests in the field of civil engineering typically involve scale models. While avoiding overly small scales, shake table tests can accurately reproduce actual seismic damage [4]. For model testing for simulating high-speed train operational safety on bridges under earthquake conditions, the feasibility of the test hinges on two aspects: first, controlling key design parameters to achieve a small-scale model while ensuring its laboratory practicality; and second, guaranteeing that the wheel–rail interactions reproduced under laboratory conditions strictly adhere to similarity laws. The former requires the scaling ratio to be as small as possible, while the latter demands the scaling ratio to be sufficiently large, making the precise selection of appropriate similarity ratio parameters critically important.

Based on the basic parameters of the existing shake table equipment, the train and bridge prototypes, and other experimental conditions in the High-Speed Railway Multi-Functional Shake Table Laboratory of Central South University, the shake table test model developed according to the similitude design parameters listed in Table 3.1 is illustrated in Fig. 3.3. Important similarity ratio parameters are analyzed below.

(1) Geometric similarity ratio

The time taken for a high-speed train traveling at its minimum speed of 250 km/h to pass through a typical 11 m × 32 m simply supported girder bridge on high-speed railways is approximately 0.4608 s. However, the strong-motion duration of an actual earthquake is typically 10–30 s. Therefore, to monitor the entire process of seismic interactions between the train and bridge and assess derailment risks, it would be necessary to simulate approximately 22 spans of simply supported girder bridges. If the train decelerates during an earthquake, the travel distance would be slightly reduced—temporarily assumed to be half (i.e., 11 spans of 32 m simply supported gilder bridges). To design a testing system for simulating train operations on bridges under earthquake conditions, the first step is to construct a bridge structural model of this length. Considering the actual performance of the shake table array system (4 × 4 m dimensions, 30*t* load capacity) and the 55-m channel spacing, the geometric similarity ratio for the experimental system should be 1∶35-1∶7. However, to ensure that the length of the bridge–train model under seismic conditions retains the critical dimensions of the relevant equipment, the geometric similarity ratio is selected as 1 10.

(2) Acceleration similarity ratio

In order to reduce the weight of the model in conventional shake table tests for civil engineering structures, the acceleration similarity ratio is often increased. While this approach introduces some degree of gravity distortion, it remains generally acceptable. However, for vehicle–bridge interaction tests under seismic excitation, the impact of gravity distortion on operational safety is significantly more critical than its effect on the seismic

Table 3.1 Similitude design parameters for shake table tests of high-speed railway continuous girder bridges.

Physical Properties	Physical Parameters	Similarity Constant	Notes
Geometric properties	Length s^l	1/10	High-precision requirements for rail and model fabrication (0.1 mm tolerance level)
Physical properties	Strain s^ε	1	
	Equivalent elastic modulus s^E	1/2	Available materials determine the similarity ratio range of 0.3-1.0
	Equivalent stress s^σ	1/2	
	Mass density s^ρ	5	Counterweighting is applied to primary structural members, excluding the track and wheels themselves, due to fabrication constraints
	Mass s^m	1/200	Considering the single-span bridge mass of approximately 2000 *t* of the prototype, the model mass of 15 *t* (single-span plus half-span) satisfies the shake table performance requirements
Load capacity	Concentrated force s^F	1/200	
	Line load s^q	1/20	
	Surface load s^p	1/2	
	Moment s^M	1/2000	
Dynamic performance	Damping coefficient s^c	$\sqrt{10}/200$	
	Period s^T	$1/\sqrt{10}$	
	Frequency s^f	$\sqrt{10}$	The shake table's frequency range of ≤50 Hz enables reproduction of the 10-20-Hz frequency components in the prototype ground motion
	Velocity s^v	$1/\sqrt{10}$	Laboratory testing requires the capability to achieve a speed of 20 m/s
	Acceleration s^a	1	Control test

Figure 3.3 Shake table test model of high-speed railway continuous girder bridge.

performance of purely civil engineering structures. Therefore, the acceleration similarity ratio must be maintained at 1:1 to eliminate gravity distortion and more accurately simulate operational safety conditions.

(3) Material property similarity ratio

In practical model tests, the structural materials generally consist of multiple components, leading to nonuniformity in the stress similarity constant. Even for single-material structures (e.g., steel buildings), it is challenging to achieve similarity in the full stress–strain curve between the model and prototype materials. Consequently, a trial-and-adjustment strategy must be adopted during actual design. Considering the material proportioning of the concrete, steel reinforcement, and rails system for train operations on bridges under earthquake conditions and the preliminary selection of brass as the model material, the material property similarity ratio (elastic modulus scaling ratio) is set at 1:2. Since strength and elastic modulus are scaled synchronously in similarity design, special attention should be paid to the selection of model materials.

(4) Track fabrication precision

As mentioned earlier, the model test adopted a geometric similarity ratio of 1 10. Under this scaling factor, the dimensions of track irregularities are proportionally reduced. Considering that short-wave track irregularities in high-speed railways are typically controlled within 3 mm, the scale model equivalent is 0.3 mm. The fabrication tolerance must be reduced by one order of magnitude, requiring the fabrication accuracy of short-wave track irregularities to reach the 0.03 mm level. Achieving such microscale precision for track irregularities is challenging, yet it remains feasible with CNC machining technology. For long-wave track irregularities at the 10-mm magnitude level, the required fabrication precision of 0.1 mm is relatively easier to achieve. As track irregularities serve as one of the primary excitation sources in the system, their fabrication accuracy significantly impacts the test results. Precision machining of track profiles represents one of the challenges in model testing. To achieve such high-precision reproduction of track irregularities, the experimental setup requires connecting four platforms through a high-rigidity chassis, upon which bridge and track structures are fabricated to conduct seismic performance tests of train operations on bridges under earthquake conditions.

(5) Wheel–rail relationship

The simulation of wheel–rail interaction constitutes the critical focus in the train-bridge coupling system model test. The current simulation strategy for wheel-rail interaction preserves the actual wheel tread and rail profile geometries without modification. Both model rails and wheelsets employ prototype metallic materials, thereby maintaining identical wheel-rail contact characteristics by default. To maintain identical wheel-rail contact profiles, the scaling ratio must not be excessively reduced while ensuring sufficient manufacturing precision. In the wheel-rail contact geometry relationship, the rail head profile dimensions are typically smaller than those of the wheel tread. For instance, in China's high-speed railway systems, the standard CHN 60 rail features a 73-mm head width, while the matching LMA-type wheel tread measures 135 mm, indicating the rail profile plays the decisive role. Under a 1 10 geometric scaling ratio, the scaled rail head dimension becomes 7.3 mm, which can be readily achieved with current CNC machining capabilities. Consequently, this configuration ensures the wheel–rail interaction fully meets experimental requirements while maintaining geometric similarity.

(6) Load-bearing capacity

For a 1 10 scale model without considering density variations, the weight scales to $(1/10)^2$ magnitude. According to the Post-tensioned Concrete Simply Supported Box Girder for 350 km/h Ballastless Track Passenger Dedicated Lines (Double-track) (Tongqiao [2008] 2322A-II), a 32-m simply supported box girder has a self-mass of 836.8 t, with the secondary dead load calculated at 1.4 t/m. This yields approximately 881.6 t per single-span girder, resulting in a total mass of 9697.6 t for an 11-span bridge. After scaling, the model's total mass becomes 97.0 t. Given that the four-platform test system has a total load-bearing capacity of 120 t, the surplus capacity during actual testing can accommodate auxiliary equipment weight, thereby fully meeting the experimental system's load-bearing requirements.

(7) Test speed

Under a geometric scale ratio of 1 10 and an acceleration scale ratio of 1 1, the velocity scale ratio is determined to be $1/\sqrt{10}$ through dimensional analysis theory. At this time, the prototype train's operational speed of 250 km/h corresponds to 21.96 m/s in the scale test model. Considering the scaled single-vehicle model weighs approximately 320 kg, the minimum motor power required to accelerate it to the target test speed is calculated as 78 kW. Commercially available motors meeting this specification can be sourced, confirming the laboratory's capability to fulfill the velocity requirements for train operational tests on bridges under earthquake conditions.

(8) Determination of other scaling ratios

After the geometric scaling ratio, material property ratio, and acceleration similarity ratio are established, all other scaling ratios can be derived from these three ratios. Under the conditions of a geometric scaling ratio of 1 10, material property ratio of 1:2, and acceleration similarity ratio of 1 1, the corresponding model similitude design can be derived as shown in Table 3.1. It should be noted that in the similarity design from prototype vehicle to scaled model, the dual-suspension structure must remain unchanged while strictly maintaining dynamic performance similarity.

3.3 Architecture of an Experimental System for High-Speed Train Operation on Bridges

3.3.1 Model Vehicle Design

The design of the model vehicle is particularly crucial throughout the entire testing process. In similitude design, the primary focus is on ensuring dynamic performance similarity of the scale-model vehicle. The design of the test model vehicle is as follows.

The prototype vehicle model is CRH 380A [5], and the model vehicle adopts a 1 10 geometric similarity ratio for similitude design, with other similitude design parameters shown in Table 3.1. Given the numerous dynamic parameters of the prototype vehicle, dynamic parameters in various directions may not simultaneously meet requirements for similitude design. To address conflicting dynamic parameters, the similitude design should prioritize similarity in vertical (z) dynamic parameters, followed by lateral (y) parameters, and then longitudinal (x) parameters. Spring parameters take precedence over damping parameters. The geometric dimensions of the bogie of the scaled model after design are illustrated in Fig. 3.4, while the dynamic parameters of the CRH 380A scale model vehicle are listed in Table 3.2.

During similitude design, air springs are simplified by replacing them with rubber springs while ensuring equivalent initial stiffness under full load conditions as much as possible. During model fabrication, C45 steel plates and Q235 steel materials are employed. Bolts are procured from commercial suppliers as required, while springs and dampers are custom-manufactured. Wheelsets are directly cut from decommissioned high-speed railway wheelsets to maintain material consistency. The completed model vehicle achieves a total mass of 108 kg. During model vehicle fabrication, the machining accuracy must be maintained at the 0.1-mm tolerance level. The completed bogie assembly should demonstrate reliable performance, withstanding repeated testing cycles without structural compromise. Before and after testing, all four wheelsets must remain grounded in a static state to prevent single or multiple wheelsets from lifting off. This ensures that no significant distortion or other deformation occurs in the wheelsets or bogies during the test.

3.3.2 Equivalent Design of a Bridge Model

In this experiment, for structural components such as girders, track slabs, and base plates, concrete is replaced with steel structures following the design principle of equivalent bending stiffness; the Cement Asphalt (CA) mortar layer between the track slab and base plate, as well as the two-geotextile–one-membrane composite layer between the base plate and the beam, is designed to achieve equivalent frictional behavior.

The fundamental principles for design and fabrication of the scale model of the track–bridge system are as follows:

① During train operational tests, the bridge girder must remain in the elastic working state without local torsional buckling or other deformations, demonstrating sufficient stiffness.

② The rail cross-section must strictly maintain geometric similarity and material similarity.

Based on the above principles, the track-bridge system of a 32m simply supported bridge of the prototype structure is scaled down for testing. The layout of the scale track-bridge model, as shown in Fig. 3.5, includes the girder, track slab, base plate, rail-bearing platform, CA mortar layer, sliding

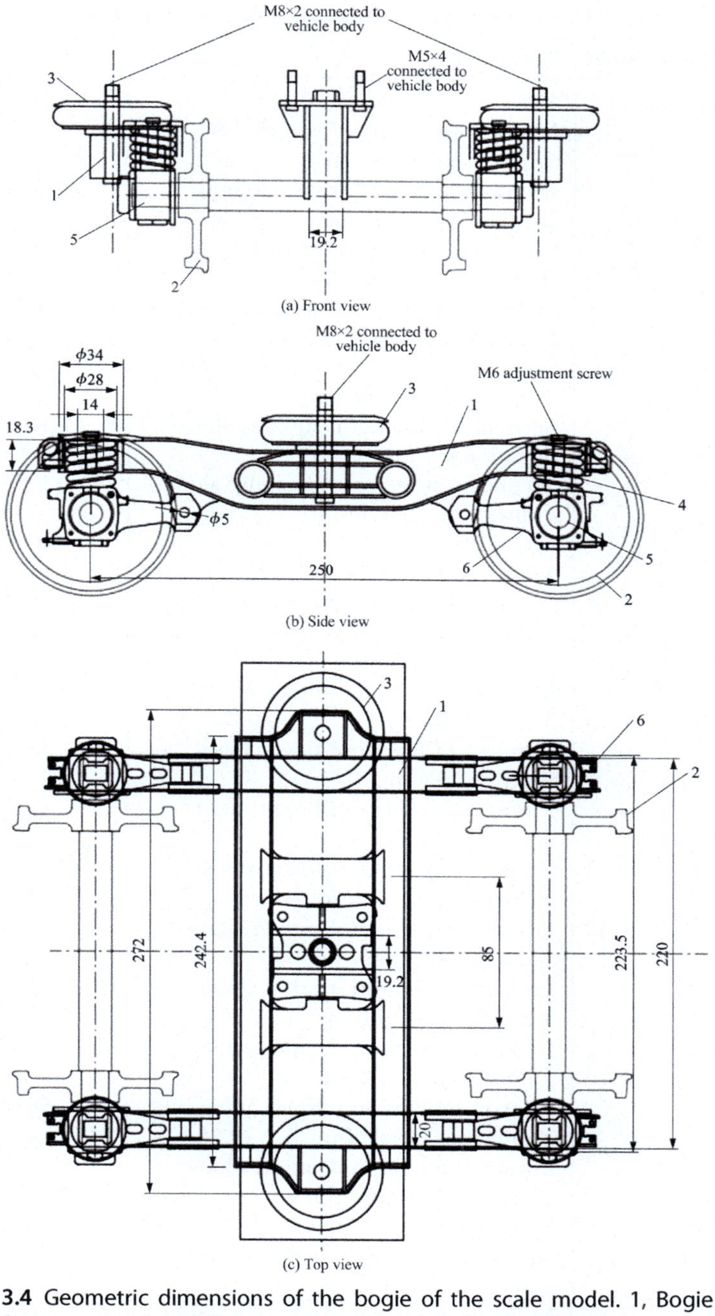

Figure 3.4 Geometric dimensions of the bogie of the scale model. 1, Bogie frame; 2, wheelset; 3, air spring; 4, frame spring; 5, vertical damper; 6, axle box arm.

Table 3.2 Train dynamics parameters of the CRH 380A scale model.

Component name	Parameter	Value
Vehicle body	Vehicle body mass m_c (kg)	219.3
	Moment of inertia of vehicle body about x-axis J_{cx} (kg·m^2)	5.47
	Moment of inertia of vehicle body about y-axis J_{cy} (kg·m^2)	82.7
	Moment of inertia of vehicle body about z-axis J_{cz} (kg·m^2)	78.05
	Secondary suspension stiffness along x-direction k_{2x} (kN·m^{-1})	8.00
	Secondary suspension damping along x-direction c_{2x} (kN·s·m^{-1})	0.08
	Secondary suspension stiffness along y-direction k_{2y} (kN·m^{-1})	8.00
	Secondary suspension damping along y-direction c_{2y} (kN·s·m^{-1})	0.93
	Secondary suspension stiffness along z-direction k_{2z} (kN·m^{-1})	9.50
	Secondary suspension damping along z-direction c_{2z} (kN·s·m^{-1})	0.63
Bogie	Bogie mass m_b (kg)	12.0
	Moment of inertia of bogie about x-axis J_{bx} (kg·m^2)	0.10
	Moment of inertia of bogie about y-axis J_{by} (kg·m^2)	0.07
	Moment of inertia of bogie about z-axis J_{bz} (kg·m^2)	0.12
	Primary suspension stiffness along x-direction k_{1x} (kN·m^{-1})	734.00
	Primary suspension damping along x-direction c_{1x} (kN·s·m^{-1})	0
	Primary suspension stiffness along y-direction k_{1y} (kN·m^{-1})	323.50
	Primary suspension damping along y-direction c_{1y} (kN·s·m^{-1})	0
	Primary suspension stiffness along z-direction k_{1z} (kN·m^{-1})	58.80
	Primary suspension damping along z-direction c_{1z} (kN·s·m^{-1})	0.10
Wheelset	Wheelset mass m_ω (kg)	12.0
	Moment of inertia of wheelset about x-axis $J_{\omega x}$ (kg·m^2)	0.10
	Moment of inertia of wheelset about y-axis $J_{\omega y}$ (kg·m^2)	0.07
	Moment of inertia of wheelset about z-axis $J_{\omega z}$ (kg·m^2)	0.12

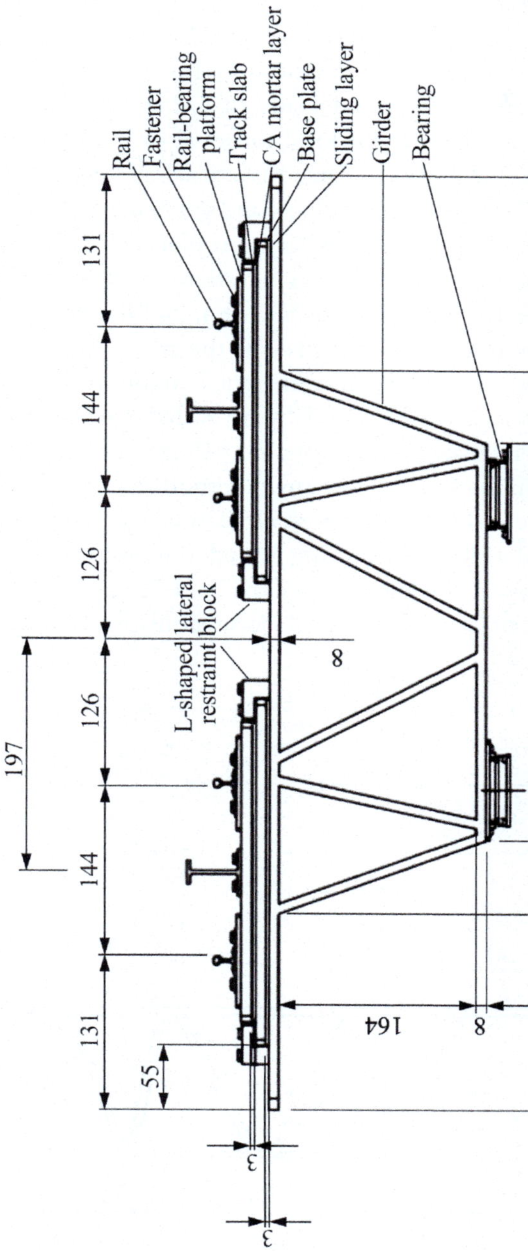

Figure 3.5 Structural layout of track-bridge model.

layer, rail, fastener, and bearing. The dimensions and materials of each component are described below.

The scaled single-span simply supported girder has a length of 3260 mm, with cross-sectional dimensions as shown in Fig. 3.6. Stiffening support plates, all 8mm thick and made of Q235 steel, are provided at intermediate segments to prevent local torsional or buckling deformation of the girder deck and bottom plate. In actual high-speed railway engineering, piers exhibit significant stiffness and generally remain undamaged under seismic loads [6,7], particularly for short piers. Therefore, the piers in this experiment are rigidly designed, as illustrated in Fig. 3.7. Steel plates with a thickness of 10 mm, width of 500 mm, and length of 800 mm are welded to the top and bottom surfaces of the piers. The cross-sectional dimensions are designed to ensure sufficient bending stiffness, and Q235 steel is used as the material. Track slabs are scaled down with equivalent flexural stiffness. The scaled dimensions are 645 mm in length, 255 mm in width, and 9 mm in thickness. They are longitudinally connected to ensure continuity and are made of Q235 steel. Base plates are designed with equivalent flexural stiffness, featuring scaled dimensions of 295 mm in width and 8.5 mm in thickness. They are laid longitudinally to ensure continuity and are made of Q235 steel. After similarity-based scaling, rail-

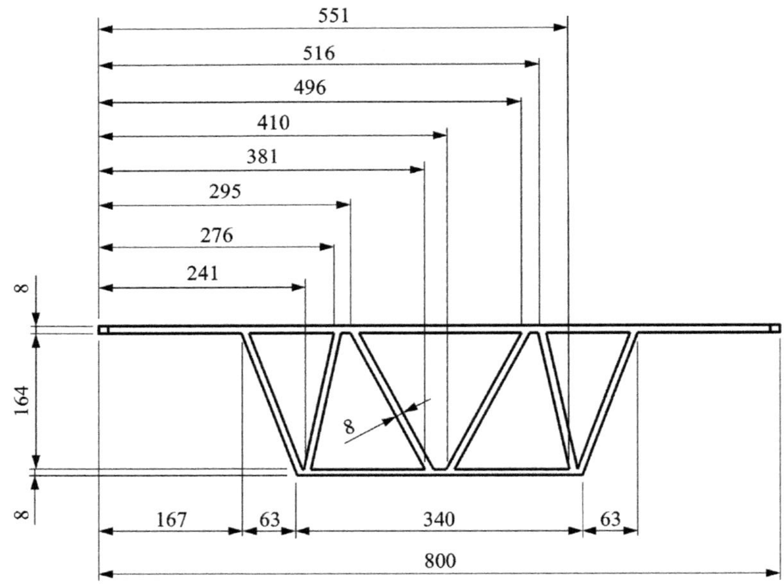

Figure 3.6 Cross-sectional dimensions of bridge structural model.

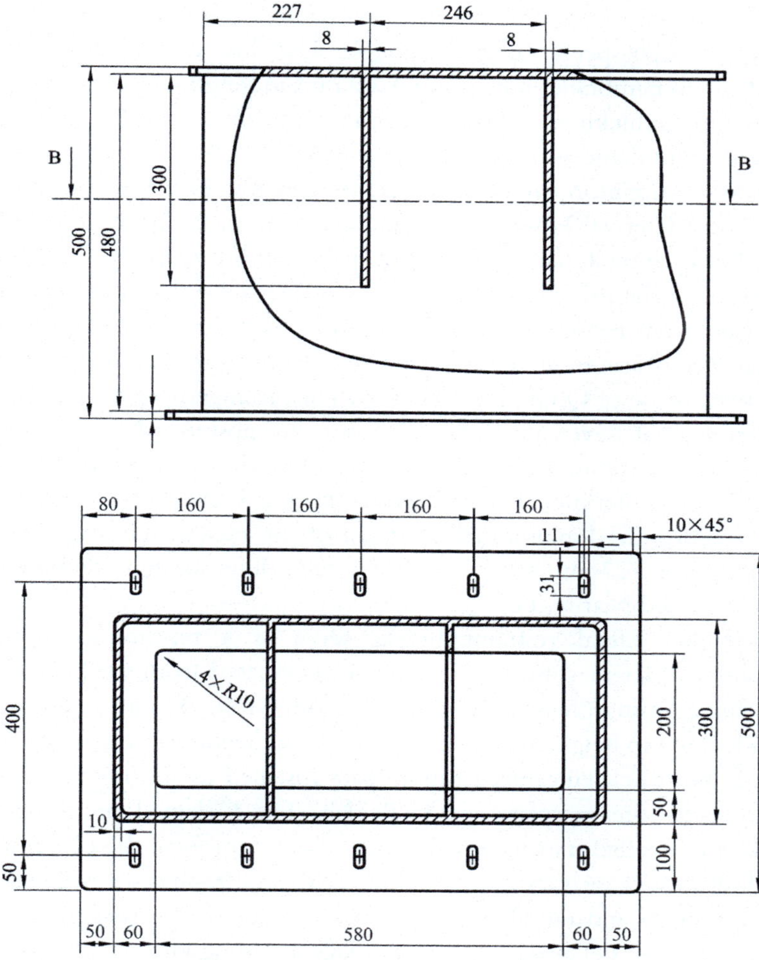

Figure 3.7 Cross-sectional dimensions of pier structural model.

bearing platforms are simulated using steel plates measuring 29 mm in length, 80 mm in width, and 4 mm in thickness. These plates are bolted to the track slabs. Each track slab is equipped with 10 pairs of rail-bearing platforms, with a longitudinal spacing of 65 mm between their centerlines. The CA mortar layer is designed by filling BASF polyurethane adhesive (Conipur M867F) between the top surface of the base plate and the bottom surface of the track slab, with a thickness of 3 mm and width of 255 mm. The sliding layer is simulated by applying BASF polyurethane adhesive (Conipur M867F) to the "two-geotextile–one-membrane" layer of actual high-speed railway bridges. The low-density polyethylene (LDPE) geomembrane, measuring 3260 mm in length, 295 mm in width, and 3 mm in thickness, is bonded to the upper surface of the girder using BASF polyurethane adhesive (Conipur M867F).

The shear reinforcement is arranged as shown in Fig. 3.8, utilizing studs with a diameter of 2 mm and length of 20.5 mm, fabricated from Q420 steel. For each track slab, two rows of shear reinforcement are arranged on both sides at the bridge joint, with two bars per row (i.e., one bar directly under each rail-bearing platform centerline). Shear grooves are positioned above fixed bearings, utilizing bolts with a diameter of 2 mm and length of 20.5 mm, made of Q345 steel. Two rows are arranged per track slab, with 14 bars in total (seven per row). Two bars are positioned directly on the track slab centerline, while six bars are symmetrically distributed on each side of the centerline. The transverse spacing between adjacent bars is 42 mm along the longitudinal direction of the bridge. The two rows of bars are spaced 27 mm apart, with their centerlines located 188 mm away from the fixed bearing end.

L-shaped lateral restraint blocks serve as a positioning restraint, requiring sufficient stiffness. Steel blocks fabricated from Q235 steel and featuring 27 mm in length, 10 mm in bottom width, 30 mm in top width, and 21.5 mm in height are used. The blocks are welded to the girder soffit, with 1-mm-thick vulcanized rubber pads installed on all three surfaces in contact with the baseplate and track slab. The L-shaped lateral restraint blocks are installed starting at a distance of 800 mm from the centerline of the shear groove near the fixed bearing, with an 800-mm spacing between adjacent blocks and a total of 12 restraint blocks per single-span girder. The rail structure model employs CHN 60 rails scaled down to 1 10, as shown in Fig. 3.9. The prototype rail materials are fabricated using CNC machining.

Fasteners are designed using vulcanized rubber pads [15 mm (transverse) × 22 mm (longitudinal) × 1 mm (thick)] beneath the rail, secured by elastic steel clips [10 mm (transverse) ×15 mm (longitudinal) × 1 mm (thick)]. The steel clips, rail-bearing platform, and track slab are bolted together, as shown in Fig. 3.10.

The bearing arrangement of the single-span bridge is shown in Fig. 3.11, while the bearing cross-section is illustrated in Fig. 3.12. The bearing incorporates a compression rubber plate (4 mm thick) to accommodate beam-end rotations, and a PTFE plate (0.5 mm thick) to facilitate

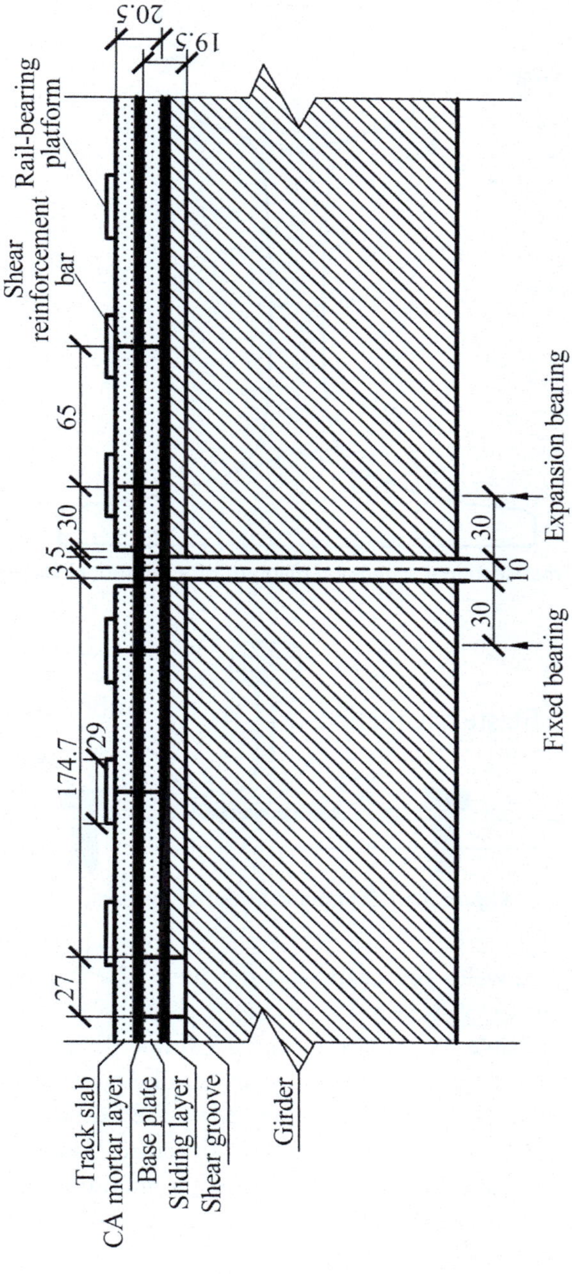

Figure 3.8 Schematic layout of shear reinforcement and shear grooves.

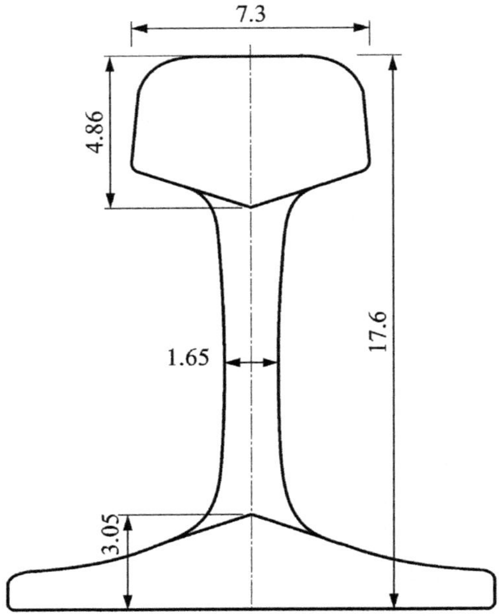

Figure 3.9 Cross-sectional schematic diagram of a rail model (scaled to 1:10).

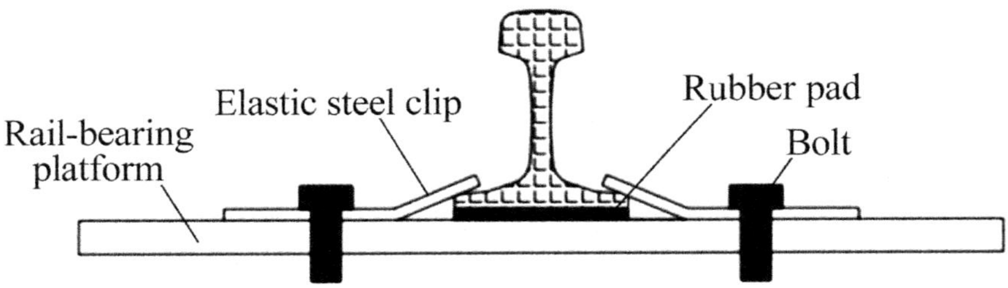

Figure 3.10 Schematic layout of fastener.

relative sliding between the beam and the pier. The bearing housing is constructed from Q235 steel.

As shown in Table 3.1, regarding the similitude design coefficients for the experimental system, the mass density similarity ratio s^{ρ} exceeds 1, demonstrating that the mass density required by the experimental system is greater than that of the actual prototype structure. Consequently, it is necessary to enhance the simulated mass by incorporating additional counterweights. The distribution principle of counterweight mass requires that the total mass of the model (including additional counterweight mass)

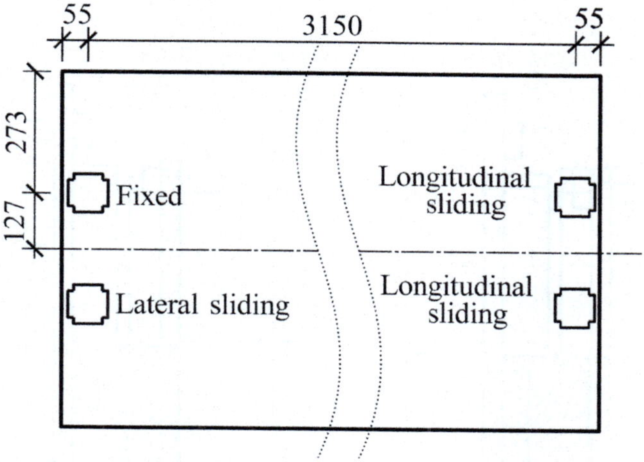

Figure 3.11 Schematic layout of the bearing for a single-span bridge.

must satisfy the mass similarity relationship with the total mass of the prototype structure. For the train model, counterweight mass is applied to the vehicle body using canvas bags filled with iron sand. For the simply supported girder, additional counterweights are applied beneath the girder model using welded steel boxes filled with lead blocks. The mass of a physical single-span simply supported girder model is 564.56 kg, as detailed in Table 3.3. According to the Post-tensioned Prestressed Concrete Simply Supported Box Girder for 350 km/h Ballastless Track Passenger Dedicated Lines (Double-track) (Tongqiao [2008] 2322A-II), a 32m simply supported box girder has a self-weight of 836.8 t, with secondary dead load calculated at 1.4 t/m. This yields approximately 881.6 t per single-span girder. After applying the mass similarity ratio, the required simulated mass is calculated as 4408 kg, indicating that the counterweight mass to be applied to the single-span simply supported girder model is 3843.44 kg. The mass of the model vehicle is 108 kg, while the simulated mass required is 280.3 kg, so an additional 172.3 kg of counterweight mass is needed.

3.3.3 Experimental System Composition

The entire experimental system consists of five parts: acceleration section, test section, deceleration section (their layout is shown in Fig. 3.13), safety protection devices, and test equipment. The composition and functions of each section are described in the following.

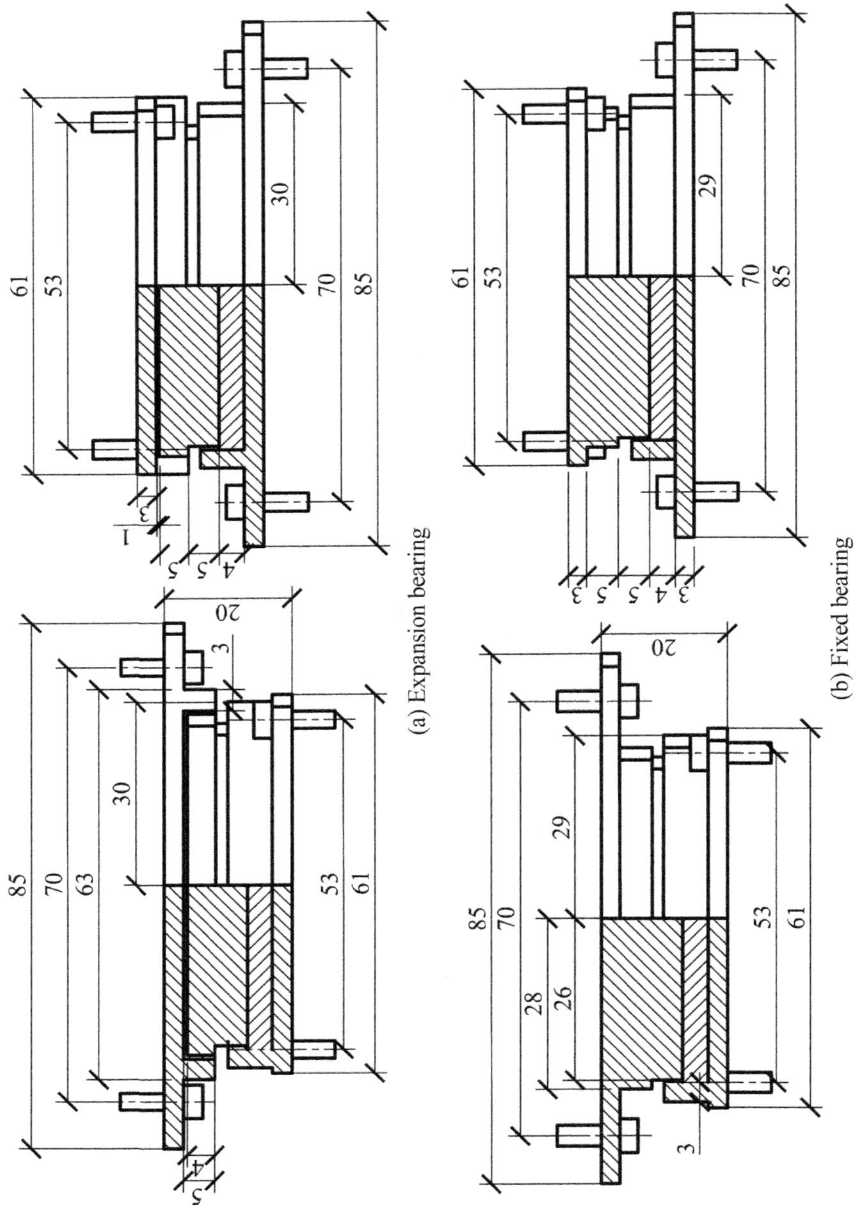

Figure 3.12 Cross-sectional schematic diagram of bearing.

Table 3.3 Mass calculation of a single-span simply supported girder model.

Model Component	Material	Density (g · mm^{-3})	Cross-sectional area (mm^2)	Length (mm)	Mass (kg)
Girder	Q235	0.01	11,935.70	3260.00	305.45
Track slab	Q235	0.01	2295.00	6450.00	116.20
Base plate	Q235	0.01	2507.50	6520.00	128.34
Rail-bearing platform	Q235	0.01	9280.00	200.00	14.57
		Gross weight			564.56

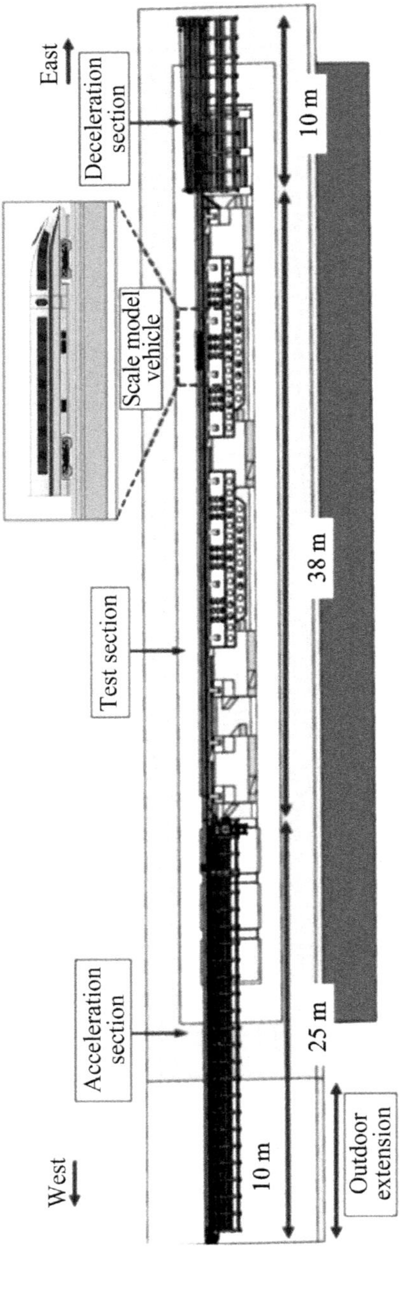

Figure 3.13 Schematic layout of the experimental system for train operations on bridges under earthquake conditions.

3.3.3.1 Acceleration section

The acceleration section is designed to propel the model vehicle to the required test velocity, driven by a Siemens 1PH 8228 servo motor with a rated speed of 1150 r/min, a rated torque of 1611 N · m, and a rated power of 194 kW. Due to a significant scale ratio, passive acceleration is employed. A servo motor drives the push mechanism via a synchronous belt, which in turn propels the model vehicle to accelerate. Upon reaching the required test velocity (maximum design speed: 20 m/s), the servo motor decelerates, allowing the model vehicle to traverse the test section under its own inertia. The acceleration mechanism of the experimental system is illustrated in Fig. 3.14. A transition mechanism is installed between the acceleration section and the test section, as illustrated in Fig. 3.14. Upon the transition of the model vehicle from the acceleration section to the test section, the connecting rail immediately disengages. Subsequently, seismic ground motion is applied to the test section, providing seismic excitation to the running train. Before the model vehicle

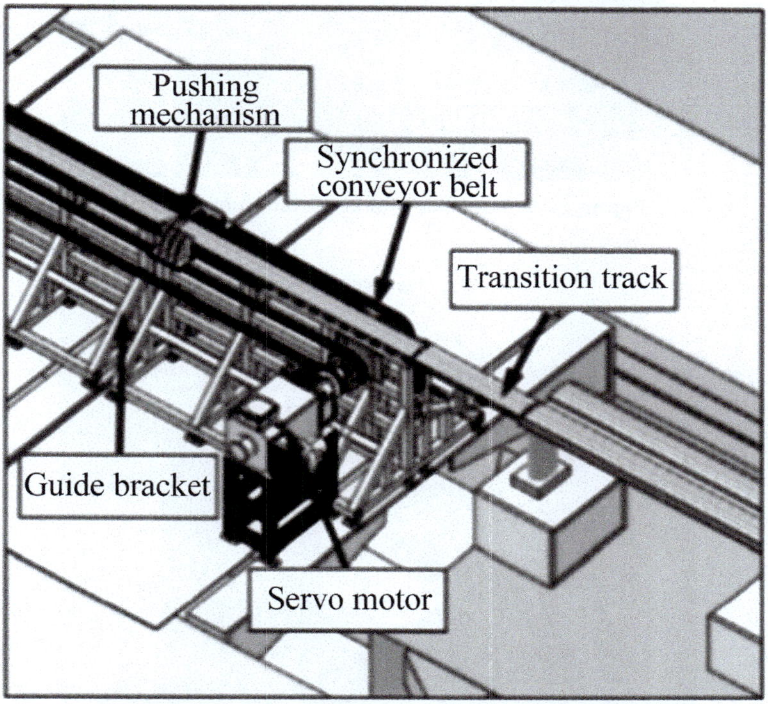

Figure 3.14 Acceleration mechanism of the experimental system.

reaches the bridge, the shake table in the test section remains stationary to ensure precise alignment between the acceleration track and the test track for smooth train passage. Actual views of the acceleration mechanism and transition mechanism are provided in Figs. 3.15 and 3.16, respectively. The test requires precise measurement of the bridge-crossing time of the model vehicle with millisecond-level accuracy, while the servo motor in the transition section must maintain sufficiently high sensitivity.

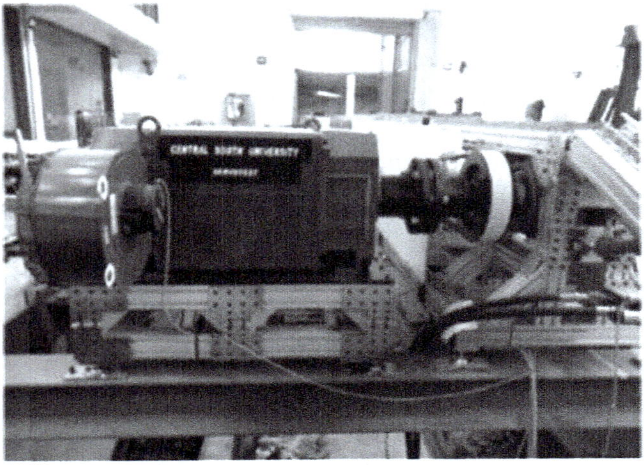

Figure 3.15 Actual view of acceleration mechanism.

Figure 3.16 Actual view of transition mechanism.

3.3.3.2 Test section

The test section consists of a four-array shaking table system and bridge specimens, with a total length of 38 m. It includes two transition tracks and 11 spans of simply supported bridges, each measuring 3.26 m in length. These girders are distributed across four shake tables, with 1 span mounted on each shake table. The spacing between Shake Table T1 and Shake Table T2 is 6.52 m, accommodating one span. The spacing between Shake Table T2 and Shake Table T3 and between Shake Table T3 and Shake Table T4 (each accommodating three spans) is 13.04 m. All components are supported by rigid load-bearing beams with sufficient stiffness. The layout of the bridge spans of the test section is shown in Fig. 3.17. It should be noted that to ensure the success of the final dynamic test without a preparatory phase, the experimental system should be equipped with an acceleration and deceleration system for model vehicle acceleration tests. The platform of the array system should be supported by movable fixture clamps, with the array system positioned at its neutral position. Before the test begins, the control system acquires the current platform position data, takes over platform position control, and retracts the movable supports to initiate testing. The specimen bridge is connected to the acceleration track via a movable transition rail, enabling independent vibrations of the specimen bridge structure. After acceleration, the model vehicle traverses the movable transition rail while transmitting the test initiation signal to the shake table control system.

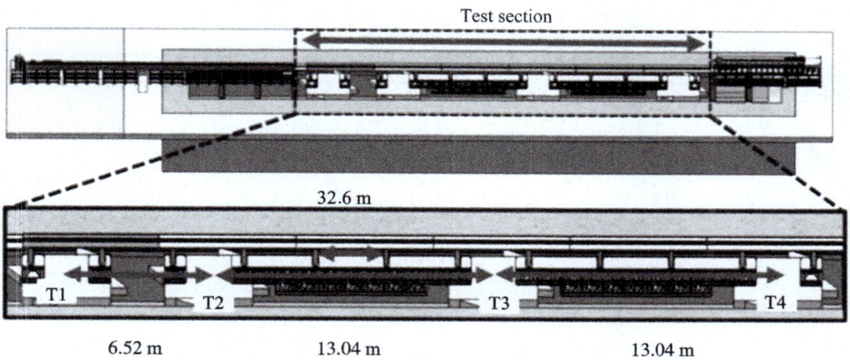

Figure 3.17 Layout of bridge spans of the test section for the train operating system.

3.3.3.3 Deceleration section

Given the high value of the model vehicle, a safe recovery procedure is implemented for its reuse. Upon passing through the test section, the model vehicle enters a deceleration net capture device mounted on linear guide rails. The deceleration net is connected to a pneumatic actuator via steel strands routed around fixed pulleys. The pneumatic actuator drives a hydraulic pump, and the hydraulic control system regulates the pressure to achieve rapid and efficient deceleration. The deceleration device of the experimental system is shown in Fig. 3.18.

To mitigate the impact of the 30-mm height difference between the test section and the deceleration section on the bogie and wheelsets, foam padding is installed beneath the acrylic panels in the deceleration section. When the model vehicle returns from the deceleration section back to the test section, a tapered transition plate is incorporated at the interface between the deceleration and test sections to facilitate the model vehicle's return to the test section and align with the track, as illustrated in Fig. 3.19.

Three nylon arcuate sliders (as shown in Fig. 3.20) are installed near the front and mid-body sections, with their undersurface positioned 2 mm above the wheel's lower edge. When the model vehicle is running on the track, the sliders remain suspended without contact; when the model vehicle enters the deceleration section from the test section, the sliders absorb transient impact loads, thereby providing bogie protection.

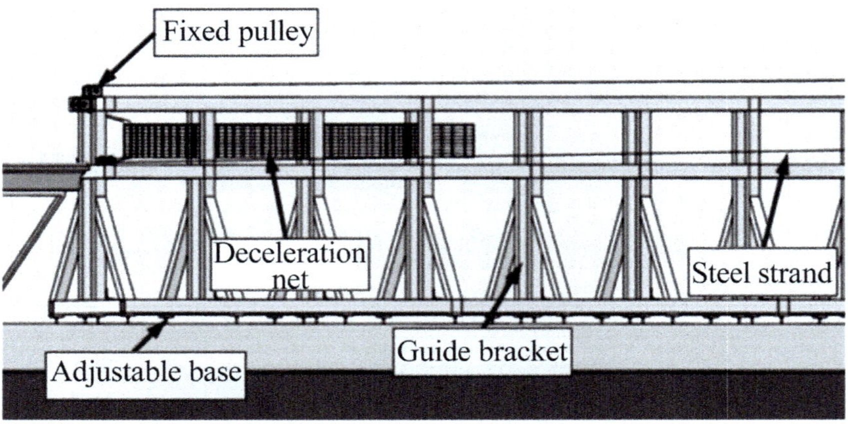

Figure 3.18 Deceleration device of experimental system.

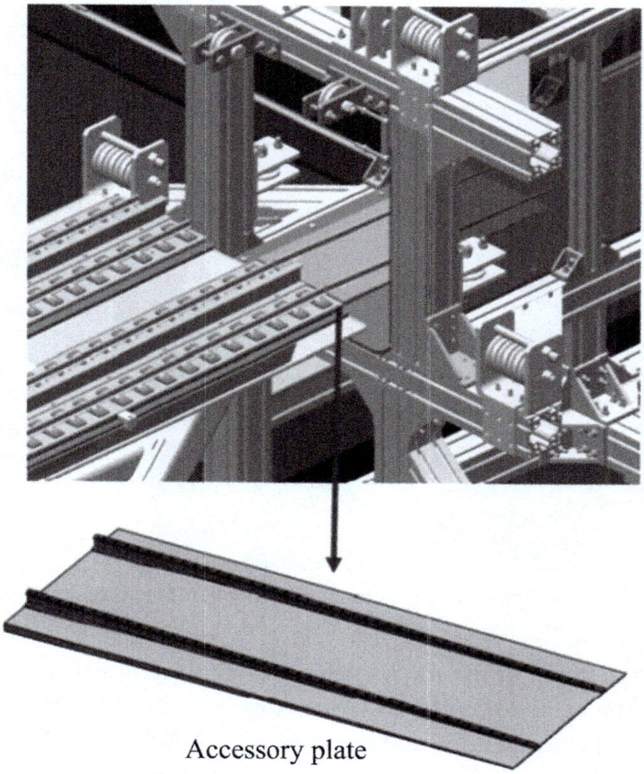

Figure 3.19 Location of accessory plate.

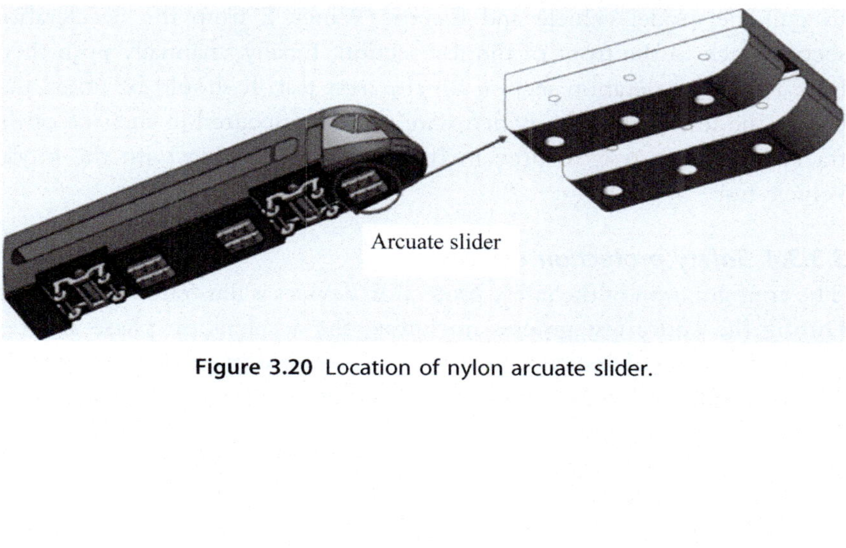

Figure 3.20 Location of nylon arcuate slider.

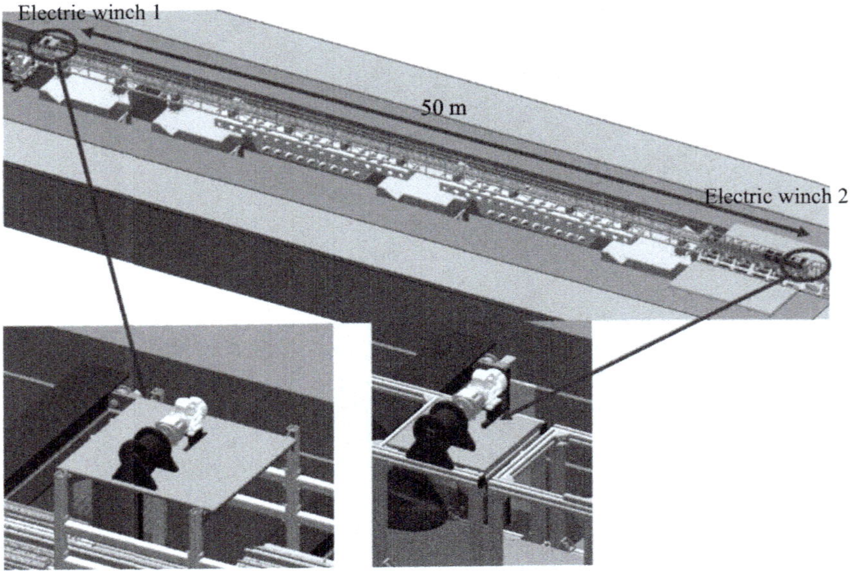

Figure 3.21 Electric winch location diagram.

After a test run of the model vehicle is completed, the system must be reset for the next test run. To reset, first add Electric Winch 1 at the junction of the acceleration and test sections, and Electric Winch 2 in front of the hydraulic cylinder in the deceleration section. Both winches are equipped with 50-m-long steel cables, and their couplers remain permanently connected, as shown in Fig. 3.21. Start Electric Winch 2 to hook Electric Winch 1 to the deceleration section. Then, start Electric Winch 1 to pull the model vehicle and Electric Winch 2 from the deceleration section back to the front of the test section. Finally, manually push them back to the acceleration section for the next test. It should be noted that before the test begins, the winch wire ropes are secured to the side of the track, not under test, to prevent them from interfering with the model vehicle test.

3.3.3.4 Safety protection devices

The configuration of the safety protection devices is illustrated in Fig. 3.22. During the entire test process, including the acceleration phase and test phase, if any derailment occurs, the test shall be immediately terminated. Therefore, the following design considerations have been implemented.

Figure 3.22 Layout of safety protection devices.

① The vehicle body is equipped with attitude sensors. The test will be terminated when the model vehicle is tilting or about to tilt.
② Fiber Bragg Grating (FBG) sensors are installed on both sides of the track. When the model vehicle derails, an alarm signal is triggered to immediately terminate the test.
③ The model vehicle is equipped with wireless cameras for visual anomaly detection as a supplementary monitoring measure.
④ Derailment protection devices are installed to remain clear of the central guard rail during normal operation. In the event of derailment, the U-shaped latch engages with the guard rail, preventing the model vehicle from falling off the experimental system and damaging testing equipment such as the shake table.

3.4 Testing of an Experimental System for High-Speed Train Operation on Bridges

This study numerically validates a scaled shake table testing methodology for high-speed train-bridge systems. The preceding theoretical analysis has demonstrated the feasibility of the tests. Here, numerical simulation further verifies that scaled test results can be effectively extrapolated to predict the prototype's dynamic response. The computational program employed in this section is a numerical framework for train-bridge interaction analysis developed in collaboration with Prof. Gu Quan et al., based on OpenSees [8]. This framework facilitates co-simulation by coupling OpenSees, which performs the finite element analysis of the bridge structure, with a MATLAB program that simulates the vehicle dynamics. The coupled system achieves co-simulation through real-time data exchange, and the Client-Server computing control technology is adopted.

3.4.1 Test Conditions

This study involves tests under various operational and seismic scenarios with parameters, including varying peak ground accelerations, train speeds, and loading directions. This section presents and analyzes results from three representative conditions: model vehicle operation at 5 m/s without seismic excitation, and under vertical Taft wave excitations of 0.1 g and 0.2 g PGA, respectively.

Figure 3.23 Photograph of the physical device for track irregularity measurement.

3.4.2 Track Irregularity Measurement

Track irregularities are a primary source of excitation in the train-track-bridge system. The system's dynamic response is highly sensitive to this input, as the use of different irregularity samples in numerical simulations can yield substantial variations in results. Therefore, the measurement accuracy of track irregularities determines the accuracy of the dynamic response in subsequent numerical simulations. The measured irregularity data serve as the foundational input for numerical verification and calibration of the experimental system. The physical measurement device for track irregularities is shown in Fig. 3.23.

3.4.2.1 Fundamental Principles of the Chord Offset Method

In this test, the chord offset method is employed to detect track irregularities, with the fundamental principles as follows.

The track irregularity to be measured can be discretized into a sequence, $y = \{y[i] | i = 1, 2, \cdots, N + 1\}$, $N+1\}$, where $N = l/\Delta x$, with l being the length of the target track section and Δx representing the sampling interval. For a given model vehicle, the chord length L is divided into two asymmetric segments L_1 and L_2 (where $L_1 \neq L_2$). The step size ΔL_s is defined as the incremental movement of the chord during measurement, with L being an integer multiple of ΔL_s. As illustrated in Figs. 3.24 and 3.25, the definition implies $\Delta x = \Delta L_s$.

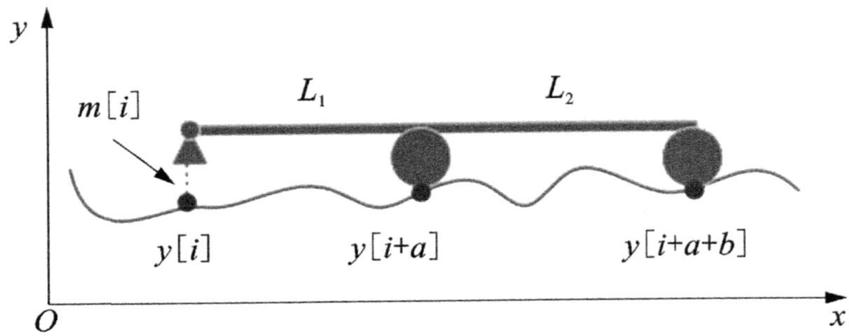

Figure 3.24 Geometric schematic of a model vehicle for track measurement.

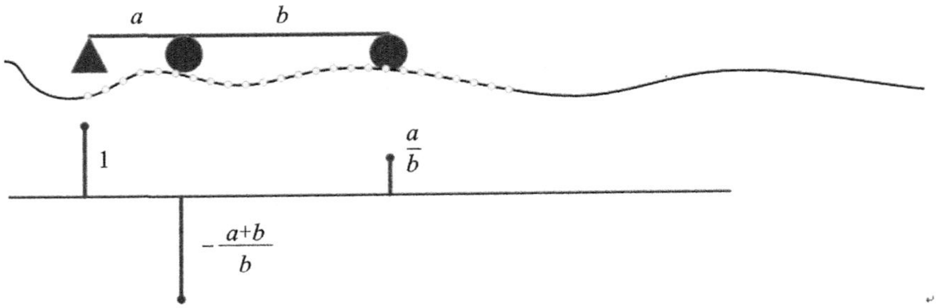

Figure 3.25 Sampling process diagram of chord offset measurement.

$$n = \frac{L}{\Delta L_s}, n \in \mathbb{N}^+ \tag{3.1}$$

Based on this definition, the order number, configuration, and minimum measurable wavelength (*MMW*) can be determined. Let n denote the number of segments when the chord length L is divided by the sampling interval ΔL_s, and let the array (a, b) represent the ratios of the two chord segments to ΔL_s, satisfying $a + b = n$.

$$a = \frac{L_1}{\Delta L_s}; b = \frac{L_2}{\Delta L_s} \tag{3.2}$$

According to the Nyquist–Shannon sampling theorem, the *MMW* for the chord offset method can be determined as:

$$MMW = 2 \cdot \Delta L_s \tag{3.3}$$

The measurement process can be modeled as a discrete impulse response filtering operation h_d applied to the track irregularity profile y, where m denotes the measured output obtained via the chord offset method.

The relationship between the chord measurement value $m[i]$ and track irregularity $y[i]$ can be expressed as:

$$m[i] = y[i] - \frac{a+b}{b} y[i+a] + \frac{a}{b} y[i+a+b] \tag{3.4}$$

The measurement process by the chord offset method is illustrated in Fig. 3.26, where laser sensors measure the normal distance between the reference chord and the rail surface. The resulting measurement sequence is defined as $m = \{m[i] | i = 1, 2, \cdots, N - n + 1\}$, where $m[i]$ is given by:

$$m[i] = y[i] - \lambda y[i+a] + \overline{\lambda} y[i+a+b]; i = 1, 2, \cdots, N - n + 1 \tag{3.5}$$

where $\lambda = -a + b/b, \overline{\lambda} = a/b$. The equation can be reformulated in matrix form as:

$$m\boldsymbol{M}y \tag{3.6}$$

where $\boldsymbol{M}$ is the measurement matrix with dimensions of $N - n + 1$ by $N + 1$, representing the mapping between the true value of track irregularity and the measured value obtained by the chord offset method. The expression for $\boldsymbol{M}$ is given by:

$$\boldsymbol{M} = \begin{bmatrix} 1 & \cdots & \lambda & \cdots & \overline{\lambda} & \cdots & 0 \\ 0 & 1 & \cdots & \lambda & \cdots & \overline{\lambda} & \cdots \\ & & \ddots & & & \ddots & \\ \cdots & 1 & \cdots & \lambda & \cdots & \overline{\lambda} & 0 \\ 0 & \cdots & 1 & \cdots & \lambda & \cdots & \overline{\lambda} \end{bmatrix} \tag{3.7}$$

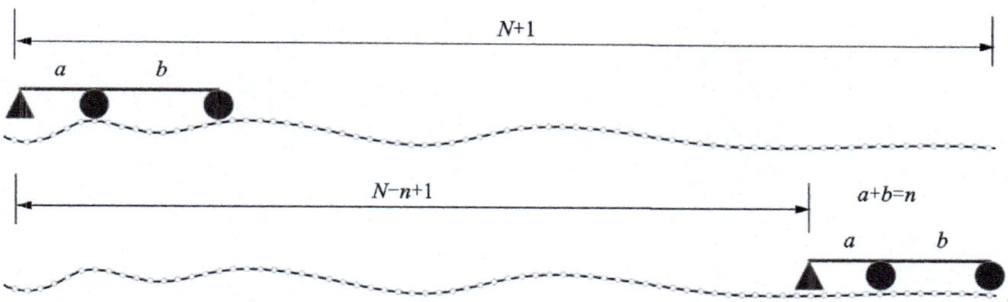

Figure 3.26 Schematic diagram of measurement by the chord offset method.

As shown by Eq. (3.7), the measurement matrix $\boldsymbol{M}$ constitutes an underdetermined matrix (rows < columns), where the number of equations is fewer than the number of unknowns. This matrix equation admits infinitely many solutions.

3.4.2.2 The Inversion Method Based on the Gradient Descent Algorithm

To address infinitely many solutions of the matrix equation, this section proposes a gradient-descent-based method for solving the system of linear equations. The underdetermined matrix equation $\boldsymbol{Ax} = \boldsymbol{b}$ is solved. Where $\boldsymbol{A}_{(\mathrm{m}\times\mathrm{n})}$ denotes an $m\times n$ coefficient matrix with $m < n$; $\boldsymbol{x}$ represents the $n\times 1$ solution vector of the matrix equation; and $\boldsymbol{b}$ is an $m\times 1$ column vector. How to solve the matrix equation $\boldsymbol{Ax} = \boldsymbol{b}$ is essentially a linear regression problem, which can be understood from the following perspectives:

$$\begin{bmatrix} a_{11} & a_{12} & \cdots & a_{1n} \\ a_{21} & a_{22} & \cdots & a_{2n} \\ \vdots & \vdots & \ddots & \vdots \\ a_{m1} & a_{m2} & \cdots & a_{mn} \end{bmatrix} \cdot \begin{bmatrix} x_1 \\ x_2 \\ \vdots \\ x_{n-1} \\ x_n \end{bmatrix} = \begin{bmatrix} b_1 \\ b_2 \\ \vdots \\ b_m \end{bmatrix} \tag{3.8}$$

The track irregularity vector $\boldsymbol{x}$ to be computed is treated as a weight column vector, while each row of the absorption matrix $\boldsymbol{A}$ serves as an input row vector. The input row vectors can be defined as $\boldsymbol{a} \in [a_1, a_2, \cdots, a_m] \in \boldsymbol{R}^{1\times n}$, and the result vector $\boldsymbol{b}$ as the output column vector. Then the relationship between the input quantity $\boldsymbol{a}$ and output quantity $\boldsymbol{b}$ in this problem can be expressed as:

$$\boldsymbol{b}_i = \sum_{l=1}^{n} \boldsymbol{a}_{il} \times \boldsymbol{x}_l \tag{3.9}$$

The input vector $\boldsymbol{a}_{\mathrm{i}}$ and its corresponding output $\boldsymbol{b}_{\mathrm{i}}$ can form a training sample, with the complete training set denoted as $\{(\boldsymbol{a}_i, \boldsymbol{b}_i), i = 1, 2, \cdots, m\}$. Define the cost function and gradient descent equation for the problem to iteratively update the weight column vector.

Define the cost function as follows:

$$\begin{aligned} J(\boldsymbol{x}) &= \sum_{i=1}^{m}\left(\sum_{l=1}^{n} \boldsymbol{a}_{i,l} \times \boldsymbol{x}_l - \boldsymbol{b}^{(i)}\right)^2 + \boldsymbol{\lambda}\sum_{l=1}^{n} \boldsymbol{x}_l \times \boldsymbol{x}_l \\ &= (\boldsymbol{Ax} - \boldsymbol{b})^{\mathrm{T}} \cdot (\boldsymbol{Ax} - \boldsymbol{b}) + \boldsymbol{\lambda x}^{\mathrm{T}}\boldsymbol{x} \end{aligned} \tag{3.10}$$

The weight column vector $\boldsymbol{x}$ here corresponds to the unknown vector to be solved in the matrix equation. When the weight column vector converges to a steady state, the solution vector of the aforementioned underdetermined system of equations is obtained. It should be noted that since the underdetermined system of equations admits infinitely many solutions, a constraint on the magnitude of the solution (i.e., a regularization term) is introduced into the cost function. By tuning the regularization coefficient λ, a balance is achieved between the accuracy of the solution and its magnitude. Computing the partial derivatives of the cost function $J(x)$ involves matrix differential calculus, where special attention must be paid to the associated operational rules.

$$\begin{aligned} \frac{\partial J(\boldsymbol{x})}{\partial \boldsymbol{x}} &= -2\boldsymbol{A}^{\mathrm{T}}\boldsymbol{b} + 2\boldsymbol{A}^{\mathrm{T}}\boldsymbol{Ax} + 2\boldsymbol{\lambda x} \\ \boldsymbol{x} &:= x - \alpha \cdot \frac{\partial J(x)}{\partial x} \end{aligned} \tag{3.11}$$

where α denotes the learning rate, whose adjustment directly controls the convergence speed of the iterative process.

After a sufficient number of weight updates through iterative refinement, the solution to the underdetermined matrix equation can be obtained using the gradient descent-based method. Since there exist infinitely many solutions, the initial weight values should be set within a reasonable range based on the actual physical meaning of track irregularity. Fig 3.27 presents a full-bridge measurement case (sampling interval: 60 mm) as an example.

The weight column vector is initialized with random values uniformly distributed in [–2, 2]. Hyperparameter settings: After multiple validation runs, the maximum viable learning rate is set to $\alpha = 0.06$ to ensure gradient descent convergence, while the regularization coefficient is fixed at $\lambda = 0.01$.

As observed in Figs. 3.27A and B, the weight values rapidly converge from random initialization to a stable solution, and the weights exhibit clear linear convergence before entering an iterative refinement phase

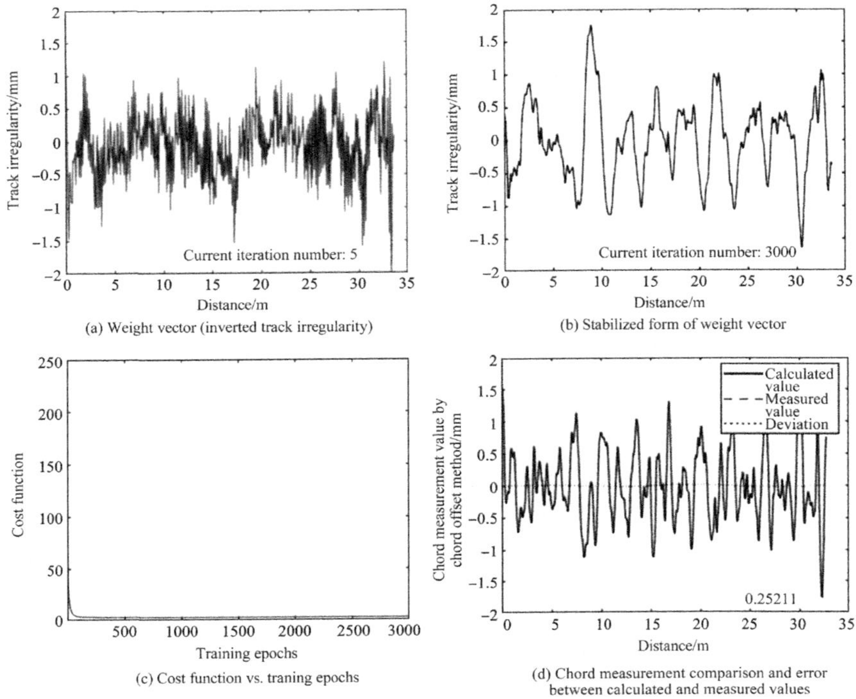

Figure 3.27 Numerical example analysis.

through gradient descent, where high-frequency oscillation signals induced by local curve initialization gradually diminish, ultimately converging to the optimal solution.

Fig 3.27C validates the computational accuracy of this method. In Fig. 3.27D, the solid black line represents the chord measurement values obtained through resampling and calculation of the inverted track irregularity (which correspond to actual measurements from the model vehicle), and the dotted black line indicates the error between these two sets of values. The computational results show slight deviations from the measured values, indicating nonperfect alignment between the two datasets. This occurs because in the gradient descent method with the regularization term $J(x) = (\boldsymbol{Ax} - \boldsymbol{b})^{\mathrm{T}} \cdot (\boldsymbol{Ax} - \boldsymbol{b}) + \boldsymbol{\lambda} x^{\mathrm{T}} \boldsymbol{x}$, the two components of the cost function cannot simultaneously reach their optima, necessitating a trade-off between these competing objectives through careful adjustment of the regularization coefficient λ.

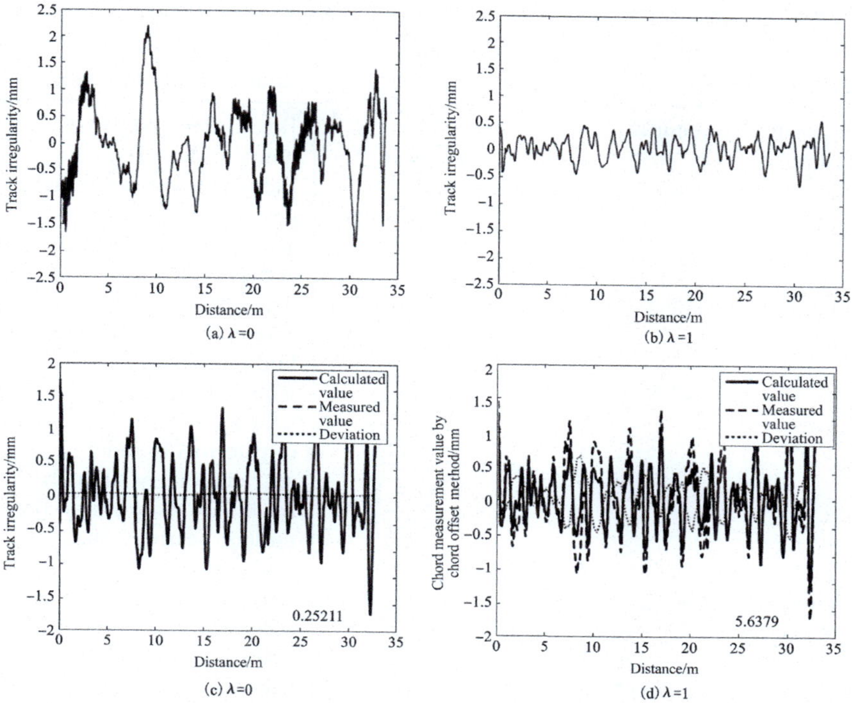

Figure 3.28 The impact of different regularization coefficients on computational results (track irregularity solved by the 001-ML method).

The regularization coefficient λ should neither be too large nor too small, as illustrated in Fig. 3.28. If the regularization coefficient λ is too small, which results in the maximum accuracy, it introduces a large number of oscillation signals that contradict the physical reality of track irregularity. If the regularization coefficient λ is too large, it over-constrains the solution magnitude, resulting in the minimum accuracy. The regularization parameter λ should be reasonably selected.

3.4.3 Experimental Data Acquisition

The experimental measurement plan was designed to capture the key dynamic responses of the coupled system. This plan encompassed the displacements and accelerations of the model vehicle's components (vehicle body, bogies, and wheelsets) and the track–bridge system's elements (rails, track slabs, base plates, box girders, bearings, and piers). Additionally, the track irregularity profile was measured. However, the

Table 3.4 Sensors, data acquisition equipment, and their measurement parameters.

Sensor Type	Data Acquisition Equipment	Measurement Parameter
Laser displacement sensor	IMC test and measurement GmbH	Box girder midspan and pier-top displacements
SDI accelerometer	IMC test and measurement GmbH	Pier-top acceleration, box girder midspan acceleration, track slab acceleration
High-speed camera	High-speed camera (VIC-3D)	The displacements and accelerations of the vehicle body, bogies, and wheelsets of the model vehicle, as well as the displacements and accelerations of the rails, track slabs, base plates, box girders, bearings, and bridge piers
Wireless accelerometer	SensorConnect system	Triaxial vehicle body acceleration of the model vehicle
IP67-rated fiber optic sensor	Vehicle control system	Model vehicle speed measurement
Accelerometer	Shake table control system	Shake table surface acceleration

model's small scale, combined with the continuous motion of the entire model vehicle, posed significant deployment challenges for conventional sensors in certain areas. Therefore, multiple types of sensors were employed to measure throughout the experiment to conduct measurements. The sensors and data acquisition equipment used are listed in Table 3.4. The specific measurement scheme and sensor placement are detailed as follows.

3.4.3.1 Conventional Sensor Arrangement for a Model Track–Bridge System

The measurement range of the high-speed camera is limited by its placement position and imaging resolution. In the current experiment, its effective measurement range is approximately 2 m. To observe the seismic-induced train responses at different locations within the entire train operating system during the experiment, additional conventional measurement points need to be deployed. Consequently, laser displacement transducers and Silicon Designs, Inc (SDI) accelerometers were deployed on the track–bridge system as illustrated in Fig. 3.29. The experimental system consists of an 11-span simply supported girder model, sequentially numbered from ① to ⑪ from west to east. Among them, Spans ①, ③, ⑦, and ⑪ are located on four shake tables. High-speed cameras are deployed near Span ⑥. To facilitate comparative analysis of experimental data, priority instrumentation is implemented for Spans ⑤, ⑥, and ⑦. Span ①, the first bridge segment subjected to vehicular loads, is instrumented with some measurement points to evaluate the bridge–vehicle interaction testing system. In Fig. 3.29, D denotes the displacement monitoring point numbering, with laser displacement sensors exclusively deployed for vertical (z-axis) measurements. In the system, A denotes the acceleration monitoring point numbering, with y (lateral) and z (vertical) axes. A9 and A10 designate the wireless acceleration sensors installed on the model vehicle body.

3.4.3.2 Layout of High-Speed Camera Measurement Points for the Model Track–Bridge System

To minimize the boundary effects of the simply supported girder, the measurement points should be positioned as far as possible from the end span of the bridge model. Research by Feng Yulin demonstrates that bridges with 5-6 approach spans exhibit consistent natural frequency characteristics [9]. Consequently, high-speed cameras are positioned adjacent to Span ⑥. Given the effective observation range of high-speed cameras of approximately 2 m, the bridge responses near piers P5 and P6 are monitored. Continuous speckle pattern measurement points are arranged on the girder web of Span ⑥, as well as on piers P5 and P6. Small discrete speckle pattern measurement points are deployed along the edges of Span ⑥ and adjacent spans, on the base plate, track slab, and rails. The schematic layout and actual view are shown in Figs. 3.30 and 3.31, respectively. Notably, during the experimental process, the bridge-crossing time of the model vehicle becomes unpredictable due to variable vehicle speed. To reduce the image data volume, the sampling frequency of the high-speed camera was adjusted based on the vehicle speed.

3.4.3.3 Placement of Measurement Points on the Model Vehicle

Instrumenting the moving model vehicle posed a significant challenge, as conventional wired sensors are difficult to wire and prone to signal

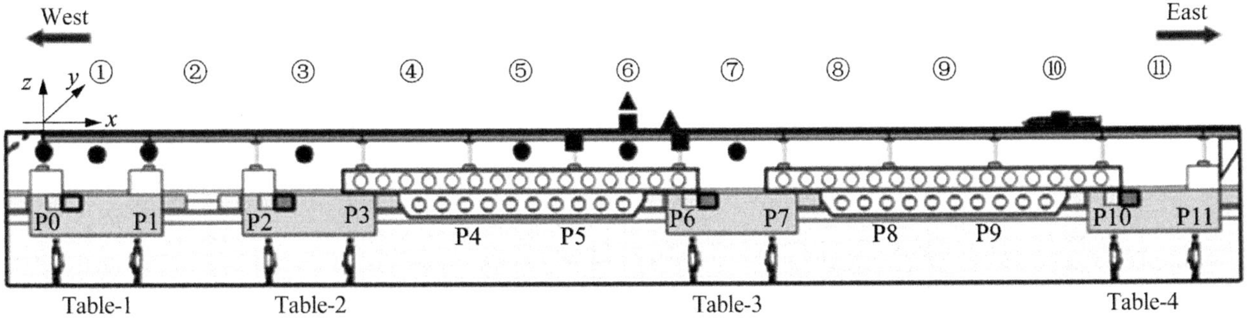

							A3	A5	A1		Bridge *y*-axis displacement sensor ID	A9	Sensor location index table
D1	D2	D3		D4		D5		D6		D7	Bridge *z*-axis displacement sensor ID		
							A4	A6	A2		Bridge *z*-axis acceleration sensor ID	A10	
								A7	A8		Track slab *z*-axis acceleration sensor ID		

● — Vertical displacement ■ — *yz*-biaxial accelerometer ▲ — *z*-axis accelerometer

Figure 3.29 Layout of acceleration and displacement measurement points for the experimental system for train operations on bridges under earthquake conditions.

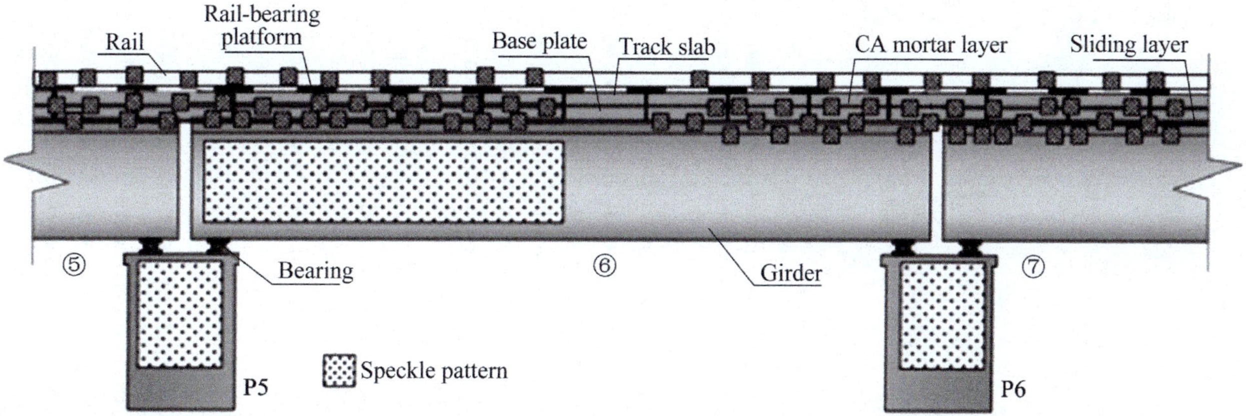

Figure 3.30 Schematic layout of high-speed camera measurement points for the experimental system for train operations on bridges under earthquake conditions.

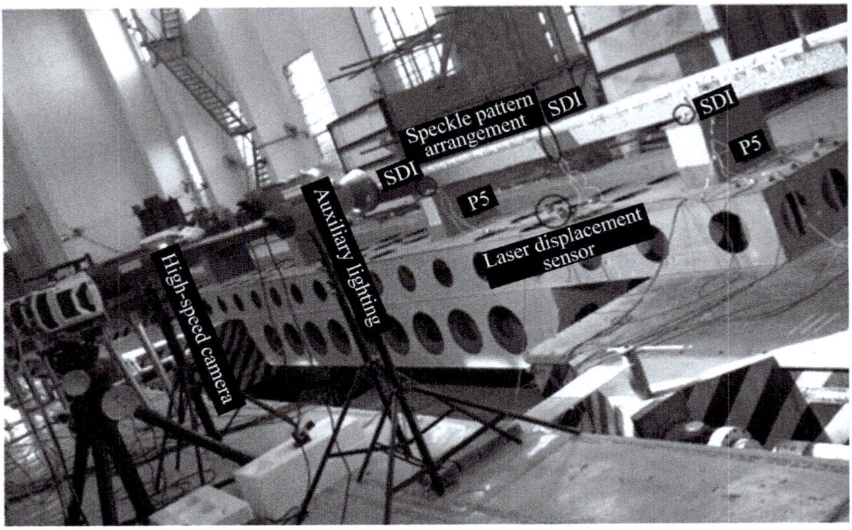

Figure 3.31 Actual view of high-speed camera measurement points for the experimental system for train operations on bridges under earthquake conditions.

instability. Therefore, this study employed wireless measurement schemes, which include both a wireless triaxial accelerometer and a high-speed camera measurement scheme. Due to their size constraints, wireless triaxial accelerometers are only installed at the midpoint of the model vehicle at a single measurement point, as shown in Fig. 3.32. The rectangular frame indicates the sensor location, with its protective cover positioned beneath it.

The speckle pattern arrangement for the high-speed camera is as follows: the middle sections of the model vehicle body, the upper body sections above the left and right bogies, the middle sections of the left and right bogies, and the axle boxes of the model wheelsets. The schematic layout and actual view of these measurement points are illustrated in Figs. 3.33 and 3.34, respectively.

3.4.4 Comparative Analysis of Test and Simulation

Based on the spatial moving elements, a full-bridge FE model for the experimental system is established with OpenSees nonlinear FE analysis software. The model is then validated against experimental test data to verify the correctness and applicability of the developed moving elements.

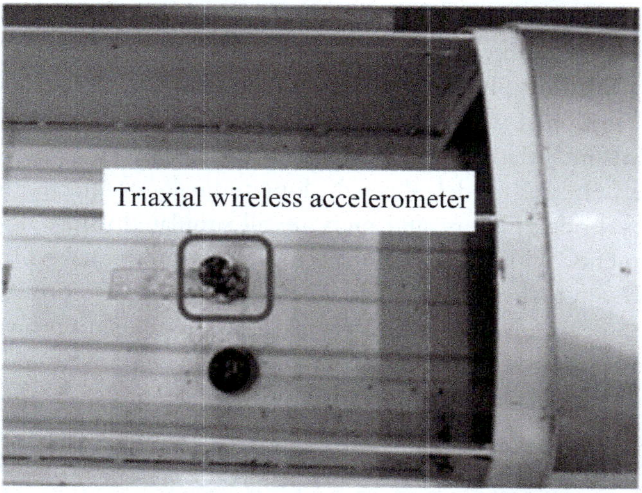

Figure 3.32 Layout of wireless accelerometers on the model vehicle body.

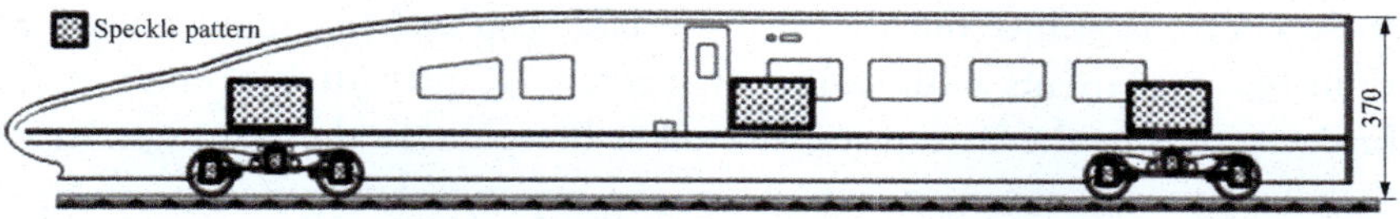

Figure 3.33 Schematic layout of high-speed camera measurement points on the model train.

Figure 3.34 Actual view of high-speed camera measurement points on the model train.

3.4.4.1 FE Model of the Experimental System

Only the test section of the experimental system is modeled and analyzed. An 11 × 32 m simply supported beam track–bridge finite element (FE) model is developed with OpenSees. The side view and front view of the established finite element (FE) model are shown in Figs. 3.35 and 3.36, respectively. In the model, the rails, longitudinally connected track slabs, base plates, bridge girders, and piers are all simulated using 3D nonlinear beam elements, with an element spacing of 65 mm, equal to the fastener spacing. Given that the experimental system is primarily constructed from steel, the Steel02 material is directly adopted for simulation, using prototype parameters from the experimental system. The rail fasteners, sliding layer, CA mortar layer, shear reinforcement, and shear grooves are modeled as spring-damper systems, implemented through zero-length elements in OpenSees with an ideal elastoplastic material model. Specific parameters are listed in Table 3.5. As the bridge girder and piers are simulated using spatial beam elements, the element nodes are the center points of the component sections. There exists a coordinate offset between these centroid points and those of the baseplate and bearings. Therefore, rigid beam elements with sufficient stiffness are provided to connect the bridge girder to the base plate, as well as the piers to the bearings. Given that the primary observation focus of the experimental system is on Span ⑥, the adjacent spans can be treated as boundary conditions, thus eliminating the need for additional boundary condition processing during modeling. The first 10 natural frequencies of the established single-span simply supported girder and the train operating system FE model are presented in Table 3.6.

Under non-seismic and Taft vertical earthquake excitations, this section validates the numerical model by comparing simulation results between experimental measurements and the aforementioned simulation model data at a vehicle speed of 5 m/s. Key response parameters, including vehicle body acceleration, midspan displacement of Span ⑥, track slab displacement, and rail displacement, are evaluated. During the test, the vehicle collides with the deceleration section after exiting the bridge. Since the simulation platform based on the moving element method cannot replicate this scenario, only the test data from the vehicle traversing the bridge are used for comparison. The detailed comparative results are presented as follows.

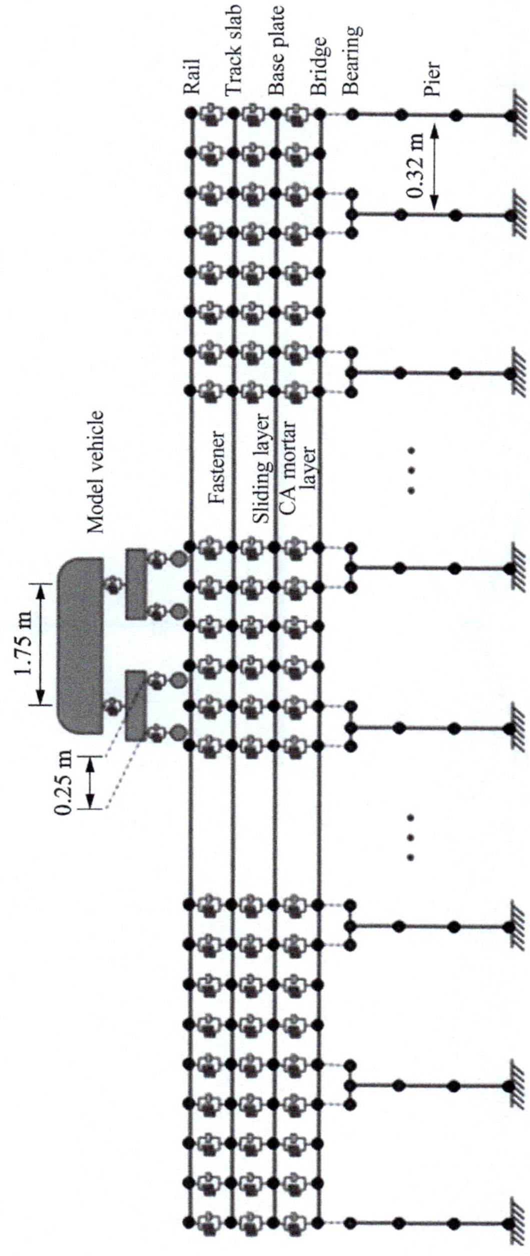

Figure 3.35 Side view of the full-bridge FE model.

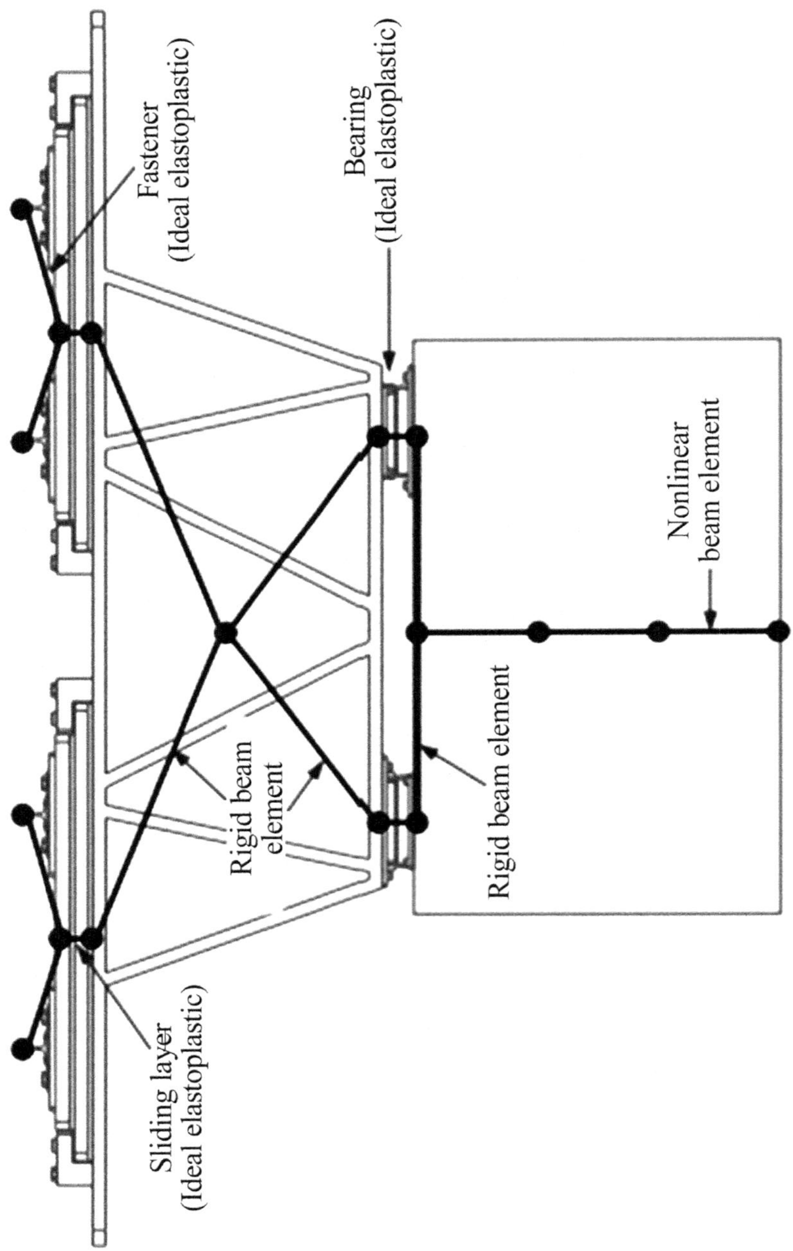

Figure 3.36 Front view of the full-bridge FE model.

Table 3.5 Force–displacement model of connection components.

Component Name	Force F_y (kN)	Displacement D_y (mm)	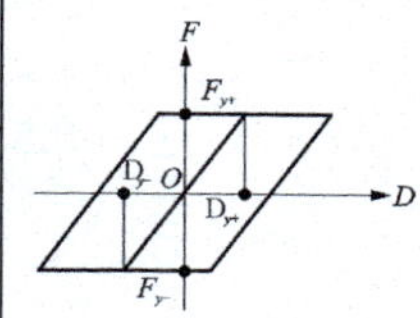
Sliding layer	1.50	0.05	
CA mortar layer	2.06	0.05	
Bearing	573.35	0.30	

Table 3.6 First 10 natural frequencies of a single-span simply supported girder and train operating system FE model.

Single-Span Simply Supported Girder Model		Train Operating-System Model	
Mode Order	**Natural Frequency (Hz)**	**Mode Order**	**Natural Frequency (Hz)**
1	3.25	1	2.94
2	4.33	2	3.10
3	5.36	3	3.50
4	9.98	4	3.66
5	10.50	5	4.21
6	10.99	6	4.89
7	14.35	7	4.90
8	20.08	8	5.45
9	22.30	9	5.66
10	25.67	10	10.23

3.4.4.2 Experimental Data Comparison

(1) Comparison of vertical and lateral accelerations of the vehicle body

According to Figs. 3.37 and 3.38, the numerical simulation results from the OpenSees platform show good agreement with the experimental measurements, and the overall vibration trends are consistent. When the model vehicle enters the first two spans of the test section from the acceleration section, unstable vibration characteristics are observed in the vehicle body acceleration. Notably, lateral acceleration exhibits unilateral deviation, likely attributable to directional impact between the wheelsets and transition rails during bridge entry, which imparts initial lateral velocity to the vehicle body. This results in a deviation in the acceleration of the first two spans. The numerical simulation, however, does not consider the acceleration process, so there is no such deviation phenomenon. The vertical acceleration of the vehicle body remains within approximately 0.4 m/s^2 under both non-seismic conditions and Taft earthquake excitations of 0.1 *g* and 0.2 *g*.

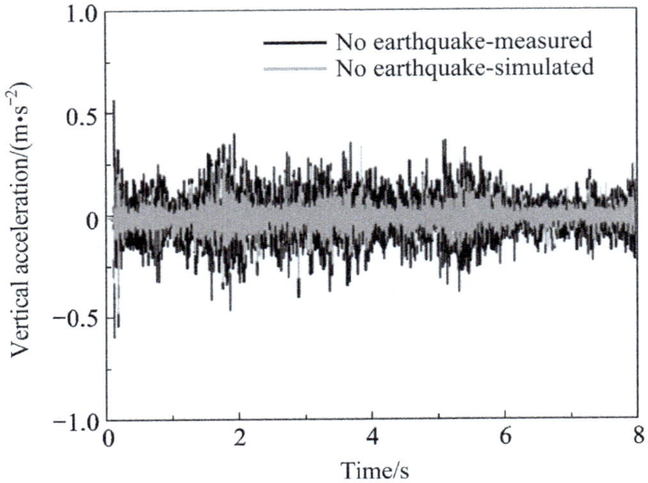

Figure 3.37 Vertical acceleration 1 of the vehicle body.

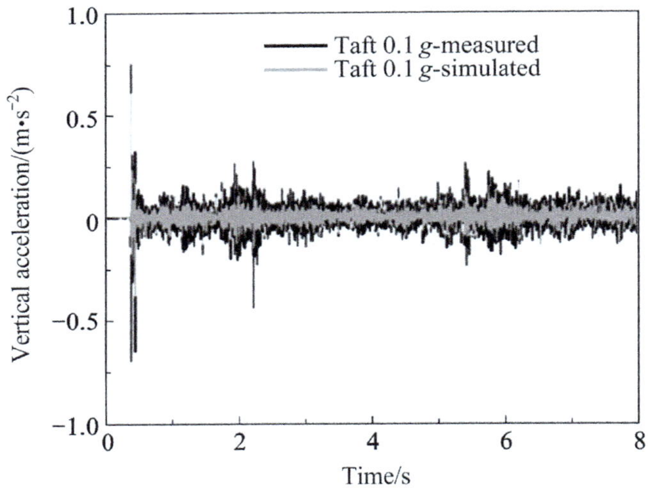

Figure 3.38 Vertical acceleration 2 of the vehicle body.

The lateral acceleration is comparatively lower, remaining consistently below 0.2 m/s^2. Experimental results show a negligible influence on the vibration response of the vehicle body. Both vertical and lateral accelerations from numerical simulations are lower than the measured values, primarily attributable to the following two factors.

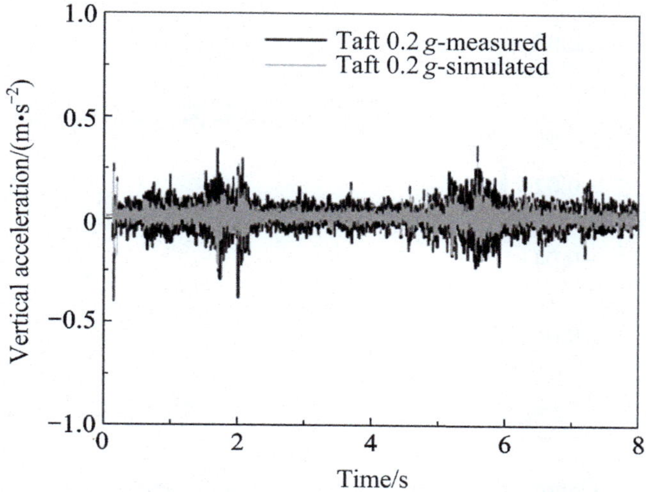

Figure 3.39 Vertical acceleration 3 of the vehicle body.

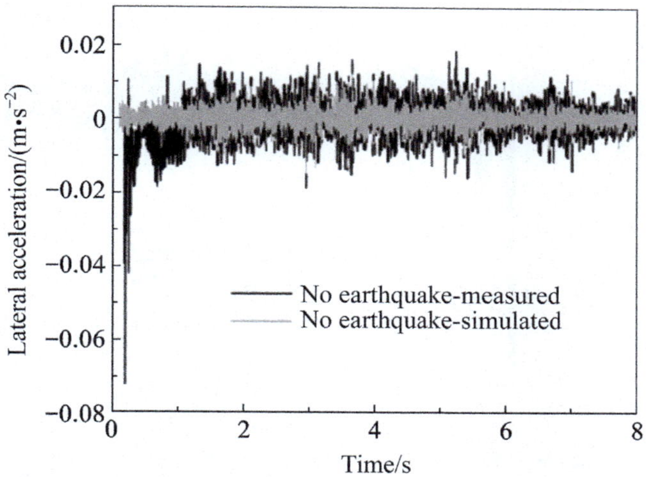

Figure 3.40 Lateral acceleration 1 of the vehicle body.

① The connection stiffness between wheelsets and bogies, as well as between bogies and the vehicle body of the model vehicle, is higher than typical values. This is primarily due to the small size of springs, which makes stiffness control challenging. In addition, as this is the first bridge–vehicle interaction test under earthquake conditions, excessive connection stiffness is adopted to prevent the model vehicle from derailing from the shake table under earthquake conditions.

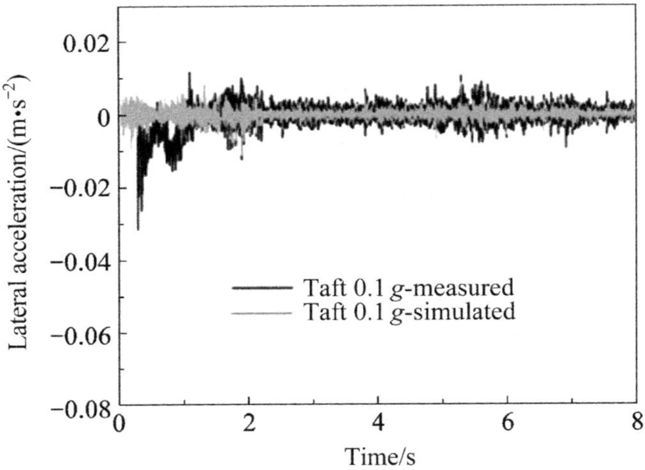

Figure 3.41 Lateral acceleration 2 of the vehicle body.

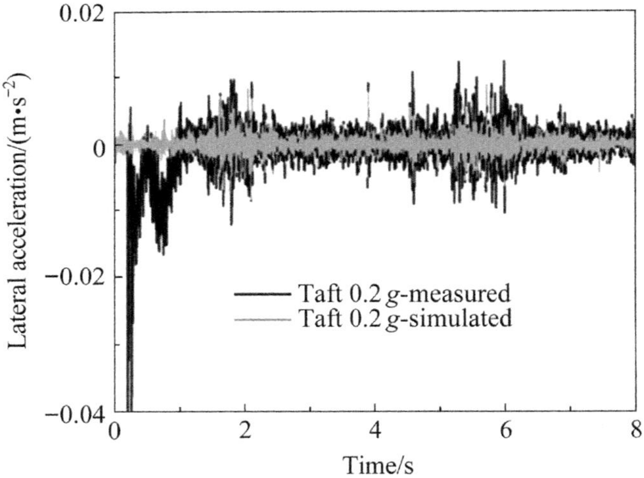

Figure 3.42 Lateral acceleration 3 of the vehicle body.

② During the experiment, no additional counterweight is applied to the body of the model vehicle, resulting in a relatively lighter vehicle body mass.

(2) Displacement comparison of the bridge at midspan

In the experiment, the displacement response of the sixth span of the simply supported girder is measured using a high-speed camera. The camera operates at a frame rate of 1000 Hz under a vehicle speed of 5 m/s. The acquisition area and typical deformation patterns are

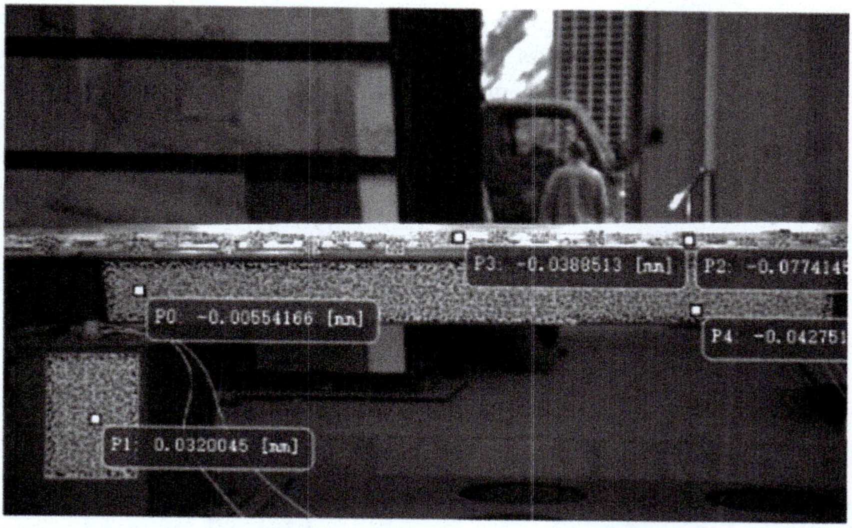

Figure 3.43 Deformation patterns of the sixth span of the simply supported girder captured by the high-speed camera at the initial moment.

shown in Figs. 3.43 and 3.44. To compare with the numerical simulation results from the OpenSees platform, the displacement responses of the bridge, track slab, and rail at the midspan position are extracted from the images generated by the high-speed camera, with the following comparisons conducted.

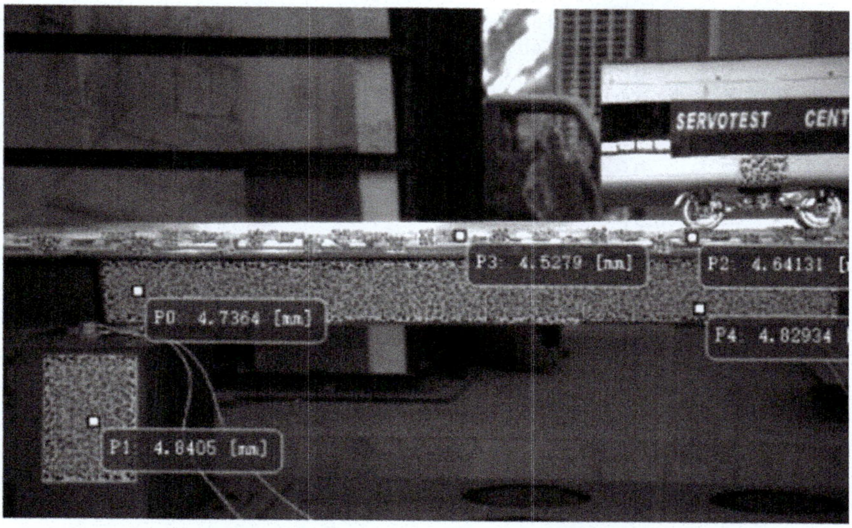

Figure 3.44 Deformation patterns of the sixth span of the simply supported girder under train loading captured by the high-speed camera.

As shown in Figs. 3.45 and 3.46, the numerical simulation results based on the OpenSees platform exhibit good agreement with the experimental measurements both vertically and laterally. When the model vehicle passes through the sixth span of the bridge, the vertical displacement response of the bridge exhibits a double-peak

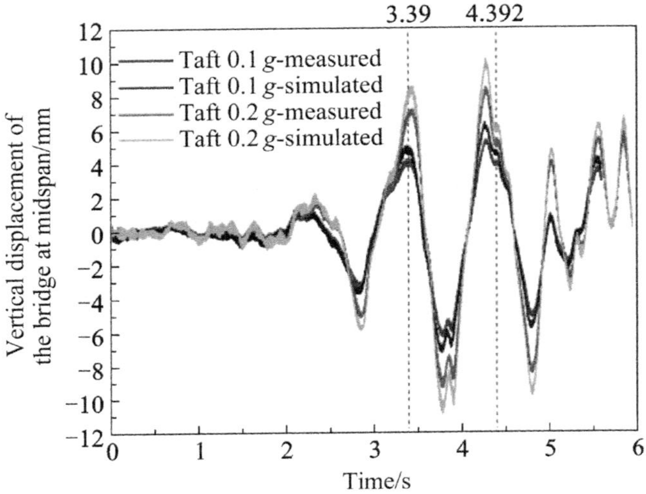

Figure 3.45 Vertical displacement of the bridge at midspan.

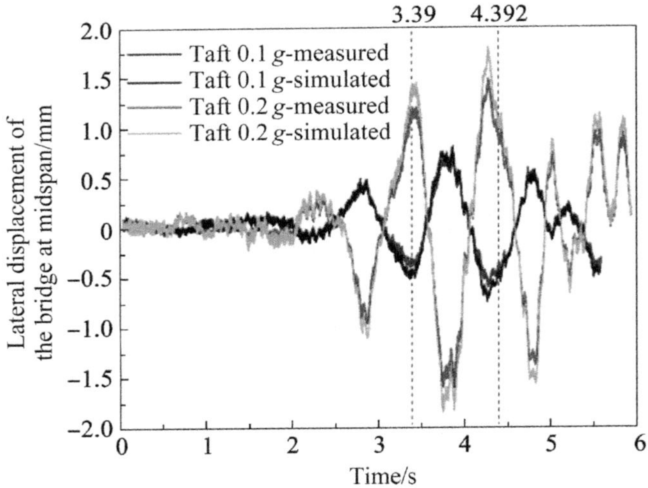

Figure 3.46 Lateral displacement of the bridge at midspan.

amplification waveform, which aligns with the field-measured waveform. This demonstrates that the indoor vehicle–bridge interaction experiment can reflect the train–bridge coupling vibration characteristics to a certain extent. Under the Taft 0.1 *g* excitation, the measured maximum vertical displacement of the bridge is approximately 10 mm, while under the Taft 0.2 *g* excitation, it is about 6.5 mm, with the simulated results are around 11 and 7 mm, respectively. The maximum horizontal displacements are relatively small: the measured values are approximately 0.7 and 1.6 mm, while the simulated values are about 0.75 and 1.8 mm, respectively. The observation that horizontal displacements are smaller than vertical displacements aligns with the conclusion that lateral stiffness typically exceeds vertical stiffness in actual engineering. The simulation values generally align with the measured data in trend, but exhibit consistently higher amplitudes. This discrepancy stems from the absence of counterweights (90 kg) on the model vehicle during physical testing, whereas the simulation incorporates counterweights (280 kg), thus increasing overall loading on the model vehicle and midspan displacement responses.

(3) Comparison of relative displacements of the track slab at midspan

Figs. 3.47 and 3.48 present the vertical and lateral displacement responses of the track slab and the rail, which generally align with midspan bridge displacements. The measured displacement differences

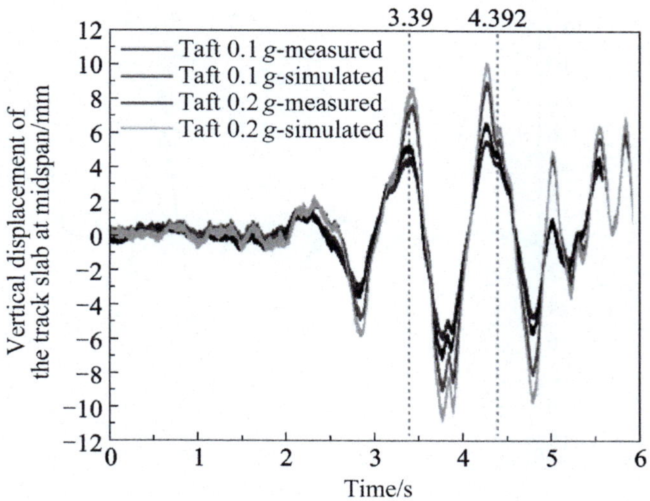

Figure 3.47 Vertical displacement of the track slab at mid-span.

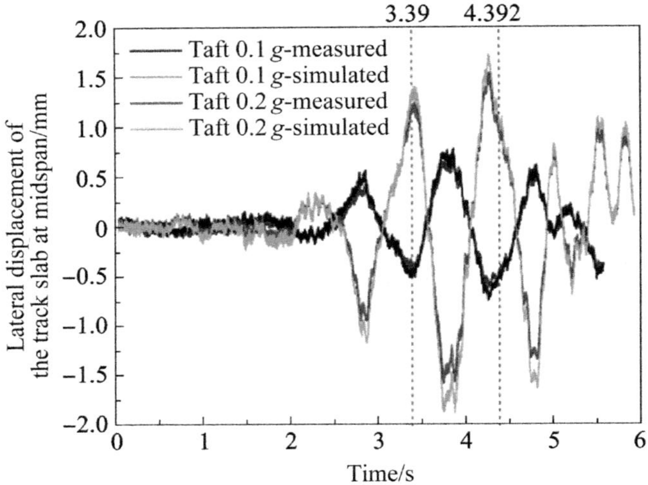

Figure 3.48 Lateral displacement of the track slab at mid-span.

between the rail and the track slab and between the track slab and the bridge in the vertical and lateral directions are illustrated in Figs. 3.49–3.54, respectively. The relative vertical displacement difference is approximately 0.5 mm, and the lateral displacement difference is approximately 0.1 mm. Figs. 3.51 and 3.52 show that the vertical displacement of the rail is less than that of the track slab, suggesting

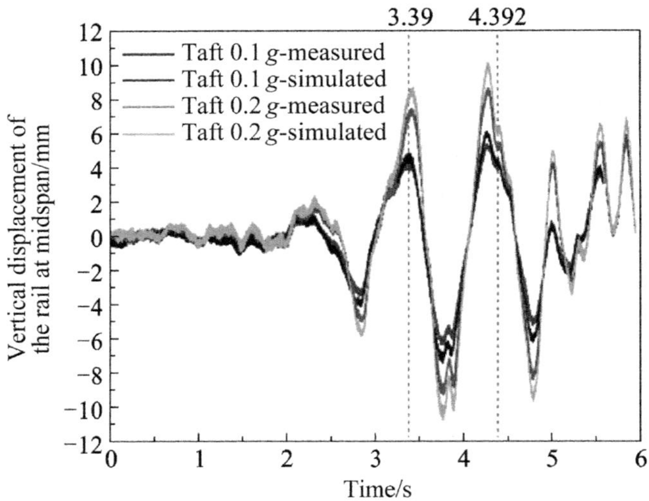

Figure 3.49 Vertical displacement of the rail at mid-span.

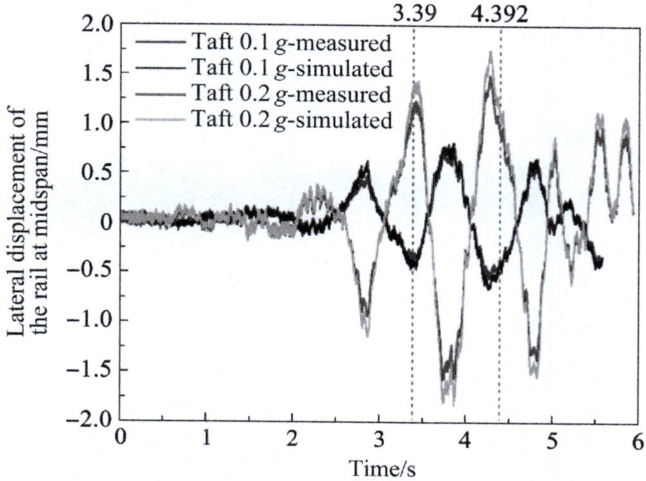

Figure 3.50 Lateral displacement of the rail at mid-span.

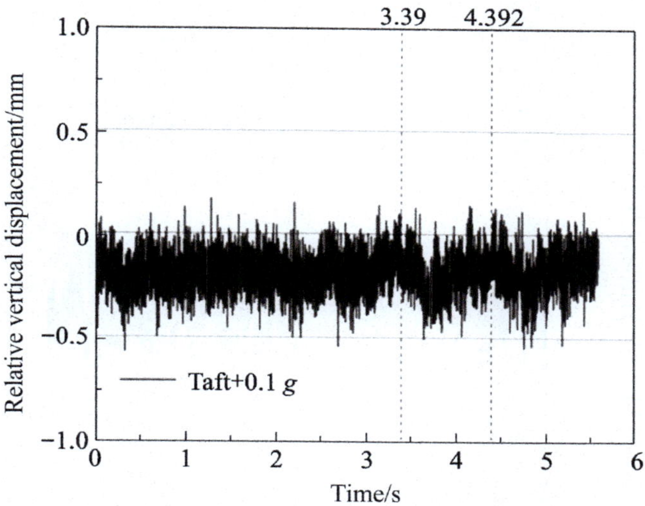

Figure 3.51 Vertical displacement difference between the rail and track slab at mid-span.

potential void formation at the interface. The lateral displacement of the rail is greater than that of the track slab, demonstrating the existence of positive-directional relative movement of the rail. Similarly, as shown in Figs. 3.53 and 3.54, the vertical displacement of the track

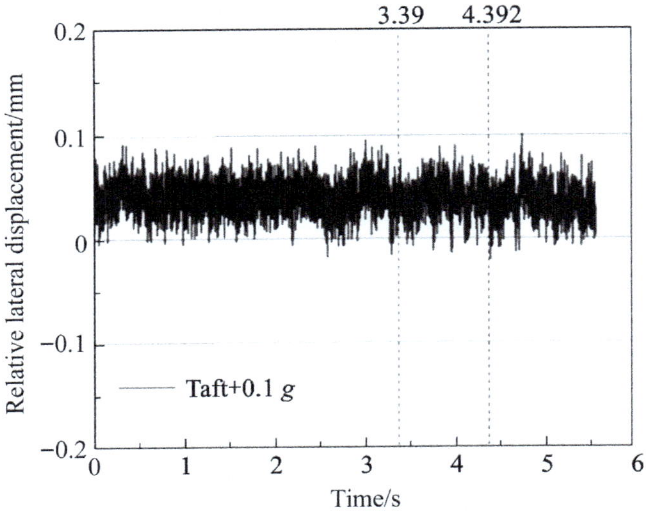

Figure 3.52 Lateral displacement difference between the rail and track slab at mid-span.

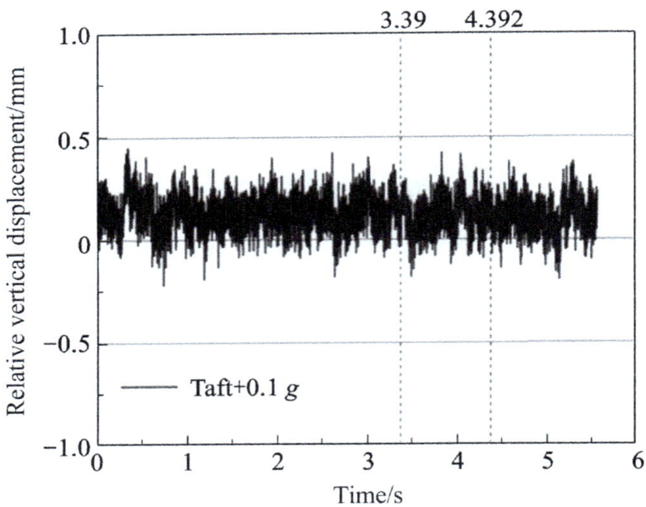

Figure 3.53 Vertical displacement difference between the track slab and bridge at mid-span.

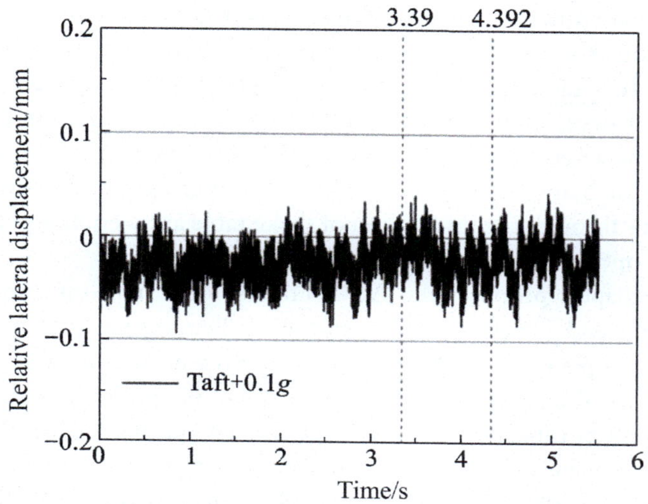

Figure 3.54 Lateral displacement difference between the track slab and bridge at mid-span.

slab is greater than that of the bridge, possibly due to insufficient bonding between them during model assembly, while the lateral displacement of the track slab is greater than that of the bridge, demonstrating the existence of positive-directional relative movement of the track slab. Comprehensive analysis demonstrates that the OpenSees simulation platform based on the moving element model can accurately simulate train-track-bridge coupling vibration under seismic conditions. This is a practical dynamic analysis method for train–bridge systems.

3.5 Summary

This chapter proposes similitude design parameters for the model tests of the experimental system for high-speed train operations on bridges. Based on the equipment performance and fabrication techniques of the shake table array system, this section presents a construction scheme for the experimental system for high-speed train operations on bridges under earthquake conditions, and analyzes the reliability of the prototype structure characteristics reproduced by the train–bridge interaction tests via theoretical analysis and numerical simulations. The research conclusions are summarized as follows.

(1) The experimental system for high-speed train operations on bridges under earthquake conditions proposed in this chapter consists of an acceleration section, a test section, a deceleration section, safety protection devices, and test equipment. This experimental system can reflect the characteristics and patterns of the prototype structure, achieving experimental testing of train operations on bridges under seismic conditions through the integration of shake table array equipment with the train operation system.

(2) The chapter proposes the ABC Theory for a train operating system testing, and indirectly addresses the challenge that scale model tests of train operations on bridges under seismic excitation cannot directly replicate the dynamic behavior of full-scale prototypes due to the inherent complexity of wheel–rail contact mechanics. By employing physical scale models to calibrate numerical simulation platforms and techniques, the calibrated numerical simulation is used to model the actual train-track-bridge system, thereby obtaining the dynamic response and safety evaluation results of the system.

(3) For the similitude design of the model for the high-speed train operations on bridges under an earthquake experimental system, key similitude design parameters to be considered include the geometric similarity ratio, acceleration similarity ratio, and material property similarity ratio. Whenever conditions permit, a high geometric similarity ratio should be prioritized to enhance the accuracy of model tests, while accuracy loss due to undersized components should be prevented. The acceleration similarity ratio should be maintained at 1.0 to prevent gravity distortion. The material property similarity ratio should be maintained within a reasonable range to allow the use of equivalent substitute materials, while fine-tuning this ratio can further optimize the design.

(4) Based on the established experimental platform, the first safety experiment of high-speed train operations on bridges under earthquake conditions was conducted. A detailed experimental scheme was developed, including model fabrication, weight application, test condition design, and data acquisition schemes. The dynamic test data of a train–bridge interaction under the coupling effect of earthquake and train were obtained.

(5) Building on the physical scale model for train operations on bridges, the previously established OpenSees train-track-bridge system simulation analysis platform was employed to develop an OpenSees FE model for indoor scaled testing of high-speed train operational safety on bridges under seismic excitation. By comparing the FE model with the experimental data, the accuracy of the OpenSees train-track-bridge system simulation analysis platform was validated, demonstrating that the OpenSees simulation platform based on the moving element model can accurately simulate the train-track-bridge coupling vibration under earthquake conditions.

References

[1] He WL. Dynamic finite element response analysis of subway station seismic performance based on soil-structure interaction. Changsha: Central South University Press; 2011 [In Chinese].
[2] Liu HY. Dynamic response and operational safety evaluation of high-speed railway long-span continuous steel truss bridges under seismic loads based on the moving element model. Changsha: Central South University; 2019 [In Chinese].
[3] Ding SS. Study on the key technology of the high-speed train carbody. Beijing: Beijing Jiaotong University; 2016 [In Chinese].
[4] Wang YH, Cheng WR, Lu F, et al. Development of the shaking table. Earthq Resist Eng Retrofit 2007;5:53–56+67 [In Chinese].
[5] Zhai WM. Vehicle-track coupled dynamics-theory and applications. Beijing: Science Press; 2020.
[6] Jiang LZ, Shao GQ, Wang H, et al. Experimental study on seismic performance of hollow piers with rounded rectangular cross section in high-speed railways. Eng Mech 2014;31(3):72–82 [In Chinese].
[7] Jiang LZ, Shao GQ, Jiang JJ, et al. Experimental study on seismic performance of solid piers with rounded rectangular cross section in high-speed railways. China Civ Eng J 2013;46(03):86–95 [In Chinese].
[8] Gu Q, Li WQ, Guo W, Liu YD, Jiang LZ, Yu ZW. A novel two-dimensional wheel-rail interaction element and its implementation in OpenSees. J China Railw Soc 2018;40(04):98–105 [In Chinese].
[9] Feng YL. Lateral irregularity of CRTS II ballastless track on high-speed railway continuous girder bridges accounting for seismic-induced damage. Changsha: Central South University; 2019 [In Chinese].

CHAPTER 4

Real-Time Hybrid Simulation Experiments of High-Speed Train Operation on Bridges

Contents

High-Speed Railway Train–Track–Bridge Systems
ISBN 978-0-443-49310-2
https://doi.org/10.1016/B978-0-443-49310-2.00004-1

4.1 Overview

High train speeds intensify the train-bridge dynamic interaction, posing significant challenges to operational safety and stability. Traditional experimental methods, such as the pseudo-static test, shake table test, and pseudo-dynamic test, cannot satisfactorily reproduce dynamic responses in real-time experimental conditions.

To address these limitations, this study employs real-time hybrid simulation (RTHS), an advanced experimental method that combines numerical simulation with physical testing to evaluate the dynamic performance of structures. Its basic principle is illustrated in Fig. 4.1. RTHS is applied to high-speed train-bridge coupling vibration research, where the bridge serves as the numerical substructure and the train as the experimental substructure. The real-time data interaction between the two substructures facilitates full-scale vehicle operation simulation within a limited experimental space. Under constrained budgets and limited space, it yields more accurate responses than pure numerical simulations, providing more valuable references for the research and design of high-speed train-bridge interactions. RTHS can authentically and accurately capture specimen performance and responses, including inertial forces, damping forces, and restoring forces, enabling in-depth research on complex components. Particularly for velocity-dependent components, RTHS currently stands as the most effective and precise experimental method. Furthermore, as the experimental substructure is part of the components of the entire structure, RTHS features low requirements for test facilities and equipment and high cost-effectiveness. From this perspective, the RTHS methodology demonstrates substantial potential for future development.

Currently, research on train-bridge coupling vibration is primarily carried out through field testing and numerical simulation. However, field testing requires the construction of actual railway lines, which are prohibitively expensive, while numerical simulation is based on multiple assumptions, often yielding results with substantial errors. In this context, RTHS technology emerges as a solution to these challenges. This chapter introduces the hybrid train-bridge testing method, and details the primary technical challenges in three aspects: hybrid simulation integration algorithms, large-scale real-time computing technology, and time-delay compensation technology for RTHS. Finally, this chapter presents the Hybrid Simulation and Analysis Experimental Platform (HSAEP).

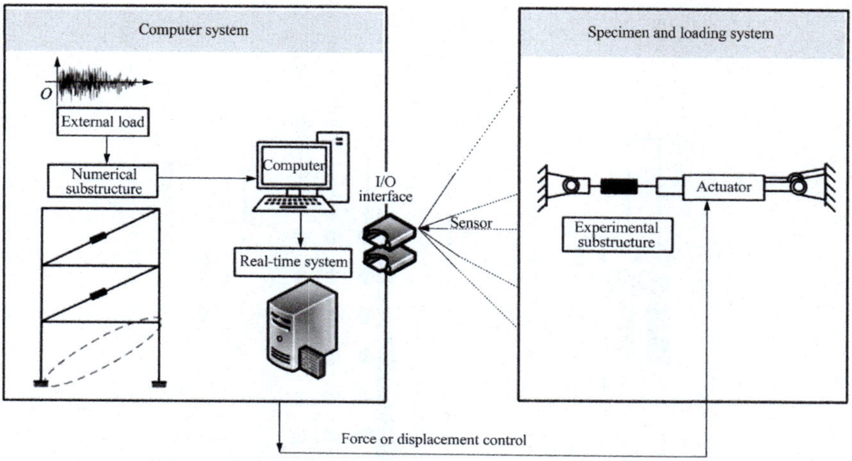

Figure 4.1 Schematic diagram of the RTHS principle.

4.2 Hybrid Simulation for High-Speed Train Operation on Bridges

4.2.1 RTHS Framework for High-Speed Train-bridge Systems

The train-track-bridge system in RTHS is divided into the numerical substructure and the experimental substructure, with its system composition and interface division shown in Fig. 4.2. The track-bridge subsystem serves as the numerical substructure, simulated in the computer using the moving load superposition algorithm for rapid solution, and then fed back to the experimental substructure. The train, serving as the experimental substructure, is fabricated into a scale model vehicle and tested on a unidirectional shake table. The equation of motion of the train-track-bridge system in RTHS can be expressed by the following equation:

$$\boldsymbol{M}_{\mathrm{N}}\ddot{\boldsymbol{x}}_{\mathrm{N}} + \boldsymbol{C}_{\mathrm{N}}\dot{\boldsymbol{x}}_{\mathrm{N}} + \boldsymbol{K}_{\mathrm{N}}\boldsymbol{x}_{\mathrm{N}} = F_{\mathrm{Train}} \tag{4.1}$$

where $\boldsymbol{M}_{\mathrm{N}}$, $\boldsymbol{C}_{\mathrm{N}}$, and $\boldsymbol{K}_{\mathrm{N}}$ represent the mass matrix, damping matrix, and stiffness matrix of the numerical substructure, respectively; $\ddot{\boldsymbol{x}}_{\mathrm{N}}$, $\dot{x}_{\mathrm{N}}$, and $\boldsymbol{x}_{\mathrm{N}}$ denote the acceleration, velocity, and displacement vectors of the numerical substructure; F_{Train} is the interface force at the boundary between the numerical and experimental substructures, representing the wheel-rail interaction force between the train and the track.

At each time step of RTHS, first, the nodal displacements of the track under train loading are computed in real time by the numerical

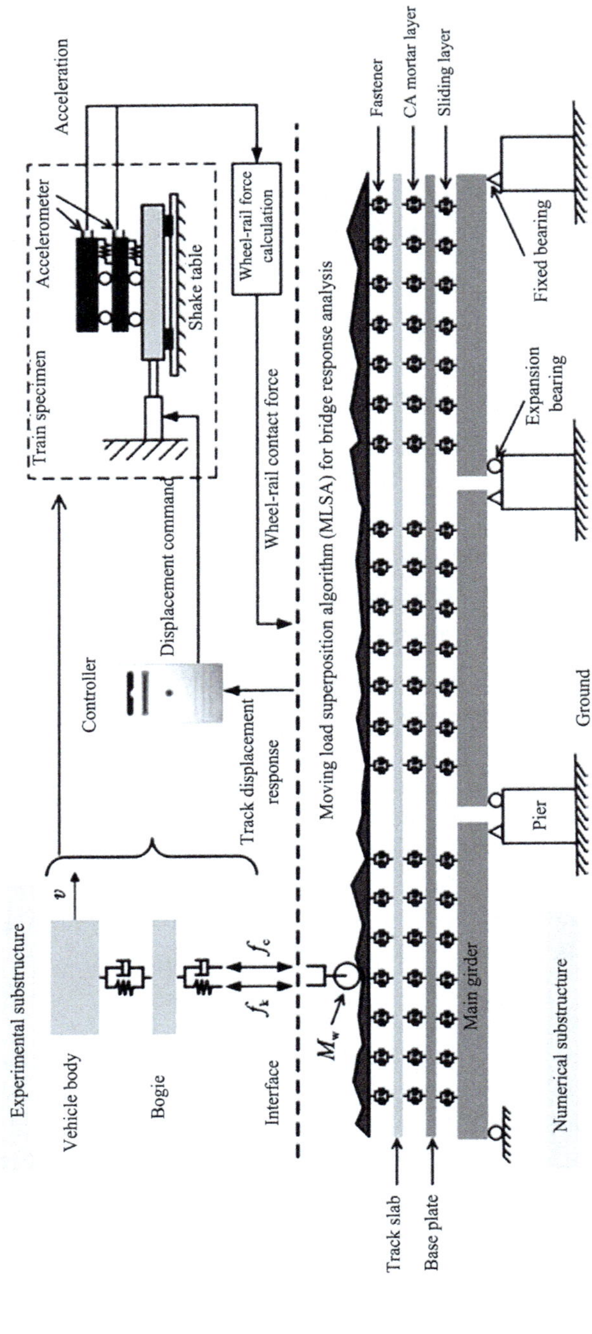

Figure 4.2 System composition and interface partitioning of the numerical and experimental substructures of RTHS experiments for high-speed train operation on bridges.

substructure; then, the track displacement responses through the real-time data transfer system are transmitted to the controller, with the controller sending command signals to the shake table to apply loading to the train; finally, the measured wheel-rail interaction forces are fed back to the numerical substructure to proceed with the next computation step. The interface between the numerical and experimental substructures keeps changing as the train moves along the track-bridge structure at speed v. This means the loading position on the numerical substructure is constantly shifting. The continuously changing boundary position between the numerical and experimental substructures represents a fundamental distinction from hybrid simulation in earthquake engineering, where the numerical–physical interface remains fixed throughout the experiment.

To reproduce the actual vibration response of trains, an RTHS framework for train-bridge coupling systems is proposed. As shown in Fig. 4.3, the division interface between the numerical and experimental substructures is defined at the wheel-rail contact point. The numerical and experimental substructures are interconnected through loading devices, sensors, and digital signal processing hardware, forming a closed-loop feedback system that solves the dynamic response of the entire structure at each time step. In the RTHS method for train-bridge coupling systems, the physical train model is installed on a shake table or other loading device, while the remaining track-bridge subsystem is simulated in real time. The vertical deformation at the wheel-rail contact point is applied to the train, the measured wheel-rail force is fed back to the numerical substructure, and then the deformation for the next time step is calculated. Rail deformation and wheel-rail forces continuously interact at the substructure division interface, forming a complete train-track-bridge system.

4.2.1.1 Numerical Substructure

Simply supported girder bridges constitute the primary bridge type in high-speed railway bridge design [1]. In the hybrid simulation of train operations on bridges, the track-bridge system is generally used as the numerical substructure. Yan et al. 's research indicates that in existing high-speed railway lines, short-span simply supported girder bridges constitute the majority of bridge structures, with the remainder consisting of continuous girder bridges and medium-to-long-span simply supported girder bridges [2]. This section will introduce the numerical substructure by taking a three-span simply supported high-speed railway bridge with the CRTS II slab track system as an example.

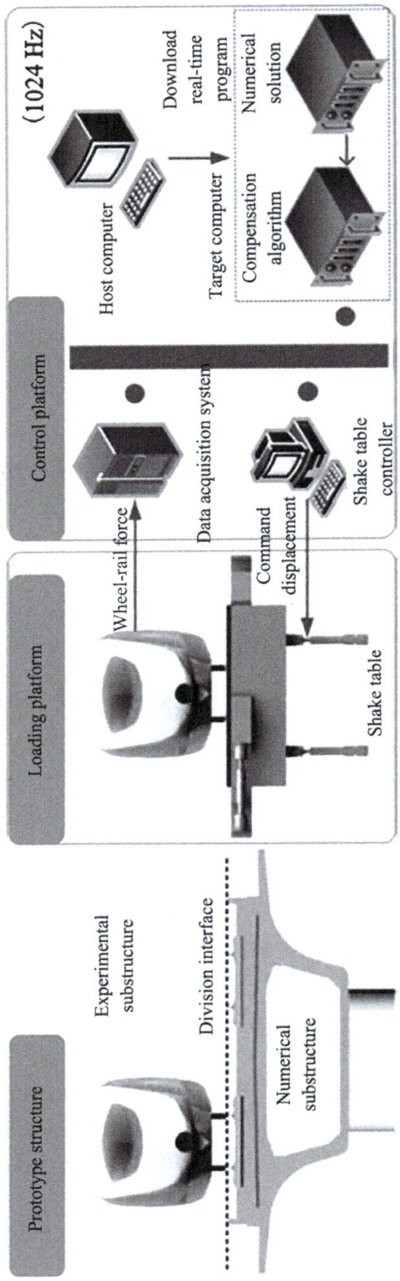

Figure 4.3 RTHS framework for train-bridge coupling systems.

The bridge structure is a three-span simply supported girder bridge. The simply supported box girder is 32.6 m long, with a calculated span length of 31.5 m. The bridge span layout is shown in Fig. 4.4A, and the total length of the bridge section is 97.8 m. The simply supported girders adopt a "fixed–free–fixed–free" bearing arrangement pattern, with the fixed bearings on the left side of the girder and the movable bearings on the right side. PZ-5000 series pot rubber bearings are employed, as shown in Fig. 4.4B. The bridge adopts double-streamlined round-ended solid piers, with pier heights of 16 and 8 m, respectively. To fully simulate the seismic response of this high-speed railway bridge, the structural model includes 50-m-long track-bridge transition sections on both sides, each comprising a base plate, a friction slab, and an end anchor. The superstructure of the simply supported girder bridge adopts the CRTS II slab ballastless track system. The track slab features a layered structure, as shown in Fig. 4.4C, comprising the track slab, CA mortar layer, base plate, two-geotextile-one-membrane sliding layer, shear reinforcement, and shear grooves. The rail uses a CHN60 profile with WJ-8C fasteners specifically designed for CRTS II slab tracks.

4.2.1.2 Experimental Substructure

In the train-bridge RTHS, the train is typically designated as the experimental substructure. Components such as springs, dampers, and suspension systems within the train are fabricated through scale-down modeling. In this section, the most fundamental train suspension system is used as an example to demonstrate the manufacturing process of the experimental substructure.

The train suspension system primarily consists of springs and dampers, designed to filter excitations between the vehicle body, bogie, and rails. Considering the symmetry of the train, a quarter-vehicle model is adopted for vertical suspension vibration research.

The equation of motion of a simplified quarter-vehicle model can be defined as follows:

$$\begin{cases} m_c\ddot{z}_c = -k_c(z_c - z_b) - c_c(\dot{z}_c - \dot{z}_b) - m_c g \\ m_b\ddot{z}_b = -k_c(z_c - z_b) - c_c(\dot{z}_c - \dot{z}_b) - k_b(z_b - z_w) - c_b(\dot{z}_b - \dot{z}_w) - m_b g \\ m_w\ddot{z}_w = F_w - k_b(z_b - z_w) - c_b(\dot{z}_w - \dot{z}_b) - m_w g \end{cases} \tag{4.2}$$

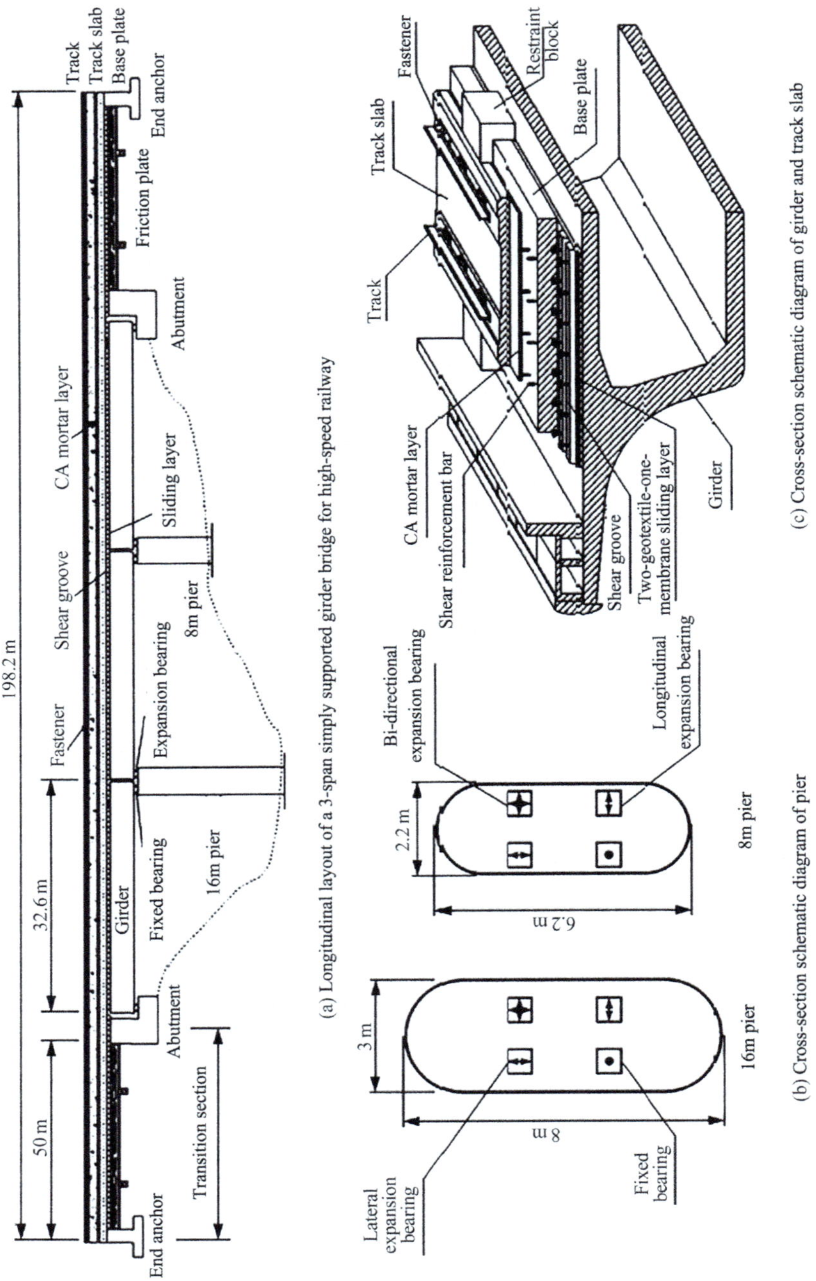

(a) Longitudinal layout of a 3-span simply supported girder bridge for high-speed railway

(b) Cross-section schematic diagram of pier

(c) Cross-section schematic diagram of girder and track slab

Figure 4.4 Schematic diagram of a high-speed railway track-bridge system.

Table 4.1 Model vehicle parameters.

Symbol	Prototype Parameters	Scale Parameters	Description
m_c	8446.5 kg	422.33 kg	Vehicle body mass
k_c	225,000 N/m	11,250 N/m	Vehicle body stiffness
c_c	10,000 N·s/m	500 N·s/m	Vehicle body damping
m_b	1028 kg	51.4 kg	Bogie mass
k_b	1,772,000 N/m	88,600 N/m	Bogie stiffness
c_b	20,000 N·s/m	1000 N·s/m	Bogie damping
m_w	1267 kg	–	Wheelset mass

where F_w denotes the wheel-rail interaction force; z and its first and second derivatives represent the displacement, velocity, and acceleration of the vehicle body, respectively; the subscripts c, b, and w correspond to the vehicle body, bogie, and wheelset, respectively.

The wheel-rail interface is treated as a rigid contact, where the vertical displacement of the wheelset is consistent with the vertical deformation of the rail at the contact point. Consequently, the wheelset displacement equals the sum of static irregularity and dynamic irregularity. According to Eq. (4.2), the wheel-rail force can be expressed as:

$$F_w = m_c\ddot{z}_c + m_b\ddot{z}_b + m_w\ddot{z}_w + (m_c + m_b + m_w)\cdot g \tag{4.3}$$

As shown in Eq. (4.3), the wheel-rail force can be obtained by combining the inertial force and the gravity of the mass block. According to this equation, the measurement method of the wheel-rail force in the experiment can be determined solely by measuring the acceleration of each mass block. The parameters of the model vehicle are calculated based on the vehicle body of the CRH380A prototype with a mass scale ratio of 0.05. The parameters of the simplified model vehicle are listed in Table 4.1.

4.2.1.3 Key Technologies

Hybrid simulation is categorized into RTHS and offline iterative hybrid simulation (OIHS). In RTHS, a feedback loop is formed between the numerical and experimental substructures, interconnected via actuators, sensors, and digital signal processing hardware, enabling the solution of equations of motion at each integration time step. In contrast, offline hybrid simulation performs global iterative computations after the completion of each time history segment.

For RTHS, numerous technical challenges need to be addressed. To meet the heightened demand for computational speed and accuracy in large-scale model real-time computing, more efficient large-scale real-time computing technologies are required. To address the time delays caused by loading devices and signal transmission systems during RTHS, real-time delay compensation techniques must be employed. To achieve better simulation and solution of numerical substructures, suitable hybrid simulation algorithms are required. These key technologies will exert a direct and profound impact on RTHS. Our research team has conducted systematic and in-depth research on these critical issues, and this chapter will elaborate on each of these key technologies in detail.

4.2.2 RTHS for High-Speed Train-bridge Systems

After the numerical and experimental substructures are built, an RTHS numerical demonstration is performed.

A moving load convolution integral method (MLCIM) analysis is performed for the three-span simply supported girder bridge subjected to post-earthquake conditions, to compute the unit impulse response matrix at wheel-rail contact points. Meanwhile, the displacement response matrix of the three-span simply supported girder bridges at midspan is calculated. MLCIM is a numerical approach based on convolution integration that accounts for the time-varying characteristics of train moving loads. Detailed explanations of MLCIM are provided in Section 4.4 of this book.

The test accounts for two working conditions: with and without rotational friction dampers (RFDs) [3] for shock absorption. The train speed is set at 350 km/h to take track irregularities into account.

4.2.2.1 RTHS Simulink Program

In the Simulink software, a hybrid simulation program for train-bridge coupling systems is established. Fig. 4.5 shows the Simulink program flowchart. The MLCIM component in this program functions to calculate the response of the track-bridge structure under wheel-rail force excitation. Meanwhile, the vehicle suspension model simulates the motion of a two-degrees-of-freedom (DOF) vehicle suspension system traversing the bridge and derives the wheel-rail interaction forces exerted by the vehicle suspension on the track-bridge structure. The vehicle suspension model is

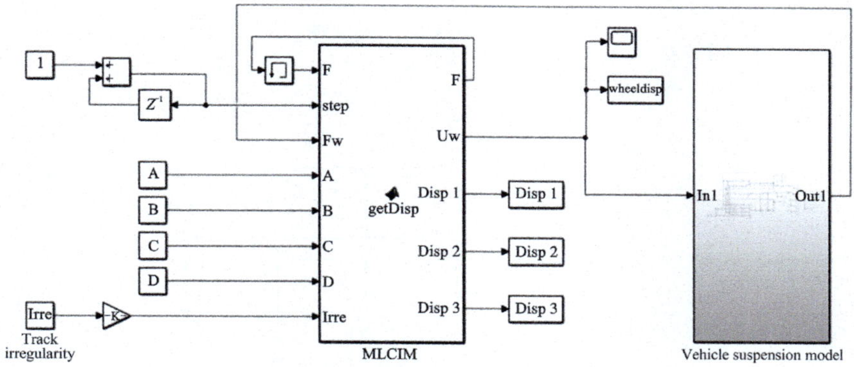

Figure 4.5 Simulink block diagram.

illustrated in Fig. 4.6. The wheelset displacement calculated from the track-bridge structure is transmitted to the shake table controller as the command. This drives the scaled vehicle model fixed on the shake table, and wheel-rail forces are computed using acceleration sensors mounted on the model mass block and then fed back into the MLCIM computational module.

Fig. 4.7 shows the structural responses of the train at 350 km/h without considering track irregularities. As shown in Fig. 4.7A, the wheelset displacement exhibits distinct periodicity when traversing the three-span simply supported girder bridge, with two peaks clearly identifiable in the response curve. Each peak value represents the train passing through a bridge pier, and each valley value represents the train passing through a bridge span, presenting the process of the train passing through a three-span simply supported girder bridge at a speed of 350 km/h. It is clearly revealed that the train response is related to the vertical stiffness of the track-bridge structure. Since the vertical stiffness of the line–bridge transition section is greater than that of the track-bridge structure, the wheelset displacement within the line–bridge transition section is relatively small. Within the bridge spans, since the vertical stiffness of the pier is greater than the vertical stiffness of the beam, the wheelset displacement near the pier is smaller than the wheelset displacement in the middle of the bridge span. When the train traverses regions with significant vertical stiffness variations, abrupt changes occur in the gradient of wheelset displacement. This is also reflected in the vertical wheel-rail contact forces, as evidenced in Fig. 4.7B, which exhibit characteristic oscillations about the static wheel load. Figs. 4.7C and D show the displacement responses of the bogie and

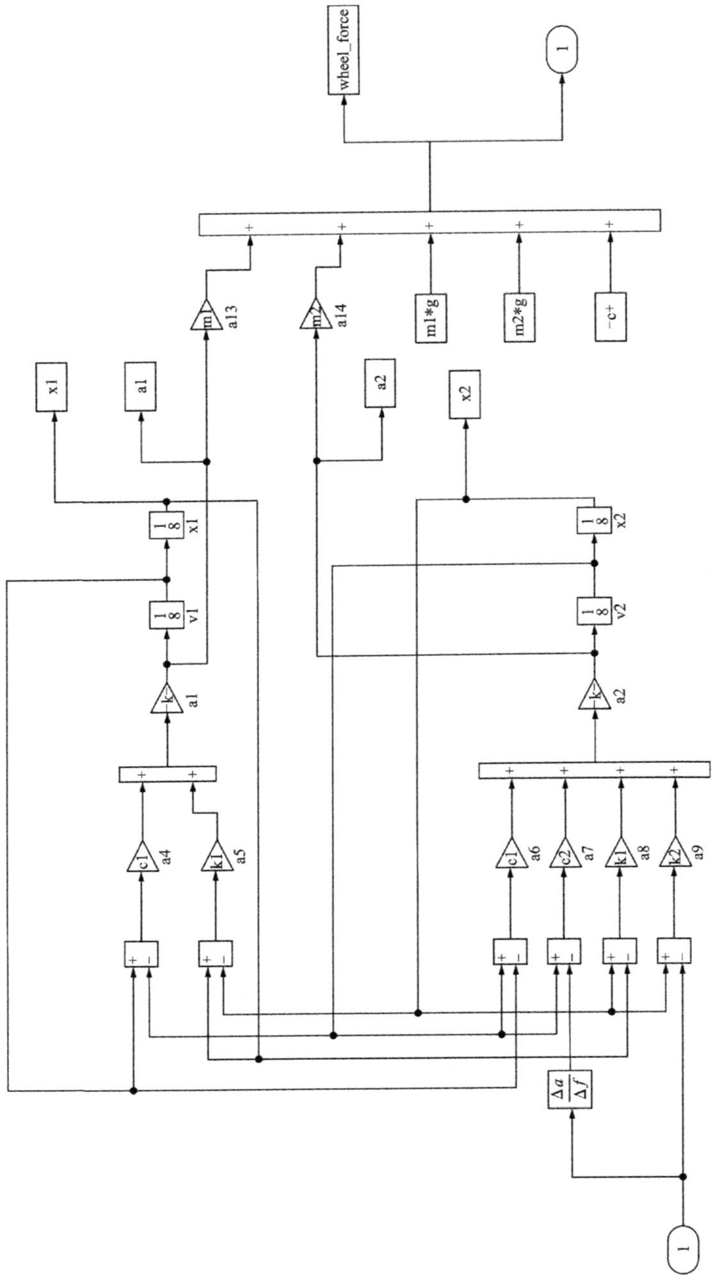

Figure 4.6 Suspension model for a vehicle body.

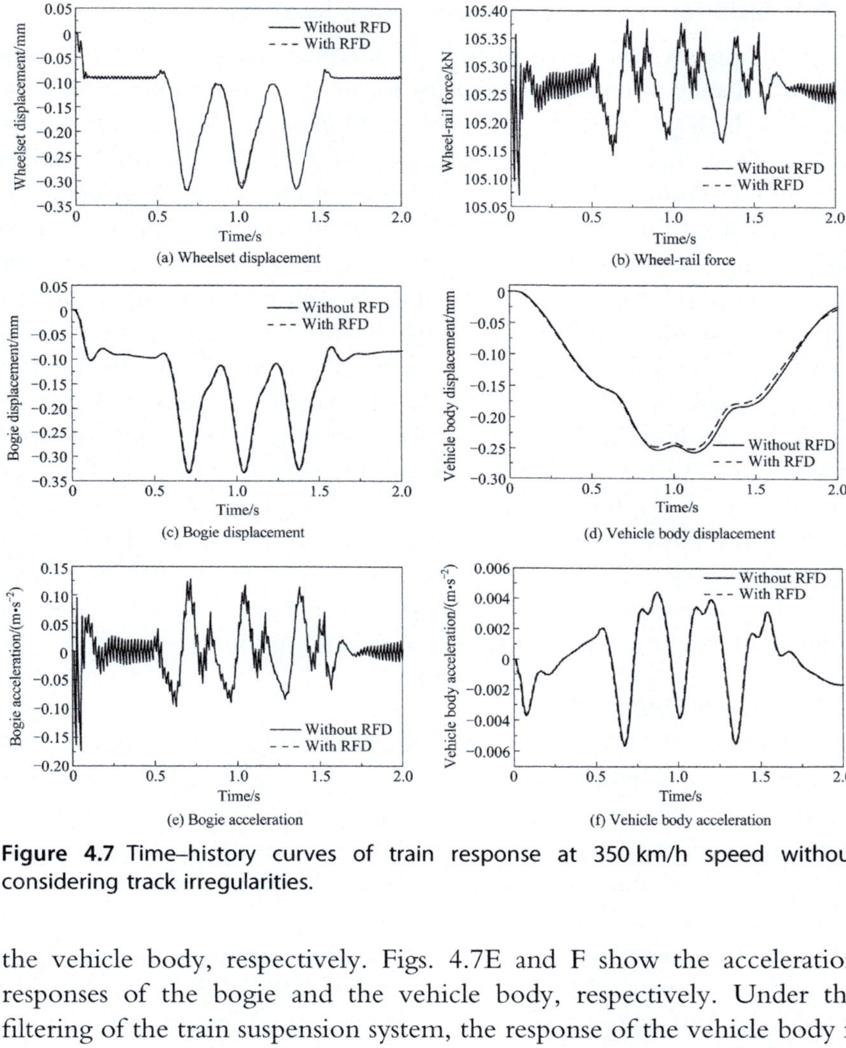

Figure 4.7 Time–history curves of train response at 350 km/h speed without considering track irregularities.

the vehicle body, respectively. Figs. 4.7E and F show the acceleration responses of the bogie and the vehicle body, respectively. Under the filtering of the train suspension system, the response of the vehicle body is smaller than that of the bogie. Fig. 4.7 compares the impact of installing RFDs on post-earthquake vehicle running performance in bridge structures. Overall, the installation of RFDs has little effect on the train displacement response. A slight reduction in displacement response is observed across wheelset, bogie, and vehicle body displacements under RFD damping control, while the acceleration responses of the wheelset and bogie are almost identical. However, a marginal decrease in vehicle body acceleration is detectable when RFDs are implemented.

4.2.3 OIHS for High-Speed Train-bridge Systems

RTHS imposes stringent real-time requirements, where the large-scale servo-hydraulic system, as an actuator for experimental substructures, demands a high-precision control system. Moreover, the computation for large-scale complex structures must be completed within milliseconds, presenting significant technical challenges. In contrast to RTHS, researchers have proposed an innovative OIHS method, which has been successfully implemented in automotive tire durability tests and nonstructural component experiments. Nonstructural components are typically equipment and parts in large buildings or automotive systems. Structurally, they resemble trains running on a bridge, but the bridge-crossing trains are mobile equipment, whereas the nonstructural components are predominantly fixed installations [4,5]. Accordingly, we have developed an OIHS system for high-speed train-track-bridge systems, and proposed two control algorithms for signal convergence between subsystems: the fixed-point iteration algorithm and the model identification convergence algorithm. These algorithms perform signal corrections from time-domain and frequency-domain perspectives, respectively. Here, we primarily introduce the fixed-point iteration algorithm.

4.2.3.1 OIHS Framework

Compared to RTHS, which simulates train-bridge interaction states at each time step, OIHS simulates actual train-running conditions on bridges by iteratively correcting force or displacement time–history response errors between numerical and experimental substructures until boundary equilibrium coordination is attained. In OIHS, the fixed-point iteration algorithm and the model identification convergence algorithm are employed for iteratively correcting response errors. The OIHS method is implemented by constructing an outer-loop controller based on the inherent inner-loop controller within a shake table system [6], with the operational principle illustrated in Fig. 4.8.

The OIHS method divides the train-track-bridge system into a physical train substructure and a numerical track-bridge substructure, with the wheel-rail contact interface serving as the partitioning boundary. The subsystems are coupled through wheel-rail interaction forces, while external excitations on the train subsystem (excluding coupling vibrations) can be defined by track irregularities. The coordinate systems of the train substructure and the track-bridge substructure maintain identical axial

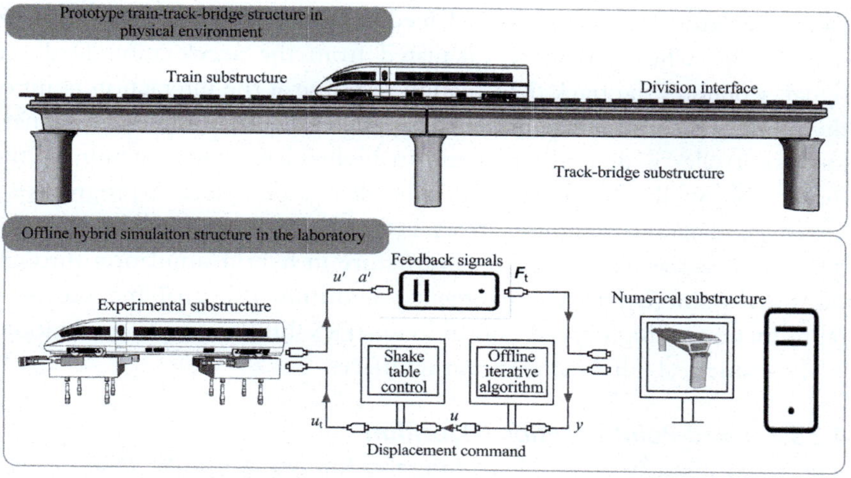

Figure 4.8 OIHS method in train-track-bridge coupling vibration analysis.

orientations and dimensional scales. Within the train substructure, each rigid body is assigned an independent coordinate system, with its origin positioned at the static equilibrium point of the respective rigid body.

$$\begin{cases} \boldsymbol{M}_{\mathrm{t}}\ddot{\boldsymbol{X}}_{\mathrm{t}} + \boldsymbol{C}_{\mathrm{t}}\dot{\boldsymbol{X}}_{\mathrm{t}} + \boldsymbol{K}_{\mathrm{t}}\boldsymbol{X}_{\mathrm{t}} = \boldsymbol{F}_{\mathrm{t}} \text{ (Train substructure)} \\ \boldsymbol{M}_{\mathrm{b}}\ddot{\boldsymbol{X}}_{\mathrm{b}} + \boldsymbol{C}_{\mathrm{b}}\dot{\boldsymbol{X}}_{\mathrm{b}} + \boldsymbol{K}_{\mathrm{b}}\boldsymbol{X}_{\mathrm{b}} = \boldsymbol{F}_{\mathrm{b}} \text{ (Track} - \text{bridge substructure)} \end{cases} \tag{4.4}$$

where $\boldsymbol{M}$, $\boldsymbol{C}$, and $\boldsymbol{K}$ represent the mass matrix, damping matrix, and stiffness matrix, respectively, which are constant matrices; $\boldsymbol{X}$ and $\boldsymbol{F}$ denote the displacement vector and force vector; the subscript t indicates the train subsystem; the subscript b represents the track-bridge subsystem; $\dot{}$ and .. correspond to the first-order and second-order time derivatives of displacement X.

In Fig. 4.8, all signals represent time histories in the Z-direction. The shake table control module serves as the inner-loop controller for reproducing the input command to the table; u denotes the displacement command generated by the offline iterative algorithm and fed into the inner-loop controller as part of the outer-loop iterative control strategy; u_{t} denotes the reproduced displacement response generated by the built-in inner-loop controller of the shake table through iterative correction with u as the target reference; u' represents the displacement response at the numerical–physical partition interface, i.e., the displacement of the shake table platform; a' is the measured feedback acceleration of the train; $\boldsymbol{F}_t$ denotes the wheel-rail force calculated from the acceleration feedback signal. y denotes the track deformation response at the interaction point of the numerical substructure under the applied wheel-rail force. The inner-loop controller is designed to perform high-fidelity reproduction of the shake table while the outer-loop controller is designed to progressively minimize dynamic response errors between the numerical bridge substructure and the physical train substructure in hybrid simulation through the fixed-point iteration convergence algorithm [7], enabling precision reproduction of high-speed railway train-track-bridge coupling vibrations at the numerical–physical partition interface.

4.2.3.2 Fixed-Point Iteration Algorithm

The fixed-point iteration algorithm, also known as simple iteration, is a mathematical operation for solving the equation $f(x) = 0$ within the interval $[a, b]$. To find the solution of the equation $f(x) = 0$, first transform the equation equivalently into the form $x = g(x)$. Then, by defining an initial value x_0, perform iterative computation using the recurrence relation $x_{k+1} = g(x_k)$ until the convergence criterion is satisfied [7].

The fixed-point iteration algorithm has specific convergence conditions: the iteration function $g(x)$ must be continuous on $[a, b]$ and satisfy the following criteria:

① When $x \in [a, b]$, $a \leqslant g(x) \leqslant \mathrm{b}$.

② There exists a positive number L such that $0 < \mathrm{L} < 1$, and $\forall x \in [a, b]$, such that $|g'(x)| \leqslant \mathrm{L}$.

Then:

① The equation $x = g(x)$ has a unique solution x^* in $[a, b]$.

② For any initial value $x_0 \in [a, b]$, the iteration $x_{k+1} = g(x_k)$ converges to x^*.

③ $\left|x_k - x^*\right| \leqslant \frac{L}{1-L}\left|x_k - x_{k-1}\right|$.

④ $\left|x_k - x^*\right| \leqslant \frac{L^k}{1-l}\left|x_1 - x_0\right|$.

During the OIHS of high-speed railway train operations on bridges, the fixed-point iteration algorithm is employed to determine the response time history that satisfies the coupling vibration performance of the train-track-bridge system. Specifically, the algorithm ensures that the displacement

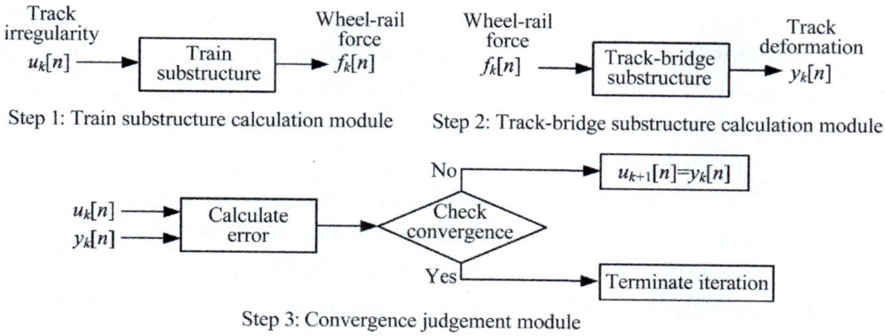

Figure 4.9 Fundamental steps of the fixed-point iteration algorithm.

error between the wheelset displacement of the train subsystem and the rail contact point displacement of the track-bridge subsystem at the partitioned boundary converges to a predefined tolerance limit. The fundamental computational procedure of the fixed-point iteration algorithm can be divided into three steps, as illustrated in Fig. 4.9. In this process, $u_k[n]$ represents the track irregularity affecting the train subsystem during the k-th iteration, $f_k[n]$ denotes the wheel-rail contact force computed in the k-th iteration, and $y_k[n]$ corresponds to the track deformation response at the wheel-rail contact point obtained in the k-th iteration.

The detailed computational process of the fixed-point iteration algorithm consists of the train substructure calculation module, the track-bridge substructure calculation module, and the convergence evaluation module.

① Train substructure calculation module: Calculate the wheel-rail forces of the train substructure. The track irregularity time–history excitation of the current iteration serves as the input signal to the train substructure. This excitation is applied to the physical vehicle model via actuators, followed by measurement of the accelerations at the vehicle body, bogies, and wheelsets. The wheel-rail force time history is then computed.

② Track–bridge substructure calculation module: Apply the wheel-rail force time history of the moving train from Step ① as the input of the track-bridge subsystem. Solve for the rail deformation response of the track-bridge system, and obtain the vertical displacement of the wheelset at the wheel-rail contact point by combining track irregularities.

③ Convergence evaluation module: Compute the error between the vertical wheel-rail displacement response obtained in Step (2) and the track irregularity excitation used in Step (1). If the error meets the predefined tolerance, terminate the iterative solution; otherwise, return to Step ① and use the wheel-rail vertical displacement response obtained in Step ② as the track irregularity excitation for the next iteration.

The error in Step ③ is quantified by the root mean square error (*RMSE*), which evaluates the iteration performance based on the deviation between the bridge responses and the system input. The equation is as follows:

$$RMSE_k = \sqrt{\frac{\sum_{i=1}^{N} [y_k(i) - u_k(i)]^2}{\sum_{i=1}^{N} [y_k(i)]^2}} \tag{4.5}$$

where, $y_k(i)$ represents the system input at the k-th iteration; $u_k(i)$ represents the system output response at the k-th iteration.

The workflow of OIHS based on fixed-point iteration is entirely different from RTHS, as illustrated in Fig. 4.10. In RTHS, the train substructure and track-bridge substructure are iteratively solved simultaneously at each time step, with convergence verification based on the dynamic response at the end of each time step. In contrast, in OIHS the two substructures are solved separately with full time–history data during each iteration cycle. Convergence is then verified by comparing the vertical displacement responses of the wheelset at the wheel-rail contact point between consecutive iterations, until the error threshold is satisfied.

4.2.3.3 Experimental Verification

The numerical–physical partition interface in the experimental system for high-speed railway train operations on bridges under earthquakes is determined to be between the train and the track, and is first studied and tested in a numerical environment.

The Simulink block diagram of the high-speed railway train-track-bridge OIHS system based on Jiang's and Ren et al. 's research [8,9] is shown in Fig. 4.11. The displacement input signal of the system acts on the train substructure, exciting the model vehicle to generate wheel-rail forces that act on the track-bridge substructure, with bridge vertical displacement responses serving as system outputs. The error between the system input and output is corrected by outer-loop fixed-point iteration, ultimately

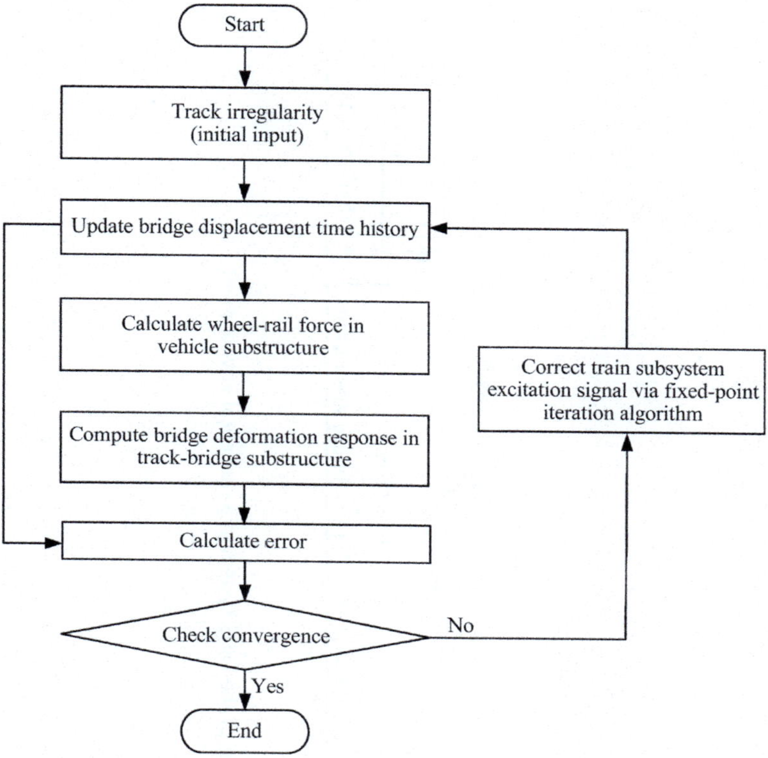

Figure 4.10 Operating procedure for OIHS based on fixed-point iteration.

achieving dynamic matching of coupling vibration in high-speed railway train-track-bridge systems.

In the simulation test, the vertical track irregularity data generated using China's low-interference spectrum are employed. When the train crosses the bridge at a speed of 300 km/h with a time step of 0.01 s, the time-domain inversion sample of the power spectral density function S(f) is calculated, yielding a maximum value of 0.861 mm.

Here, we will verify OIHS tests for high-speed train operation on bridges in the numerical environment.

(1) Simulation of high-speed train operation on bridges without track irregularities

For OIHS without track irregularities, the initial input of the system is zero. However, considering that the simplified modeling of the high-speed railway train-track-bridge system may cause certain impacts on the results, and that the establishment of the OIHS system is based on the ability of the shake table to stably reproduce vibrations without accounting for errors

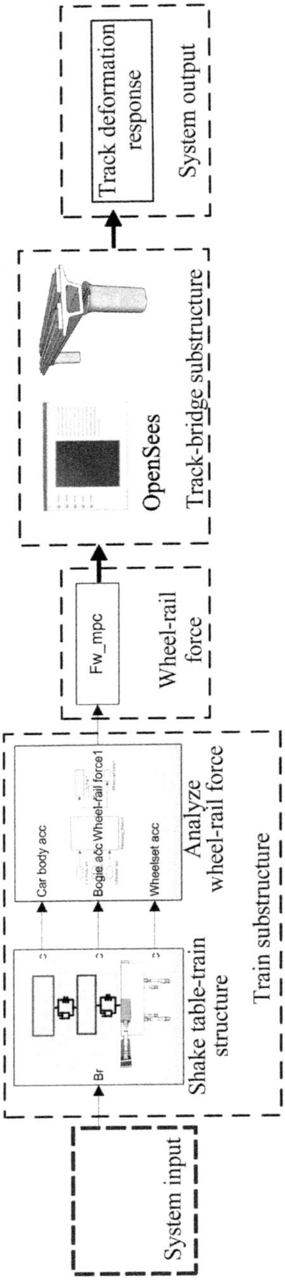

Figure 4.11 Simulink architecture of the OIHS system in a numerical environment.

caused by reduced reproduction accuracy of the shake table, the existing errors are added to the initial input of the system in the form of random white noise. The initial input of the system is applied to the train subsystem at a speed of 300 km/h to conduct OIHS. The system input signals and bridge response signals within the same iteration are recorded, compared with the standard solution of the train-bridge interaction simulation without track irregularities (Fig. 4.12), and the error between the bridge response and the standard solution is calculated. The comparison results are shown in Fig. 4.13.

Fig. 4.12A presents the time–history results of the simulation of high-speed train operation on bridges without track irregularities, while Fig. 4.12B shows the corresponding vertical displacement spectrum of the bridge. It can be observed that in the absence of track irregularities, the frequency range of the vertical displacement induced by high-speed train passage is 0.3–5 Hz.

Fig. 4.13 presents a comparison of the system input and bridge response in the simulation of train operations on bridges without track irregularities at different iteration counts, along with the error between the bridge response and the standard solution. It can be observed that as the number of iterations increases, the waveform of the bridge response gradually aligns with that of the standard solution, with the response error significantly decreasing and ultimately stabilizing at the order of 10^{-3} mm. To better illustrate the convergence effect of the OIHS method, the RMSE results of the bridge response in each iteration compared with the standard solution and the current system input are shown in Fig. 4.14.

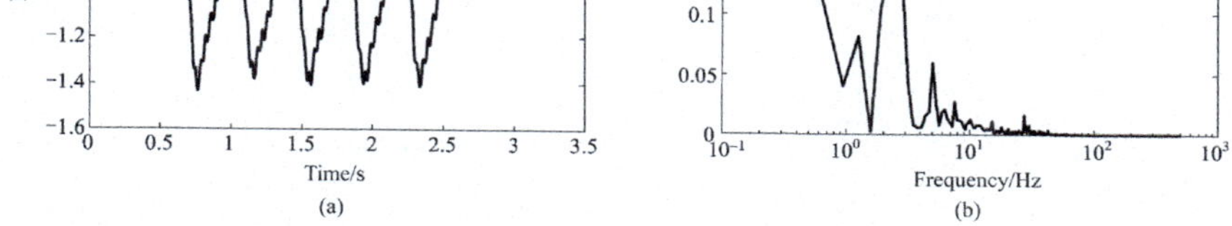

Figure 4.12 Vertical displacement response of a bridge after high-speed train operation on bridges without track irregularities.

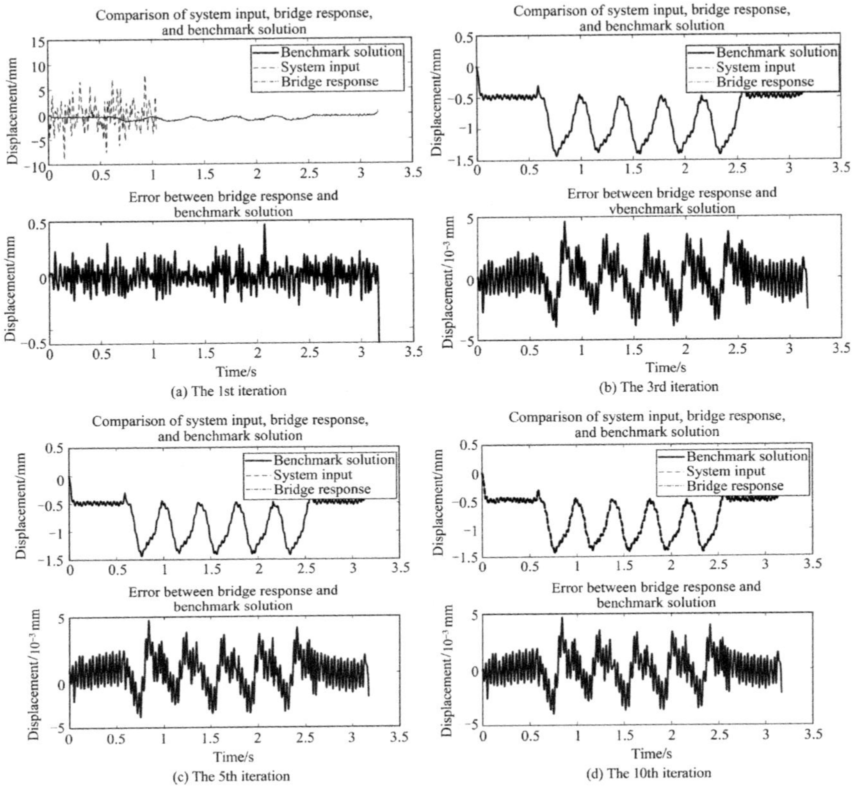

Figure 4.13 Comparison between OIHS results and standard solutions.

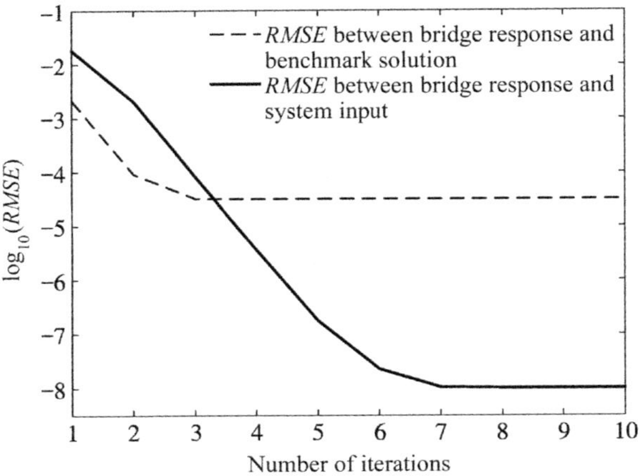

Figure 4.14 RMSE of the first 10 iterations.

The RMSE of each iteration further indicates that as the iteration count increases, the relative error between offline hybrid simulation results and train-bridge interaction simulation results gradually decreases. After 3 iterations, the error magnitude drops to 10^{-4} mm and remains stable; while the absolute error of the current system input decreases to 10^{-8} mm after 7 iterations and remains stable, meeting the convergence threshold requirement for approaching the standard solution.

(2) OIHS verification with track irregularities

The vertical track irregularity time-domain sample sequence, generated by the inverse Fourier transform method, is applied as the vertical external excitation for the high-speed railway train-bridge interaction model. Considering the errors introduced by simplified modeling and shake table reproduction accuracy, low-amplitude broadband white noise is superimposed on the track irregularity time history to simulate such effects. The initial system input is applied to the train subsystem traveling at 300 km/h, and an offline hybrid simulation is performed. The system input signals and bridge response signals from the same iteration are recorded and compared with the standard solution of the train-bridge interaction simulation with track irregularities, as shown in Fig. 4.15. Finally, the error between the bridge response and the standard solution is calculated, with the results presented in Fig. 4.16.

Fig. 4.15A shows the time–history simulation results of the high-speed railway train operations on bridges with track irregularities, while Fig. 4.15B shows the corresponding vertical displacement spectrum of the bridge. It can be observed from Fig. 4.15 that when track irregularities are

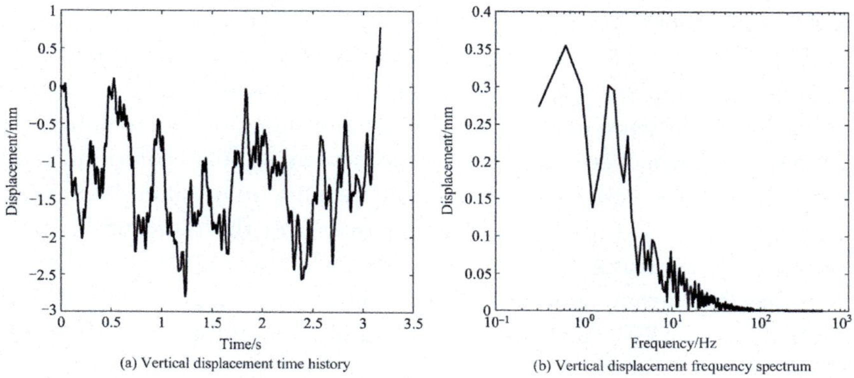

Figure 4.15 Vertical displacement response of the bridge after high-speed train operation on bridges with track irregularities.

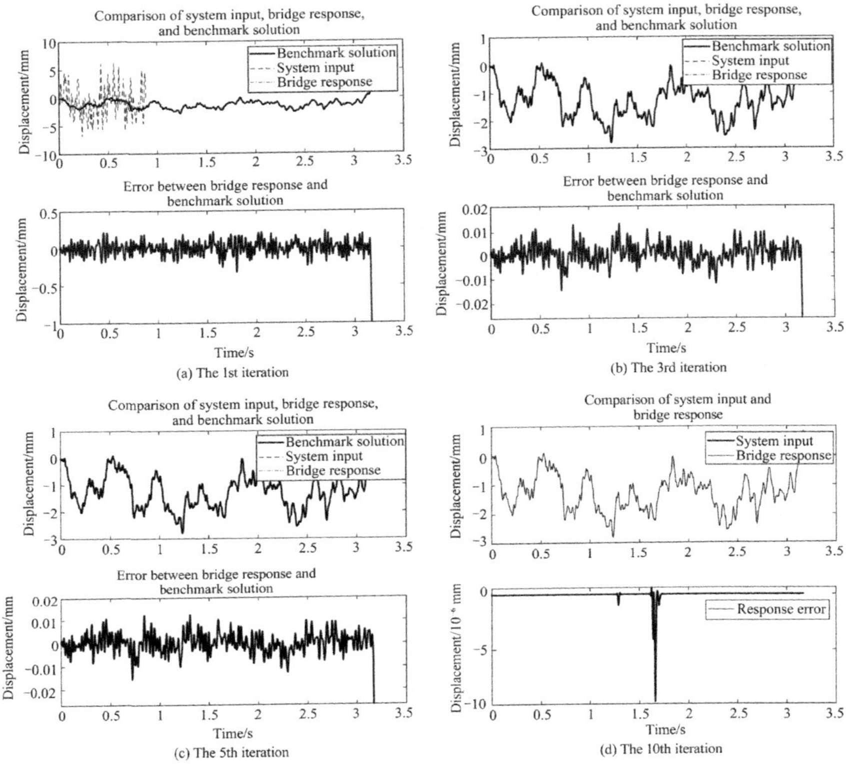

Figure 4.16 Comparison between OIHS results and standard solutions.

introduced, the dominant frequency of the vertical displacement of the bridge after train passage is 0.3–10 Hz, which corresponds to the characteristic frequencies of track irregularities.

Fig. 4.16 presents the comparison between system inputs, bridge responses, and standard solutions (i.e., simulation results of train operations on bridges with track irregularities) at different iteration steps, along with the response errors between bridge responses and standard solutions. Even when the initial system inputs are affected by white noise, after 3 iterations, the response error between bridge responses and the standard solutions significantly decreases and remains essentially at the 10^{-2} mm level. Likewise, to further demonstrate the convergence effect of the OIHS method, the RMSE values for each iteration are calculated and shown in Fig. 4.17.

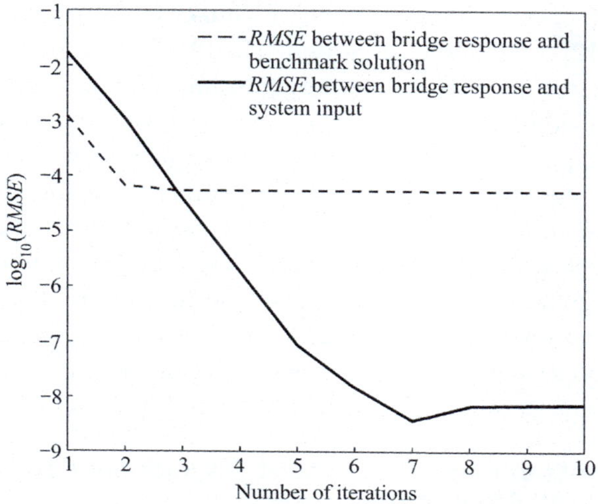

Figure 4.17 RMSE of the first 10 iterations.

The RMSE results further demonstrate that: as the iteration count increases, the relative error between OIHS results and the simulation results of train operations on bridges with track irregularities gradually decreases. After 3 iterations, the error magnitude drops to 10^{-4} mm and remains stable; while the absolute error of the current system input decreases to 10^{-8} mm after 8 iterations and remains stable, and the error is less than the threshold and meets the convergence threshold requirement for standard solutions.

4.3 RTHS Algorithms

The numerical substructure is typically simulated using Finite Element (FE) models, thus requiring an appropriate numerical integration algorithm for its solution. The numerical integration algorithm is crucial for the smooth implementation of RTHS. In RTHS, not only does the experimental process itself impose stringent requirements on the algorithm, but structural characteristics such as the mass ratio between the numerical substructure and experimental substructure can also significantly affect the performance of the algorithm. Furthermore, excessively long integration time steps and time delays both influence the stability and accuracy of the algorithmic system by amplifying the high-frequency modes of the structure. However, most existing explicit algorithms (e.g., explicit Newmark method [10], central difference method, Gui-λ family algorithms [11], CR algorithm [12], and Chang algorithm [13,14]) lack numerical damping. Due to time step discretization in the algorithm and time delays induced by experimental equipment, system errors typically introduce spurious high-frequency responses to the structure. Consequently, it is imperative to develop an algorithm with appropriate numerical damping capable of dissipating these unwanted high-frequency oscillations. To address these needs, this chapter proposes a novel family of unconditionally stable explicit integration algorithms that feature adjustable numerical damping through parametric control (capable of null damping), explicit formulations for both displacement and velocity components, and iteration-free implementation ensuring computational efficiency in RTHS.

4.3.1 Single-Step Method of Explicit Displacement and Velocity (SSMEDV) Algorithm

We have developed SSMEDV, a novel unconditionally stable explicit integration algorithm. This algorithm features explicit single-step formulations for both displacement and velocity, with its time-domain expressions given by:

$$\begin{cases} x_i = x_{i-1} + (\Delta t)\dot{x}_{i-1} + \alpha_1(\Delta t)^2\ddot{x}_{i-1} \\ \dot{x}_i = \dot{x}_{i-1} + \alpha_2\Delta t\ddot{x}_{i-1} \end{cases} \tag{4.6}$$

where, α_1 and α_2 are control parameters; x and its first and second derivatives denote the displacement, velocity, and acceleration responses of the structure, respectively, and the subscript i denotes the computation step; Δt is the time step size. For single-DOF (SDOF) structural systems, the dynamic equation can be expressed as Eq. (4.7).

$$m\ddot{x} + c\dot{x} + kx = f \tag{4.7}$$

where m, c, and k represent the mass, damping, and stiffness of the structure, respectively; x and its first and second derivatives denote the displacement, velocity and acceleration responses of the structure, respectively, and f indicates the external load.

Based on the discussion of transfer functions in the frequency domain presented in Chapter 2, the pole values of the transfer function correspond to the eigenvalues of the algorithm, which characterize the performance of the algorithm. Therefore, the z-transform can be applied to Eq. (4.6) in the frequency domain, as shown in Eq. (4.8). The transfer function expression for the SDOF system is given by:

$$\begin{cases} \dot{x}_i(z) = \dfrac{\dfrac{(z-1)\alpha_2}{(z-1)\alpha_1+\alpha_2}x_i(z)}{\Delta t} \\ \ddot{x}_i(z) = \dfrac{\dfrac{(z-1)^2}{(z-1)\alpha_1+\alpha_2}x_i(z)}{\Delta t^2} \end{cases} \tag{4.8}$$

$$G(z) = \frac{\alpha_1 \Delta t z + (\alpha_2 - \alpha_1)\Delta t^2}{m\left(z^2 + \left(\alpha_1 \Omega^2 + 2\alpha_2 \xi \Omega - 2\right) - z + (\alpha_2 - \alpha_1)\Omega^2 - 2\alpha_2 \xi \Omega + 1\right)} \tag{4.9}$$

where $\Omega = w\Delta t$, with ω denoting the natural frequency of the structure, and ξ representing the damping ratio.

The coefficients of the transfer function are listed in Table 4.2.

The SSMEDV algorithm proposed by our research team exhibits appropriate numerical damping characteristics. Although existing algorithms such as the Wilson-θ method [15], Hilber-Hughes-Taylor α method (HHT-α) [16], and Newmark method all possess numerical damping when parameter $\gamma \neq 0.5$, these algorithms are implicit and therefore not ideally suited for RTHS with real-time requirements. Therefore, the algorithm should simultaneously maintain superior computational performance. Since its initial proposal by Newmark in 1959, the Newmark method has been widely applied, particularly in engineering fields. This algorithm is governed by two parameters (γ, β) that determine its computational characteristics. When $\gamma > 0.5$, the algorithm becomes an unconditionally stable implicit formulation, while the β parameter can be adjusted to achieve optimal numerical damping. Furthermore, many contemporary algorithms are essentially refined variants of the Newmark method, including the Wilson-θ method [10], the Gui-λ algorithm [11], and the KR (2014) algorithm [17]. Building upon this foundation, this

Table 4.2 Coefficient terms of transfer functions in the novel integration algorithm.

Numerator		Denominator	
n_2	0	d_2	m
n_1	$\alpha_1 \Delta t^2$	d_1	$\alpha_1 k \Delta t^2 + \alpha_2 c \Delta t - 2m$
n_0	$(\alpha 2 - \alpha 1)\Delta t^2$	d_0	$m + (\alpha_2 - \alpha_1) k \Delta t^{3-} \alpha_2 c \Delta t$

chapter proposes an explicit formulation of the Newmark method through frequency-domain transfer function analysis, transforming the SSMEDV algorithm into an unconditionally stable single-step explicit algorithm that maintains comparable stability, accuracy, and numerical damping characteristics to the original Newmark method.

Consequently, the Newmark method is reformulated in the Z domain, with the coefficients of its transfer function listed in Table 4.3. By equating the pole values of the transfer function of the SSMEDV algorithm, the corresponding values of α_1 and α_2 can be obtained, as shown in Eq. (4.10).

$$\alpha_1 = \frac{(1+2\gamma)m}{2\beta k\Delta t^2 + 2\gamma c\Delta t + 2m}, \alpha_2 = \frac{m}{\beta k\Delta t^2 + \gamma c\Delta t2m} \tag{4.10}$$

4.3.2 Characteristics of the SSMEDV Algorithm

4.3.2.1 Stability [18]

Based on Wu Xiaoli's research findings, since Multiple Degrees of Freedom (MDOF) structures can be decoupled into superimposed SDOF systems through modal decomposition while preserving performance characteristics, the numerical properties of this algorithm can be analyzed using SDOF structures [18].

In the frequency domain, the stability of the algorithm can be determined by ensuring that the maximum value of its transfer function poles does not exceed 1. The calculation equation for the pole values of the transfer function of the algorithm is as follows:

$$A_1 Z^2 + A_2 Z + A_3 = 0 \tag{4.11}$$

where, A_1, A_2, and A_3 are the coefficient terms of the denominator of the transfer function, $A_1 = 2\beta\Omega^2 + 4\gamma\xi\Omega + 2$, $A_2 = (2\gamma\text{-}4\beta + 1)\,\Omega^2 + (4.8\gamma)\,\xi\Omega - 4$, $A_3 = (2\beta - 2\gamma + 1)\Omega^2 + (4\gamma - 4)\xi\Omega + 2$, where $\Omega = \omega\Delta t$, $\omega^2 = K/M, \xi = C/(2\omega M)$.

The two solutions of Eq. (4.11) are:

$$Z_{1,2} = \frac{-A_2 \pm \sqrt{A_2^2 - 4A_1A_3}}{2A_1}$$

In structural dynamics, the equations of motion of a system can only be expressed in vibrational form when the two principal eigenvalues are complex conjugates. Therefore, to ensure computational feasibility and algorithmic stability, the following conditions must be satisfied:

Table 4.3 Coefficient terms of transfer functions in the Newmark family of algorithms.

Numerator		Denominator	
n_2	$2\beta dt^2$	d_2	$2\beta k\Delta t2 + 2\gamma c\Delta t + 2m$
n_1	$(2\gamma - 4\beta + 1)\Delta t^2$	d_1	$(2\gamma - 4\beta + 1)k\Delta t^2 + c\Delta t(3 - 4\gamma) - 4m$
n_0	$(2\beta - 2\gamma + 1)\Delta t^2$	d_0	$(2\beta - 4\gamma + 1)k\Delta t^2 + c\Delta t(2\gamma - 2) + 2m$

$$\begin{cases} A_2^2 - 4A_1A_3 < 0 \\ |Z_{1,2}| = \dfrac{\sqrt{(-A_2)^2 + \left(-A_2^2 + 4A_1A_3\right)^2}}{2A_1} = \dfrac{A_3}{A_1} \leqslant 1 \end{cases} \tag{4.12}$$

Substituting Eq. (4.11) into Eq. (4.12) yields:

$$\begin{cases} \left[(2\gamma + 1)^2 - 16\beta\right]\Omega^2 + (8 - 16\gamma)\Omega + 16\xi^2 - 16 < 0 \\ (1 - 2\gamma)\Omega^2 - 4\xi\Omega \leq 0 \end{cases} \tag{4.13}$$

The solution of Eq. (4.13) demonstrates that the algorithm achieves unconditional stability when $\gamma \geq 1/2$ and $\beta \geqslant (2\gamma + 1)^2/16$. As illustrated in Fig. 4.18, the SSMEDV algorithm demonstrates its stability performance in pure numerical computations under various combinations of γ and β values. Fig. 4.18A shows that under the above parameter conditions, the root loci of the algorithm are all within the stability circle. Furthermore, Fig. 4.18B reveals that the stability of the algorithm improves as the γ value increases. However, when γ is fixed, even if the unconditional stability condition $\beta \geqslant (2\gamma + 1)^2/16$ is satisfied, its stability gradually degrades as the β value increases.

4.3.2.2 Accuracy

Accuracy is a critical performance metric in numerical integration algorithms, reflecting the computational accuracy of structural analysis. Employing high-precision integration algorithms ensures the authenticity

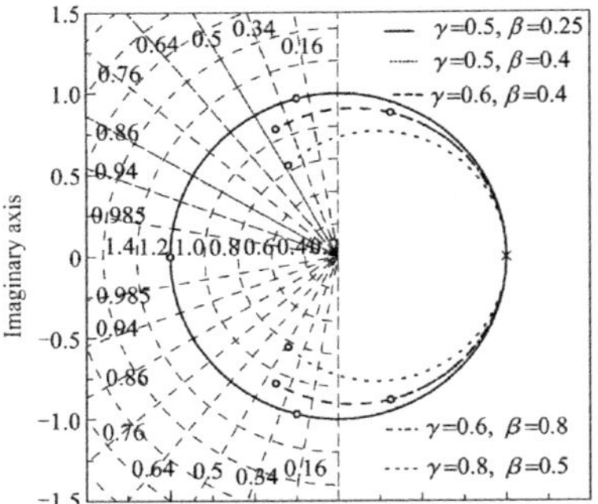

(a) Root locus plot

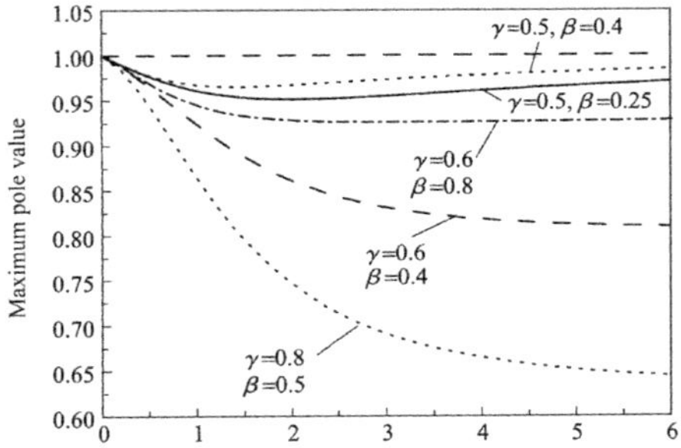

(b) Maximum pole value of algorithm

Figure 4.18 Stability performance of the SSMEDV algorithm in pure numerical computation.

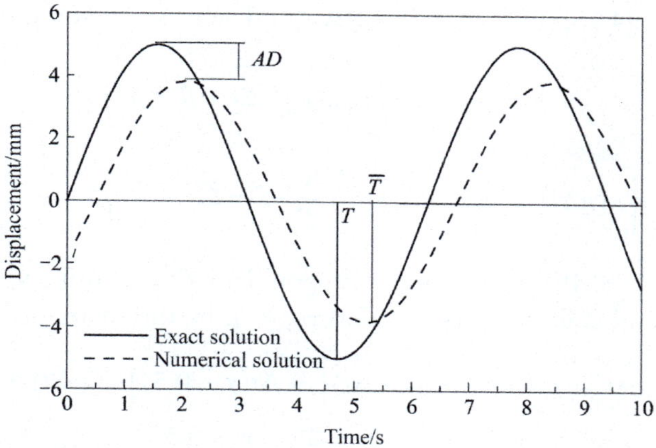

Figure 4.19 Schematic diagram of algorithm accuracy calculation [19].

and reliability of hybrid simulation results. The definition of algorithmic accuracy can be represented by Fig. 4.19, where errors primarily manifest in amplitude and period. However, as noted by Chopra in practical computations, since amplitude decay calculations depend on initial values, accuracy is commonly characterized by numerical damping ξ and periodic decay rate $\left(\overline{T}-T\right)/T$ [19].

The numerical algorithm for structural dynamic response can be expressed in finite difference formulation:

$$\sum_{j=-2}^{j=p}\left(\alpha_j m+\gamma_j c\Delta t+\beta_j k\Delta t^2\right)x_{n+j}=0 \tag{4.14}$$

where the value of p is algorithm-dependent, indicating that the current system state is dependent on the states from the previous several steps. For single-step-type algorithms such as the explicit Newmark method and the SSMEDV algorithm proposed in this section, the displacement truncation error of the algorithm is defined as follows:

$$\tau=\frac{1}{\Delta t^2}\sum_{j=-2}^{j=1}\left(\alpha_j m+\gamma_j c\Delta t+\beta_j k\Delta t^2\right)x\left((n+j)\Delta t\right) \tag{4.15}$$

For Eq. (4.15), expand $x[(n+j)\Delta t]$ in Taylor series about time nt:

$$\tau = \frac{1}{\Delta t^2}\sum_{j=2}^{1}\left(\alpha_j m + \gamma_j c\Delta t + \beta_j k\Delta t^2\right) \left\{x(n\Delta t) + j\Delta t\dot{x}(n\Delta t) + \frac{(j\Delta t)^2}{2!}\ddot{x}(n\Delta t) + \cdots\right\} \tag{4.16}$$

For the SSMEDV algorithm proposed in this section, its truncation error (expanded up to second order) is given by the following expression:

$$\tau = \left[m\ddot{x}(n\Delta t) + c\dot{x}(n\Delta t) + kx(n\Delta t)\right] + \Delta t(\gamma - 0.5)\left[c\ddot{x}(n\Delta t) + k\dot{x}(n\Delta t)\right] + \Delta t^2\left[\frac{1}{12}mx^{(4)}(n\Delta t) + \frac{1}{6}cx^{(3)}(n\Delta t) + \frac{2\beta - \gamma + 0.5}{2}kx^{(2)}(n\Delta t)\right] + O\left(\Delta t^3\right) \tag{4.17}$$

When $\gamma = 0.5$, Eq. (4.17) reduces to Eq. (4.18), resulting in second-order accuracy for the displacement component.

$$\tau = \left[m\ddot{x}(n\Delta t) + c\dot{x}(n\Delta t) + kx(n\Delta t)\right] + \Delta t^{*}0 + \Delta t^2 \left[\frac{1}{12}mx^{(4)}(n\Delta t) + \frac{1}{6}cx^{(3)}(n\Delta t) + \frac{2\beta - \gamma + 0.5}{2}kx^{(2)}(n\Delta t)\right] + O\left(\Delta t^3\right) \tag{4.18}$$

When $\gamma \neq 0.5$, Eq. (4.17) simplifies to Eq. (4.19), resulting in first-order accuracy for the displacement component.

$$\tau = \left[m\ddot{x}(n\Delta t) + c\dot{x}(n\Delta t) + kx(n\Delta t)\right] + \Delta t(\gamma - 0.5)\left[c\ddot{x}(n\Delta t) + k\dot{x}(n\Delta t)\right] + O\left(\Delta t^2\right) \tag{4.19}$$

The accuracy analysis expression indicates that since the characteristic function of the SSMEDV algorithm is identical to that of the Newmark method, the SSMEDV algorithm exhibits an accuracy relationship similar to the Newmark method, achieving second-order accuracy when $\gamma = 0.5$. This accuracy is verified through a numerical example of a single-DOF (SDOF) linear structure under free vibration. The example structure has a

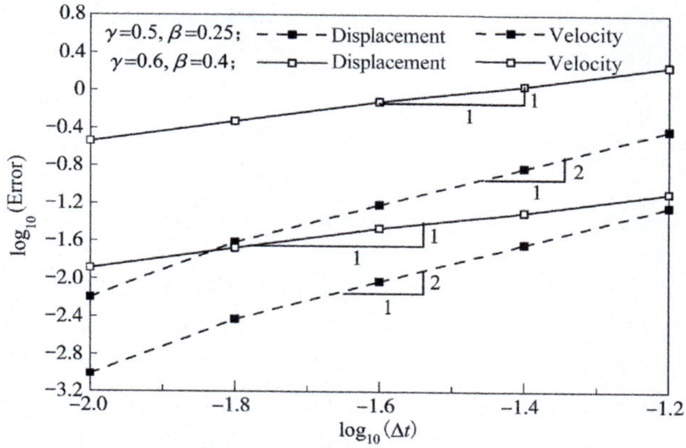

Figure 4.20 Numerical example of algorithmic displacement and velocity errors.

mass of 80 kg, a natural frequency of 12π/s, and initial conditions of $x_0 = 1$ mm and $v_0 = 0$ mm/s. Therefore, the dynamic equation of this structure is expressed in Eq. (4.20), while its theoretical dynamic response is given by Eq. (4.21). The displacement and velocity errors of the SSMEDV algorithm at $t = T$ are presented in Fig. 4.20. The results demonstrate that when $\gamma = 0.5$, the SSMEDV algorithm achieves second-order accuracy in displacement, while the velocity maintains the same order of accuracy but with lower accuracy compared to displacement. Moreover, when $[\gamma, \beta] = [0.6, 0.4]$, the SSMEDV algorithm demonstrates lower accuracy compared to the case with $\gamma = 0.5$, particularly exhibiting reduced accuracy in velocity computation, primarily due to the enhanced numerical damping effect induced by this parameter combination. In theoretical ideal calculations, numerical damping tends to reduce solution accuracy. However, in practical applications, particularly in RTHS, as inevitable system errors from testing equipment are often high-frequency errors, the existence of numerical damping can effectively reduce the influence of error factors, thereby obtaining a solution with accuracy higher than the algorithm without numerical damping.

$$m\ddot{x}_i + kx_i = 0 \tag{4.20}$$

$$x_i = \cos(\omega_i t_i) \tag{4.21}$$

Chopra demonstrates that algorithm accuracy can typically be characterized by numerical damping $(\bar{\xi})$ and periodic decay rate $\left[\left(\overline{T} - T\right)/T\right]$ [19]. When the transfer function poles of the algorithm are obtained as a pair of complex conjugates, the numerical damping and periodic decay rate are calculated as follows:

$$Z_{1,2} = \frac{-A_2 \pm i\sqrt{4A_1A_3 - A_2^2}}{2A_1} = A + Bi; \ \overline{\Omega} = \arctan(B/A) \tag{4.22}$$

$$\xi = -\ln\left(A^2 + B^2\right) / 2\overline{\Omega}; \ PE = \left(\overline{T} - T\right) / T = \left(\Omega - \overline{\Omega}\right) / \overline{\Omega} \tag{4.23}$$

Fig. 4.21 illustrates the numerical damping and periodic decay rate performance of the SSMEDV algorithm. It can be seen that when $\gamma = 0.5$, the algorithm exhibits zero numerical damping; when $\gamma > 0.5$, the algorithm introduces positive numerical damping, causing structural vibration decays even in the absence of physical damping, with the damping effect intensifying as γ increases; when $\gamma < 0.5$, negative numerical damping is introduced, leading to algorithmic instability. The periodic decay rate of the algorithm is independent of the γ coefficient, though when γ remains constant, the periodic decay rate increases continuously with the growth of β. In other words, under the condition of unconditional stability of the algorithm ($\gamma \geq 0.5$ and $\beta \geqslant (2\gamma + 1)^2/16$), when β exactly equals $(2\gamma + 1)^2/16$, the algorithm achieves its minimal periodic decay rate, corresponding to optimal accuracy.

In summary, this algorithm features excellent performance, possessing stability and accuracy equivalent to the Newmark method; by adjusting the γ and β coefficients in the algorithm, it can achieve varying stability characteristics, including zero or positive numerical damping for different accuracy requirements, making it widely applicable. Moreover, since both displacement and velocity in this algorithm are explicitly expressed through single-step formulations, it demonstrates superior advantages for RTHS.

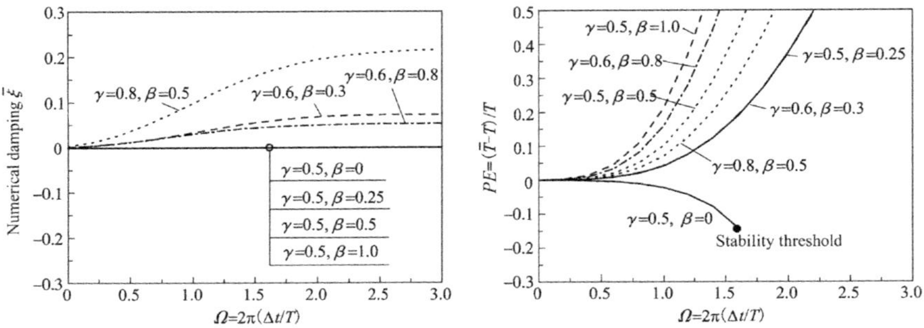

(a) Algorithmic numerical damping (b) Algorithmic period elongation ratio

Figure 4.21 Accuracy performance of the SSMEDV algorithm.

4.4 Real-Time Numerical Model Calculation Methods

In the RTHS of high-speed railway systems, due to high train speeds, sufficiently long bridge sections and refined numerical train-track-bridge coupling models are required, imposing stringent requirements on both computational accuracy and efficiency. Chen Menghui pointed out that RTHS demands high computational efficiency, as each step must be completed within an extremely limited time [20], further increasing implementation challenges. According to the research by Gu Quan et al. on high-efficiency computational methods for RTHS of high-speed railway train-track-bridge coupling systems, the moving load integration method [21] proposed in this section can address real-time computation challenges for large-scale bridge models.

The moving load integration method stores the data required for the numerical substructure calculation of the bridge prior to testing. During actual computation, it retrieves and processes data corresponding to each time step to obtain the displacement response at the current step. This section further extends the method to localized nonlinear scenarios by integrating elastoplastic numerical substructure techniques, and systematically validates the feasibility and reliability of the method through the RTHS of the scaled-down high-speed railway on shake tables.

4.4.1 Moving Load Integration Method

For linear elastic bridge systems, the time-varying wheel-rail interaction forces can be superimposed in the following manner before being applied to the bridge structure.

$$f(x,t)=\sum_{k=1}^{n} F(x_k,t_k)L_k(x,t) \tag{4.24}$$

where, x denotes the bridge position; t represents time; n is the number of moving loads (e.g., axles) whose influence ranges cover the position x at time t; the subscript k indicates the current time step; t_k stands for the k-th time step, while x_k corresponds to the wheel-rail force application position at that instant. For a train moving at velocity v, the relationship between t_k and x_k is defined as $x_k = vt_k$; $F(x_k, t_k)$ signifies the wheel-rail interaction force measured at position x_k and time t_k through RTHS; $L_k(x, t)$ represents the triangular unit impulse function acting at position x_k during time step t_k. The functional graphs of $f(x, t)$ and $L_k(x, t)$ are illustrated in Fig. 4.22.

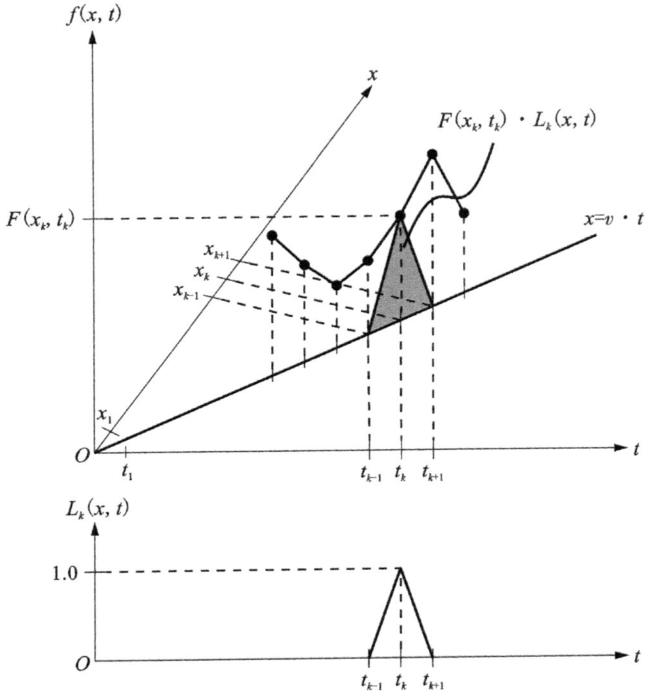

Figure 4.22 Functional graphs of $f(x, t)$ and $L_k(x, t)$.

In linear elastic structures, the displacement of a bridge at any position (x, t) can be calculated based on the principle of linear superposition.

$$u(x,t) = \sum_{k=1}^{t} F(x_k, t_k) u_k(x,t) \tag{4.25}$$

where, $u(x, t)$ represents the displacement response induced by the triangular unit impulse $L_k(x, t)$, constituting a bivariate function of position x and time t. When $k = 1$, its functional graph is shown in Fig. 4.23 $u(x, t)$ is likewise a bivariate function of position x and time t, with its functional graph shown in Fig. 4.24.

As shown in Eq. (4.25), for the superposition calculation of $u(x, t)$, computation and storage of u_k (x, t) are required. In the final calculation, they are retrieved and superimposed, with the total data volume being determined by the number of computational time steps. In the RTHS of high-speed railway train-track-bridge coupling systems, the bridge model exhibits considerable total length and the test requires an extremely small

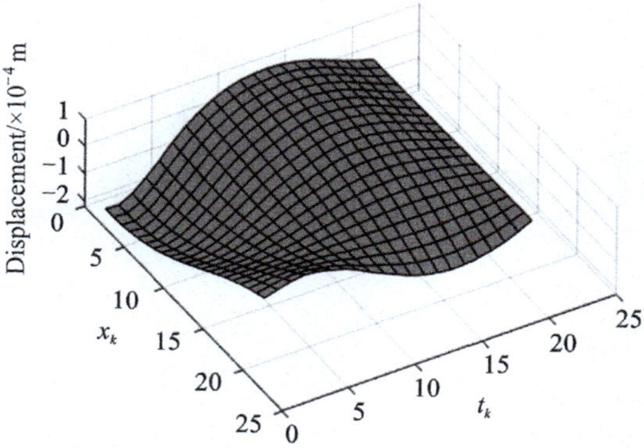

Figure 4.23 Functional graph of $u_{k=1}(x, t)$.

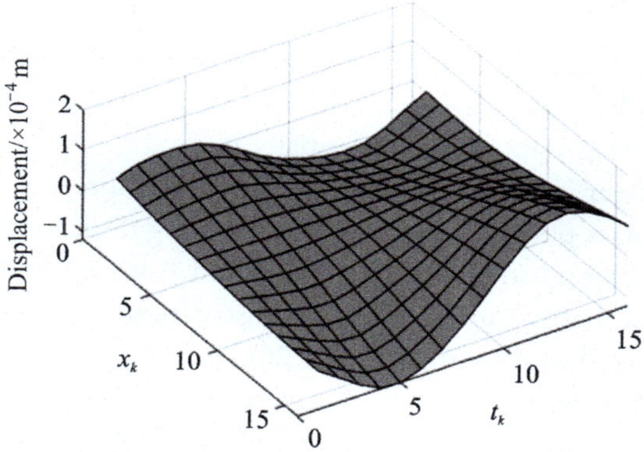

Figure 4.24 Functional graph of $u(x, t)$.

time step. The computation of $u(x, t)$ via Eq. (4.25) necessitates storing and inputting massive data volumes, resulting in excessive computation time per step that fails to meet the control requirements of RTHS.

During the RTHS of train operations on bridges, once the train speed and time step are determined, the location of the wheel-rail contact force on the bridge is uniquely defined at every time instant. Since in RTHS, only the bridge response at the location of the wheel-rail force needs to be calculated at each time step, with no need to calculate the full bridge response, it is unnecessary to store the full structural response function for a unit impulse acting at each moment. Instead, only the displacement response at the application point of the bridge at subsequent time steps after the unit impulse acts need to be stored. The data at these specific points are

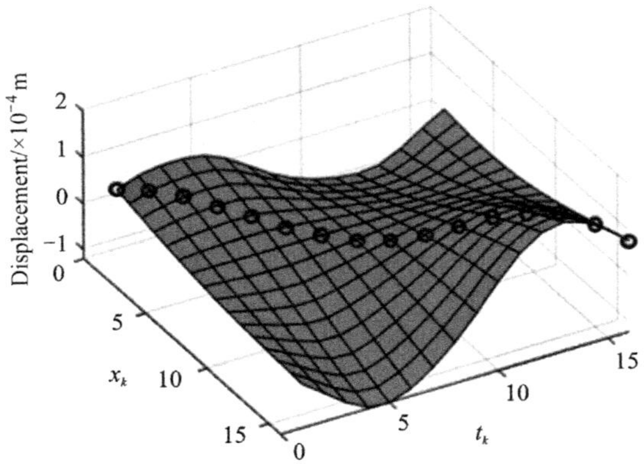

Figure 4.25 Schematic diagram of valid data points for *u*(*x*, *t*) (Note: The solid circles in the figure denote valid data points.)

extracted from a large amount of data, and only minimal valid data is stored (Fig. 4.25). On this basis, superimposed calculations are performed, rendering all other data as non-essential.

For the linear system, when a unit impulse force is applied to the rail-wheel contact position at time i, the resulting displacement response at the corresponding position at time j is denoted as $u_{ij}(t)$. In dynamic analysis, u_{ij} constitutes a matrix, and the complete structural displacement response data are organized as shown in Eq. (4.26).

For scenarios involving multiple moving loads, each load is processed using this data structure individually, and in the final calculation, all data are linearly superimposed. To enhance computational efficiency during superposition, only the unit impulse response components corresponding to the current time step from previously acting data, rather than all stored response data, are extracted.

$$\boldsymbol{U} = \begin{bmatrix} u_{11} & \cdots & u_{1n} \\ \vdots & \ddots & \vdots \\ u_{n1} & \cdots & u_{nn} \end{bmatrix} \tag{4.26}$$

Taking time step i as an example, the displacement response at a specified bridge location is calculated using the optimized data storage matrix as follows:

$$u_i = \sum_{n=1}^{i} f_n u_{n,i} \tag{4.27}$$

where, f_n represents the actual wheel-rail interaction force at the corresponding bridge location at time t_n, i.e., the magnitude of real-time wheel-rail force measured in RTHS.

As shown in Eq. (4.27), the displacement at the wheel-rail force application position for each time step is calculated by extracting the response at the required position from the global response recorded under a unit moving impulse load. These extracted responses form a data matrix, where only the corresponding data needs to be retrieved for linear superposition during computation. Therefore, for the response calculation at time step n, the SSMEDV algorithm requires only the superposition of n data points, significantly enhancing computational speed.

The moving load integration method requires pre-storage and real-time access to an $n \times n$ data matrix during computation ($n =$ total number of time steps to be calculated). Nevertheless, in RTHS, the value of n may still be relatively large, making this process time-consuming. In fact, the influence of wheel-rail forces is limited in scope, and their effects at sufficiently distant locations (e.g., beyond 3-5 spans of a bridge) can be neglected. Hence, in practical calculations, the number of columns n in matrix U can be set to a finite value. Given that the effective influence range of moving loads is well-defined and the number of rows increases continuously until the far end of the bridge, it should be noted that this method is suitable for continuous girder bridges or simply supported girder bridges with short spans, but not suitable for long-span bridges. In RTHS, the data matrix can be read in batches by loading an appropriate number of rows (e.g., n rows) at each iteration. This reduces the single-time data loading volume during computation, thereby decreasing memory usage and enhancing computational efficiency.

The aforementioned algorithm is based on linear elastic systems and the displacement superposition method. Below we, present Sun Baoyin et al.'s proposal to extend the moving load integration method to locally nonlinear systems using the elastoplastic numerical substructure approach [22,23].

4.4.2 Nonlinear Correction of the Moving Load Integration Method

For bridge structures where localized nonlinearities may occur at certain piers, an elastoplastic numerical substructure approach is adopted. The global bridge system is still analyzed using the original linear model, while nonlinear components are isolated and computed with nonlinear models [22]. The internal force difference between linear and nonlinear components is then computed and applied as nonlinear corrective forces at bridge nodes. These corrective forces, treated as additional external loads, generate displacements that are superimposed onto the linear bridge displacements. This superposition yields the true displacement response of nonlinear bridge structures, equivalently superimposing a stationary-load-induced response component onto the moving-load-induced bridge response. It is noted that the displacement response induced by nonlinear corrective forces can also be treated as a special case of the moving load superposition method. This response can be calculated with Eq. (4.27), with the only distinction being the alternative matrix formulation in Eq. (4.26), where the load positions must remain invariant.

Below is a brief introduction to the calculation method of nonlinear corrective forces in the numerical substructure method.

The equation of motion for the global structure based on the FE method is given by:

$$\boldsymbol{M}\ddot{\boldsymbol{u}}(t) + \boldsymbol{C}\dot{\boldsymbol{u}}(t) + R(\boldsymbol{u}(t)) = \boldsymbol{P}(t) \tag{4.28}$$

where, $\boldsymbol{M}$ and $\boldsymbol{C}$ denote the mass matrix and damping matrix of the structure, respectively; $\ddot{\boldsymbol{u}}(t)$, $\dot{\boldsymbol{u}}(t)$, and $\boldsymbol{u}(t)$, represent the acceleration, velocity, and displacement vectors of the corresponding note, respectively; $\boldsymbol{R}$ denotes the internal force of the structure; $\boldsymbol{P}$ is the external load vector.

Discretize Eq. (4.28) at time t by the Newmark-β method:

$$\left[\frac{1}{\beta(\Delta t)^2}M + \frac{\alpha}{\beta\Delta t}C\right]u_{n+1} = F_{n+1} - R(u_{n+1}) \tag{4.29}$$

where, α and β are introduced parameters; F_{n+1} represents the equivalent external load.

By adding the $F_{\mathrm{un}+1}$ term to both sides of Eq. (4.29) and rearranging:

$$K^{dyn}u_{n+1} = F_{n+1} + \widetilde{F}_{n+1} \tag{4.30}$$

where, K^{dyn} is the equivalent stiffness; $\widetilde{F}_{n+1}$ is the nonlinear corrective force.

$$\widetilde{F}_{n+1} = Ku_{n+1} - R(u_{n+1}) \tag{4.31}$$

In Eq. (4.31), the right-hand side represents the fully elastic main structure, while the left-hand side term $Fn+1$ denotes the equivalent external load obtained through time discretization using the Newmark-β method at time t. F is determined by the difference between the linear force of the main structure and the nonlinear force of the substructure.

In this study, the constitutive relations of nonlinear materials are directly incorporated into the program to determine the internal forces of nonlinear structural components, with the nonlinear corrective forces being obtained through Eq. (4.31).

4.4.3 Validation of the Moving Load Integration Methods in RTHS

An 11-span track-bridge system with a total length of 330 m, as shown in Fig. 4.26, is adopted as the numerical substructure model for the test. The parameter values for the track and bridge are listed in Table 4.4.

In the test, the shake table system shown in Fig. 4.27 is adopted in the simulation. It is an electro-hydraulic shake table with a platform size of 1 × 1 m, a maximum load capacity of 10 t, an effective excitation frequency range of 0.1–50.0 Hz, and a maximum output acceleration of 1.2 g. A 1/4 CRH380A high-speed train vehicle body (represented by a single wheelset) is selected as the experimental substructure. The time step for the experiment is 1/128 s, and the running speed is 230 km/h. To simplify the research, a single wheelset from one bogie of the train is isolated for investigation, as illustrated in Fig. 4.28. The parameters of the experimental substructure are listed in Table 4.5.

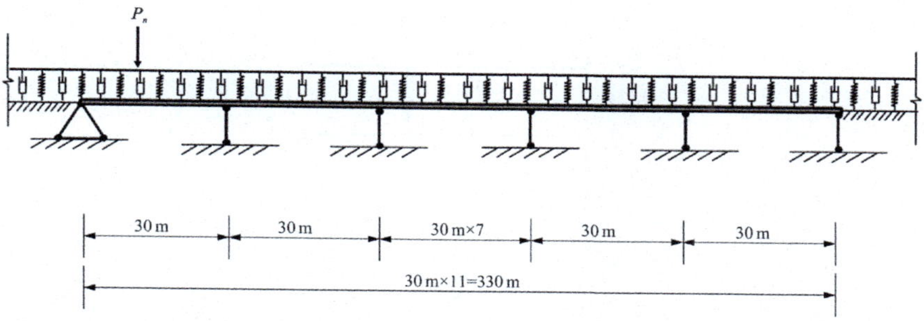

Figure 4.26 Numerical substructure model (track-bridge system).

Table 4.4 Parameters of the track-bridge system.

Parameter	Value
Rail linear mass $\overline{m}_c/(kg\cdot m^{-1})$	51.5
Rail elastic modulus E_r/Pa	2.06×10^{11}
Rail cross-sectional moment of inertia I_r/m^4	2.037×10^{-5}
Rail cross-sectional area A_r/m^2	77.45×10^{-4}
Spring stiffness $k_{rb}/(N \cdot m^{-1})$	6.58×10^7
Damping $C_{rb}/(N \cdot s \cdot m^{-1})$	3.21×10^4
Bridge linear mass $\overline{m}_b/(kg\cdot m^{-1})$	1.2×10^4
Bridge elastic modulus E_b/Pa	2.943×10^9
Bridge cross-sectional moment of inertia I_b/m^4	2.88
Bridge cross-sectional area A_b/m^2	7.94
Single-span length of bridge l/m	30
Rigid subgrade length l_0/m	400
Bridge damping ratio ξ	0.02

Figure 4.27 Shake table system.

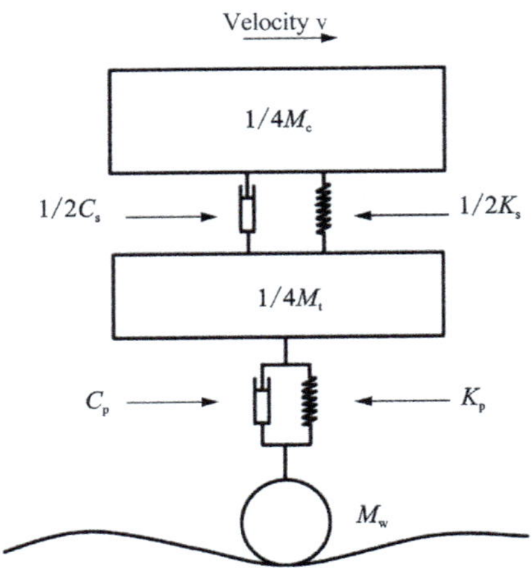

Figure 4.28 Simplified 1/4 vehicle body model.

Table 4.5 Parameters of the experimental substructure.

Substructure	Parameter	Value
Vehicle body	$\frac{1}{4}m_c/kg$	8446.5
	$\frac{1}{2}K_s/(\mathrm{N \cdot m^{-1}})$	225,000
	$\frac{1}{2}C_s/(\mathrm{N \cdot s \cdot m^{-1}})$	10,000
Bogie	m_t/kg	1028
	$K_p/(\mathrm{N \cdot m^{-1}})$	1,772,000
	$C_p/(\mathrm{N \cdot s \cdot m^{-1}})$	20,000
Wheelset	m_w/kg	1267
Total	m_{sum}/kg	10,741.5

In the table, m_c, m_t, and m_w represent the masses of the vehicle body, bogie, and wheelset, respectively; K_s and K_p denote the spring stiffness between the vehicle body and bogie, and between the bogie and wheelset, respectively; C_s and C_p represent the damping coefficients between the vehicle body and bogie, and between the bogie and wheelset, respectively; m_{sum} indicates the total mass (including the wheelset).

In the RTHS, the track-bridge system is treated as the numerical substructure, where the displacement response at the current time step is calculated using the moving load integration method; the wheelset and bogie assembly is treated as the experimental substructure, where the wheel-rail interaction force at the current time step is computed based on the acceleration measured by sensors on the shake table. To minimize the amplitude and phase errors between the displacement command signal derived from the bridge numerical substructure and the input signal received by the train substructure, the adaptive time series (ATS) method [24] proposed by DARBY is employed for time-delay compensation.

In the RTHS, the Pulsar iterative control strategy (ICS) is employed to reproduce the system response. The real-time simulation is implemented on the MATLAB/Simulink platform, and the Simulink-compiled model is executed on a separate target computer xPC-Target using the Simulink Real-Time toolbox. Simulink Real-Time supports the SCRAMNet interface. It uses SCRAMNet to transmit various signals from the Pulsar system in real time while reading external control commands into Pulsar.

4.4.3.1 Verification Under Linear Elastic Systems

In this section, the feasibility and effectiveness of applying the moving load integration method to linear elastic systems in RTHS are verified.

Prior to applying the moving load integration method, it is necessary to calculate the initial displacement of the track-bridge system induced by the dead load before train movement. Subsequently, the unit impulse load is sequentially applied to corresponding positions on the track-bridge model according to the actual computational time steps, and the full time-history displacement responses of all nodes in the track-bridge system under current loading conditions are recorded.

The simulation process is illustrated in Fig. 4.29. The displacement calculated by the moving load integration method at the i-th step is applied to the experimental substructure. The wheel-rail force at the $(i + 1)$-th step is then obtained by calculating the acceleration measured from the shake table and is subsequently applied to the numerical substructure. The displacement at the $(i + 1)$-th step is computed and loaded onto the experimental substructure again. This process is repeated iteratively until the test is completed.

Fig. 4.30 shows the variation of wheel-rail forces over time during the experimental process. A numerical model of the track-bridge system identical to the experimental setup is established in OpenSees. The wheel-rail forces recorded during the test are applied at corresponding time steps,

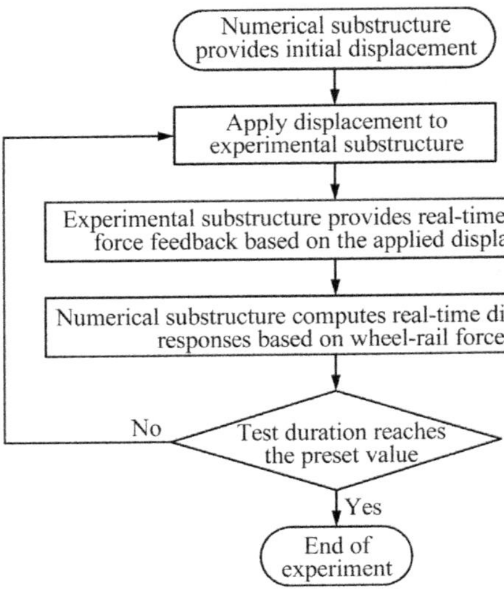

Figure 4.29 RTHS process in linear elastic systems.

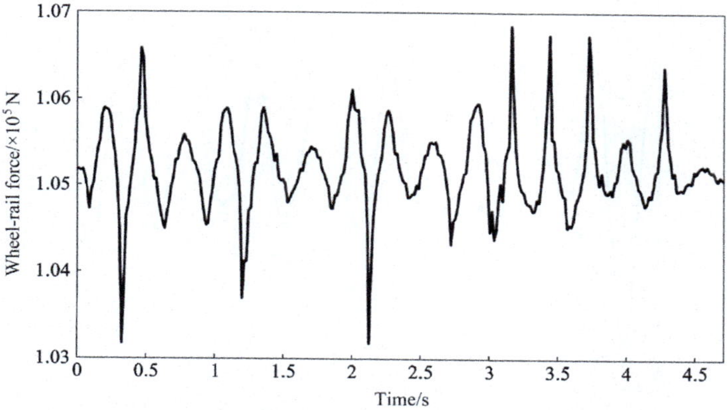

Figure 4.30 Experimental wheel-rail forces in linear elastic systems.

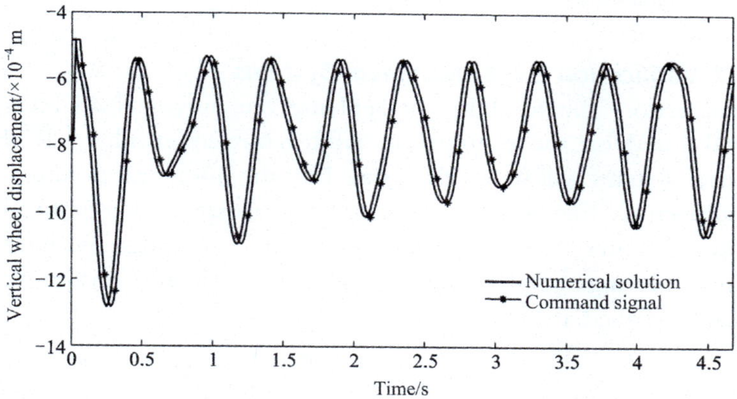

Figure 4.31 Comparison between numerical signals and command signals in linear elastic systems.

and the numerical results are then compared with the command signals recorded by the experimental substructure, as shown in Fig. 4.31. After applying the ATS method for time delay compensation, the average time delay in this test is measured as 9.9 ms. Subsequently, the time-delay-processed command signals from the test are compared with numerical solutions, with the results shown in Fig. 4.32. The obtained results verify the reliability of applying the moving load integration method to linear elastic systems in RTHS.

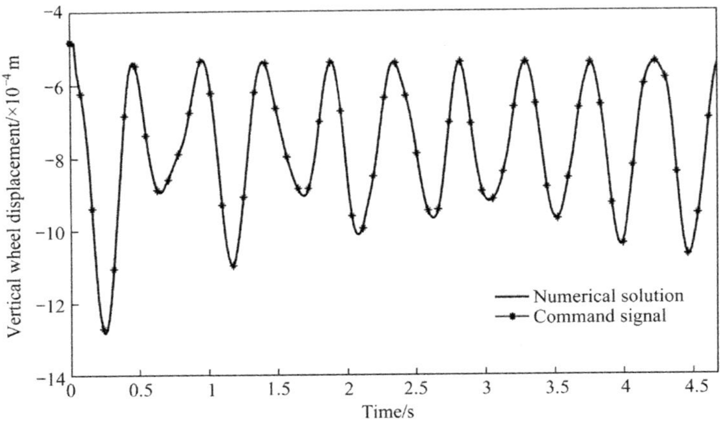

Figure 4.32 Comparison between time delay compensated numerical signals and command signals in linear elastic systems.

4.4.3.2 Verification Under Nonlinear Systems

In this section, the feasibility of applying the moving load integration method based on the elastoplastic substructure approach to RTHS of nonlinear systems will be verified. Based on the numerical example in the first subsection, this section replaces the bearings at the midspan with nonlinear bearings to simulate localized nonlinear bridge behavior. The nonlinear bearings employ Duffing materials, with the constitutive relationship described as follows:

$$\sigma = E(\varepsilon + \gamma^3) \tag{4.32}$$

where $E = 13.89$ MPa, $\gamma = -3000$, σ and ε represent stress and strain, respectively.

The computational process of this algorithm is complex when applied to nonlinear systems. Below, we provide a detailed explanation using the numerical example. First, replace the nonlinear structure in this case with the linear structure from the numerical example in the previous section, thereby converting the entire track-bridge system into a linear-elastic system. Apply unit loads sequentially to corresponding positions on the track-bridge model according to the time steps, and record the complete time-history displacement responses of all nodes in the track-bridge system under current loading conditions. Form the fundamental data matrix for the moving load integration method based on the recorded displacement responses corresponding to all loading conditions. Considering the internal force differences between nonlinear and linear systems, in addition to the fundamental data matrix, two other types of data matrices need to be compiled. The first type of matrix (Matrix A) corresponds to displacement responses at nonlinear nodes under unit load actions at the current time step. In this numerical example, both ends of the nonlinear bearing are nonlinear nodes. Therefore, it is necessary to record the complete time-history displacement responses of both nodes when unit loads are sequentially applied at corresponding positions on the track-bridge model according to the time steps. The corresponding data matrix is formed based on the recorded displacement responses of nonlinear nodes for subsequent calculation of nonlinear correction forces. The second type of matrix (Matrix B) corresponds to displacement responses at all nodes of the track-bridge model when unit loads are applied to the nonlinear nodes. In this numerical example, unit loads are sequentially applied to the nonlinear nodes according to the time steps. The complete time–history displacement responses of all nodes in the track-bridge model under the current loading condition are recorded, and the corresponding data matrix is subsequently formed.

During the actual computation, after the required data matrix is completed, the displacement response at the corresponding position of the current time step under linear conditions is first calculated based on the fundamental matrix of the moving load integration method. Then, the displacement at the nonlinear nodes for the current time step is computed according to Matrix A, and the nonlinear correction force at the nonlinear nodes is calculated by combining the constitutive relation formula of the nonlinear material in the program. Subsequently, Matrix B is used to calculate the displacement response at the load application position when the nonlinear correction force acts on the nonlinear nodes, which is then superimposed with the previously obtained displacement response. Finally, the true displacement response of the structure is obtained.

The specific RTHS process is illustrated in Fig. 4.33. The displacement calculated by the moving load integration method at the i-th step is applied to the experimental substructure. The wheel-rail force at the $(i + 1)$-th step is then computed from the acceleration measured by the shake table and subsequently applied to the numerical substructure. The numerical substructure first calculates the displacement response at the wheel-rail force application position under a linear elastic system for the current time step based on the wheel-rail force. It then computes the nonlinear correction force for the current time step, uses this force to correct the

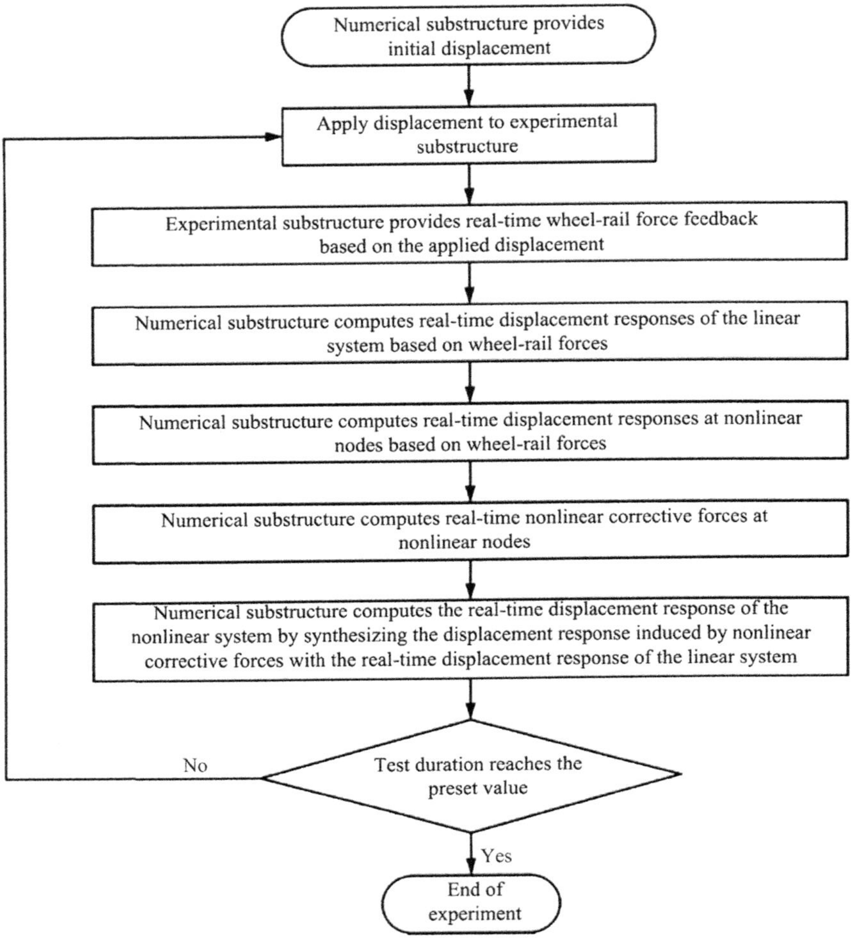

Figure 4.33 RTHS process in nonlinear systems.

displacement response obtained from the linear elastic system, and finally determines the displacement for the $(i + 1)$-th step, which is applied to the experimental substructure. This process is repeated iteratively until the test is completed.

Fig. 4.34 shows the variation of wheel-rail forces over time during the test under nonlinear system conditions. The wheel-rail forces recorded during the test are loaded into the OpenSees numerical model at corresponding time steps for calculation, and the numerical results are compared with the command signals recorded by the experimental substructure, as

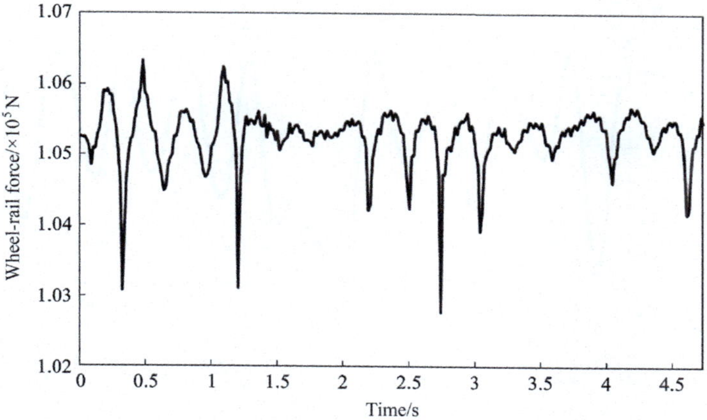

Figure 4.34 The time–history curve of wheel-rail force variation during testing in nonlinear systems.

shown in Fig. 4.35. After applying the ATS method for time-delay compensation, the average time delay in this test is 8.6 ms. The command signals are then processed according to the experimental time delay and compared with the numerical solutions, with the results presented in Fig. 4.36. The results validate the reliability of applying the moving load integration method with elastoplastic numerical substructures to RTHS under nonlinear system conditions.

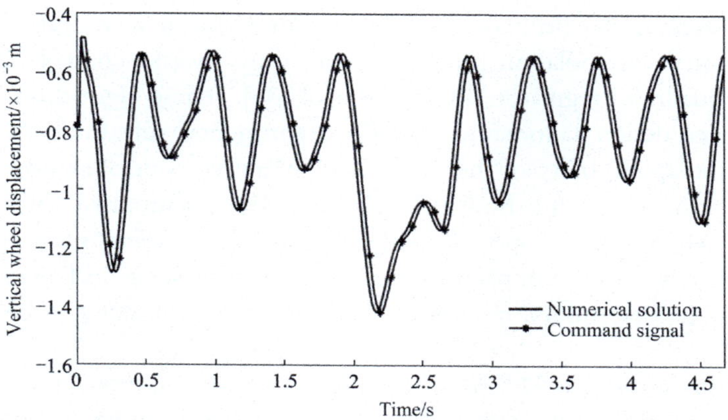

Figure 4.35 Comparison between numerical signals and command signals in nonlinear systems.

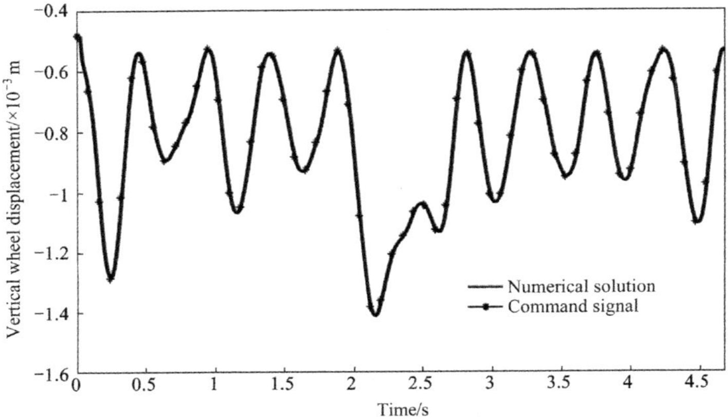

Figure 4.36 Comparison between time-delay compensated numerical signals and command signals in nonlinear systems.

4.5 Time Delay Compensation Techniques

To obtain reliable test results, the synchronization between the numerical substructure and the experimental substructure is critical. Given the characteristics of the loading system, a time delay exists between the excitation applied to the experimental substructure and the reference displacement output from the numerical substructure, which is equivalent to introducing negative damping into the driving system. When the negative damping exceeds the system damping, instability will occur. Therefore, time-delay compensation must be provided to the loading system [25].

Currently, the primary time-delay compensation techniques include polynomial extrapolation [26–28], feedforward inverse control [29], and linear quadratic regulator (LQR) control [30]. These algorithms require controller design tailored to specific experimental substructures, which limits the applicability of the controllers. Adaptive control algorithms can address this issue, with both ATS control [31] and minimal control synthesis (MCS) [32] proposed to compensate for loading system delays. However, some control parameters in adaptive time series (ATS) and MCS need to be determined empirically, increasing the difficulty in controller tuning.

In this section, RTHS is applied to bridge-vehicle interaction studies, and an adaptive composite controller [33] consisting of adaptive state feedback control (ASFC) and an interpolation prediction algorithm are

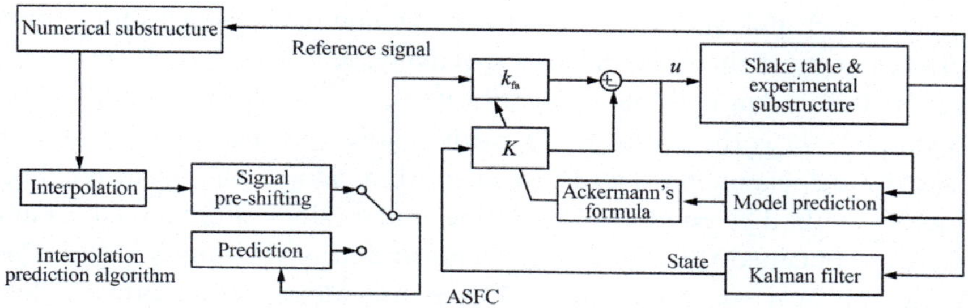

Figure 4.37 Block diagram of ASFC for RTHS.

proposed. Without requiring prior knowledge of the experimental substructure, the controller compensates for time delays in the hybrid simulation loading system, as illustrated in Fig. 4.37. The adaptability of ASFC reduces the time delay of the loading system while ensuring tracking accuracy, and its residual time delays are compensated by the interpolation prediction algorithm.

4.5.1 ASFC

Using the shake table as the loading device for the experimental substructure in the bridge-vehicle RTHS, the ASFC treats both the shake table and the experimental substructure as the controlled plant. By relocating the poles to the target poles, it adjusts the transient response characteristics of the controlled plant. Because the shake table is nonlinear and the experimental substructure model is unknown before testing, we employ online model estimation to identify the controlled plant's dynamics in real time and automatically update control parameters to enhance tracking accuracy and response speed when prior knowledge is unavailable.

4.5.1.1 Online Model Estimation

If the experimental substructure is regarded as an equivalent mass with disturbances, the controlled plant can be represented by a third-order model. Additionally, Phillips et al. [34] proposed that a third-order linear model can simulate both the shake table and the experimental substructure. Therefore, the third-order discrete model given in Eq. (4.33) is adopted as the model structure for online estimation.

$$\begin{cases} \boldsymbol{A}(z^{-1})y(k) = B(z^{-1})u(k) + v(k) \\ \boldsymbol{A}(z^{-1}) = 1 + a_1 z^{-2} + a_2 z^{-2} + a_3 a^{-3} \\ \boldsymbol{B}(z^{-1}) = b_1 z^{-1} + b_2 z^{-2} + b_3 z^{-3} \end{cases} \tag{4.33}$$

where, $u(k)$ and $y(k)$ represent the input command signal and output displacement of the shake table, respectively, a_i, $b_i(i = 1, 2, 3)$ are model parameters requiring online identification; z^{-1} denotes the difference operator; $v(k)$ is the sensor noise, with a noise variance of $R_k = E|vv^t|$. The model parameter estimation algorithm adopts the recursive least squares method given in Eq. (4.34). The initial values of the model parameters are obtained through offline identification of the empty shake table using the least squares method. The entire identification process does not require prior knowledge of the experimental substructure. The sensor noise of the output displacement of the shake table is accounted for through the gain matrix $G_{\mathrm{LS}}(k)$ [35].

$$\begin{cases} \widehat{\theta}(k) = \widehat{\theta}(k-1) + G_{\mathrm{LS}}(k)\left[y(k) - \varphi(k)^{\mathrm{T}}\widehat{\theta}(k-1)\right] \\ \widehat{\theta}(k) = [a_1, a_2, a_3, b_1, b_2, b_3]^{\mathrm{T}} \\ \varphi(k) = [-y(k-1), \cdots, -y(k-3), u(k-1), \cdots, u(k-3)]^{\mathrm{T}} \end{cases} \tag{4.34}$$

where, $\widehat{\theta}(k)$ represents the estimated parameter vector at the k-th control step, $\varphi(k)$ denotes the input–output data at the k-th step. The gain matrix $G_{\mathrm{LS}}(k) = \frac{P_{\mathrm{LS}}(k-1)\varphi(k)}{R_k + \varphi(k)^{\mathrm{T}} P_{\mathrm{LS}}(k-1)\varphi(k)}$, where the covariance matrix $P_{\mathrm{LS}}(k) = P_{\mathrm{LS}}(k-1) - G_{\mathrm{LS}}(k)\varphi(k)^{\mathrm{T}} P(k-1)$.

Before the controller design, it is necessary to convert the estimated third-order transfer function model into the controllable canonical form of the state-space equation as shown in Eq. (4.35).

$$\begin{cases} x(k+1) = \boldsymbol{F}x(k) + Gu(k) \\ y(k) = [b_3 b_2 b_1]x(k) \\ \boldsymbol{F} = \begin{bmatrix} 0 & 1 & 0 \\ 0 & 0 & 1 \\ -a_3 & -a_2 & -a_1 \end{bmatrix}, \boldsymbol{G} = \begin{bmatrix} 0 \\ 0 \\ 1 \end{bmatrix} \end{cases} \tag{4.35}$$

where, $x(k)$ represents the state of the controlled plant.

Changes in model parameters may cause fluctuations in control commands and even lead to response divergence in the controlled plant. Therefore, limiting these changes helps enhance the robustness of the controlled plant. The relative error between the current estimated model parameters and those of previous steps is selected as the criterion. When the metric is smaller than the preset tolerance, the current estimated parameters are deemed valid. The model parameters of the controlled plant are estimated at each control step. The controller design model is updated only when the estimated parameters are deemed valid. Once the model is updated, the control parameters are immediately redesigned automatically for the new model to ensure controller applicability.

4.5.1.2 State Feedback Control

State feedback control based on pole placement is adopted as the control law for ASFC. To ensure stability, target poles are placed inside the complex unit circle using the empirical transient-response formula for second-order discrete systems, and a conjugate pair is chosen as the dominant poles based on peak time and damping ratio. Dorf et al.'s research indicates that when the damping ratio is between 0.5 and 0.8 [36], the response speed and overshoot of the system can remain within reasonable ranges. The feedback control matrix $\boldsymbol{K}$ is computed using Ackermann's equation:

$$\boldsymbol{K} = [\,10\cdots00\,]\boldsymbol{P}_c^{-1}\boldsymbol{q}(\boldsymbol{F}) \tag{4.36}$$

where, the controllable matrix $\boldsymbol{P}_c = [\boldsymbol{F}^{n-1}\boldsymbol{F}^{n-2}\mathbf{G}\cdots\boldsymbol{F}\boldsymbol{G}\boldsymbol{G}]$, where $\boldsymbol{q}(\boldsymbol{F})$ is the characteristic equation of the target pole. The control command signal $u(k)$ is generated by the following equation:

$$u(k) = -\boldsymbol{K}\boldsymbol{x}(k-1) + k_{fa}r(k) \tag{4.37}$$

where, $\boldsymbol{x}(k-1)$ is the state of the controlled plant at step $k-1$, obtained through Kalman filter observation; k_{fa} is the forward amplification gain; $r(k)$ is the input reference displacement signal of the shake table.

4.5.1.3 Steady-State Error Compensation

By configuring the poles of the controlled plant, the transient response characteristics can be improved, but steady-state error is unavoidable. According to the final value theorem [36], the forward amplification gain k_{fa} in Eq. (4.38) is used to compensate for the steady-state error.

$$k_{fa} = \frac{1}{\left(\dfrac{b_1 + b_2 + b_3}{1 + a_1^* + a_2^* + a_3^*}\right)} \tag{4.38}$$

where, b_1, b_2, and b_3 are the parameters of the estimation model; a_1^*, a_2^*, and a_3^* are the parameters of the target pole eigenfunction.

4.5.2 Interpolation Prediction Algorithm

ASFC can reduce the response time of the shake table but cannot eliminate it entirely. Additionally, the control step size of the shake table is typically smaller than the analysis step size of the numerical substructure. Therefore, an interpolation prediction algorithm, composed of an interpolation algorithm and a prediction algorithm [37], is adopted as the feedforward controller outside the ASFC feedback control loop.

To synchronize the time steps of interaction signals between the numerical substructure and the experimental substructure, the displacement response of the numerical substructure is interpolated to match the same time interval as the vibration table control, which is then used as the input reference displacement signal for the shake table. Niku demonstrated that a fifth-order polynomial interpolation algorithm can ensure continuity in displacement, velocity, and acceleration [38]. Therefore, the fifth-order polynomial interpolation algorithm in Eq. (4.39) is adopted to interpolate the displacement response of the numerical substructure.

$$\left\{\begin{aligned} I(k) &= c_0 + c_1 k + c_2 k^2 + c_3 k^3 + c_4 k^4 + c_5 k^5 \\ \dot{I}(k) &= c_1 + 2c_2 k + 3c_3 k^3 + 4c_4 k^3 + 5c_5 k^4 \\ \ddot{I}(k) &= 2c_2 + 6c_3 k + 12c_4 k^2 + 20c_5 k^3 \end{aligned}\right. \tag{4.39}$$

where, $I(k)$, $\dot{I}(k)$, and $\ddot{I}(k)$ represent the interpolated displacement, velocity, and acceleration, respectively; k denotes the k-th control step. c_1, c_2, c_3, c_4, and c_5 are the coefficients of the fifth-order interpolation polynomial.

To compensate for time delays, a set of interpolated data between any two adjacent analysis steps of the numerical substructure is processed. As shown in Fig. 4.38B, the first $M - 1$ values of the interpolated signal are discarded, and the M-th value is adopted as the input reference displacement signal for the shake table at the k-th step. To maintain the continuity of the reference displacement signal, the triangular-marked data in Fig. 4.38B, i.e., the one-step-ahead predicted value of the interpolated signal set, is adopted as the reference displacement signal until step $N + 1$, where the interpolated signals from steps $N + 1$ to $N + 2$ are used as the reference displacement signal. This algorithm requires only one-step-ahead prediction to compensate for time delays exceeding one control step size, thereby reducing errors introduced by multi-step-ahead prediction.

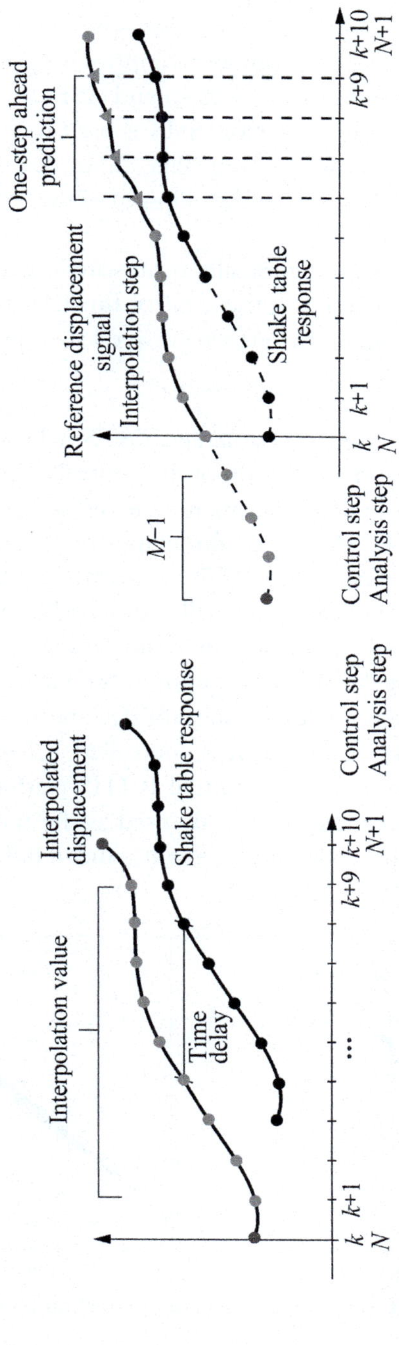

(a) Original reference signal and response

(b) Processed reference signal and response

Figure 4.38 Comparison of signals before and after interpolation prediction algorithm processing.

4.5.3 Experimental Verification

On the virtual RTHS platform for high-speed railway train-bridge coupling vibration, the adaptive composite control proposed in this section is compared with the combination of ATS and ASFC.

The synchronization subspace plot (SSP) is used to evaluate the time-delay compensation effect of the controller. The ideal SSP is a straight line with a slope of 1, and the wider the ellipsoid of the SSP, the greater the time delay.

ATS and interpolation prediction algorithms are integrated with ASFC, respectively. Eq. (4.40) is used as the control function for ATS, and parameters are estimated online through recursive least squares.

$$u = \alpha_0 r + \alpha_1 \dot{r} \tag{4.40}$$

where α_0 and α_1 are the control parameters of ATS, whose initial values are obtained through offline identification of the controlled plant under ASFC control; r is the target signal displacement; $\dot{r}$ is the target signal velocity.

As shown in Figs. 4.39A and B, the ellipse of the interpolation prediction algorithm combined with ASFC is narrower than that of ATS combined with ASFC, indicating a smaller time delay in the shake table under adaptive composite control. To evaluate the control effect, the displacement of the wheel-rail contact point in the virtual RTHS platform is compared with the response of the ideal bridge-train model. In the ideal RTHS model, the shake table model is controlled as Model 1, while the remaining parts are identical to the virtual RTHS platform, meaning the interpolated track displacement is directly used as the train excitation. By comparing the train excitation displacements under different controls in

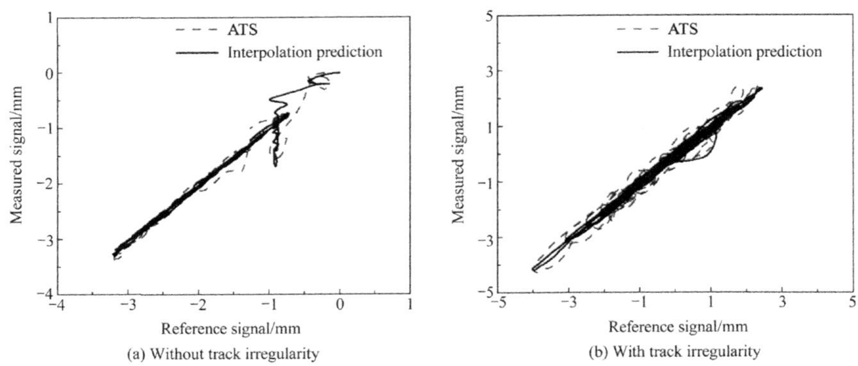

Figure 4.39 SSP and interpolation prediction of ATS.

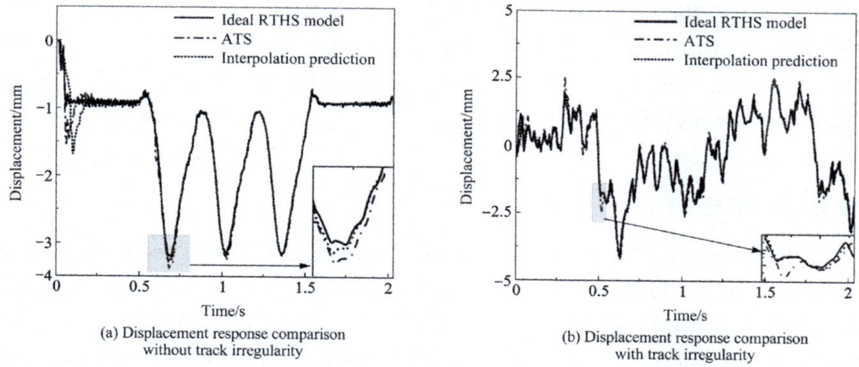

Figure 4.40 Comparison between ATS and the interpolation prediction algorithm.

Figs. 4.40A and B between the virtual RTHS platform and the ideal bridge–vehicle model, it is evident that the virtual RTHS platform under the combined control of the interpolation prediction algorithm and ASFC provides train displacement excitation closer to the ideal model.

For operating conditions without track irregularities, small-amplitude white-noise excitation can be applied offline to identify the fixed experimental substructure on the shake table. This approach compensates for insufficient excitation components during the initial test phase, thereby correcting the initial model parameters of the online estimator to reduce control errors. Fig. 4.41 demonstrates that after the initial model parameters are corrected, the excitation displacement error of the train of the virtual RTHS platform is significantly reduced under adaptive composite control. In summary, under the premise of unknown experimental

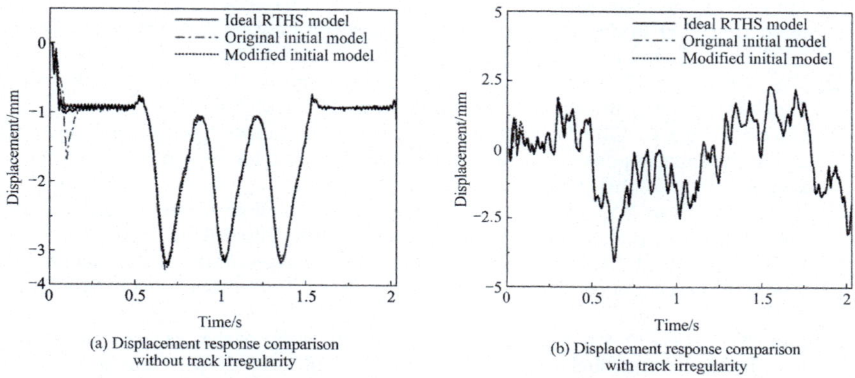

Figure 4.41 Comparison of control effects before and after initial model correction.

substructure, the adaptive composite control combining the interpolation prediction algorithm and ASFC can autonomously adjust control parameters and achieve superior shake table tracking accuracy and effective time-delay compensation.

4.6 Summary

This chapter elaborates on the overall framework and key technologies of RTHS tests of high-speed train operations on bridges, and introduces a hybrid simulation platform architecture. The train-track-bridge system is divided into the numerical substructure and the experimental substructure for the hybrid simulation for high-speed train operation on bridges. Based on differences in time-history computation, hybrid simulation can be classified into RTHS and OIHS, both of which can be divided into the numerical substructure and the experimental substructure.

Considering the characteristics of RTHS, this chapter focuses on three core technologies: To better simulate and solve the numerical substructure, the SSMEDV algorithm is proposed with its characteristics explained in detail; to meet heightened demand for computational speed and accuracy in large-scale real-time simulations, the moving load integration method is presented, fulfilling the requirements of RTHS for rapid computation; to address time delays caused by loading devices and signal transmission devices during RTHS, an ASFC is introduced, delivering superior shake table tracking accuracy and effective time-delay compensation.

References

[1] Liu JF, Liu CY. A summary of bridges on Qin-Shen passenger dedicated railway and pondering over bridge construction for high-speed railway. Railway Stand Design 2004;7:134–8. 159 [In Chinese].

[2] Yan B, Dai G-L, Hu N. Recent development of design and construction of short span high-speed railway bridges in China. Eng Struct 2015;100:707–17.

[3] Guo W, Zeng C, Gou H, et al. Rotational friction damper's performance for controlling seismic response of high speed railway bridge-track system. Comput Model Eng Sci 2019;120:491–515.

[4] Zhou HM, Shao XY, Tian YP, et al. Reproducing response spectra in shaking table tests of nonstructural components. Soil Dynam Earthq Eng 2019;127:105835.

[5] Fricke D, Frost M. Development of a full-vehicle hybrid-simulation test using hybrid system response convergence (HSRC). SAE Int J Passen Cars-Mech Sys 2012;5 (2):921–36.

[6] Luo ZB, Yang ZD, Chen L, et al. Numerical simulation of the control strategy for a large hydraulic centrifuge shaker. J Vib Eng 2015;28(1):18–26 [In Chinese].

[7] Herbert A. Fixed point equations and nonlinear eigenvalue problems in ordered banach spaces. SIAM Rev 1976;18(4):620–709.

[8] Jiang YZ, Wu PB, Zeng J, et al. Intelligent batch process method for analyzing the effect of the suspension parameters on the vibration of the suspended monorail. Adv Mech Eng 2020;12(10):16878140296642.

[9] Ren SB, Arie R, Kees K. Dynamic simulation of the maglev vehicle/guideway system. J Bridge Eng 2010;15(3):269–78.

[10] Newmark NM. A method of computation for structural dynamics. J Eng Mech ASCE 1959;85(3):67–94.

[11] Gui Y, Wang JT, Jin F, et al. Development of a family of explicit algorithms for structural dynamics with unconditional stability. Nonlinear Dyn 2014;77:1157–70.

[12] Chen C, Ricles JM. Development of direct integration algorithms using discrete control theory. J Eng Mech 2008;134(8):676–83.

[13] Chang SY. Explicit pseudodynamic algorithm with unconditional stability. J Eng Mech 2002;128(9):935–47.

[14] Chang SY. Improved explicit method for structure dynamics. J Eng Mech 2007;8 (4):329–40.

[15] Wilson EL. A computer program for the dynamic stress analysis of underground structures, division of structural engineering structural mechanics. Berkeley: University of California; 1968.

[16] Hilber HM, Hughes TJR, Taylor RL. Improved numerical dissipation for time integration algorithms in structural mechanics. Earthq Eng Struct Dynam 1977;5:283–92.

[17] Kolay C, Ricles JM. Development of a family of unconditionally stable explicit direct integration algorithms with controllable numerical energy dissipation. Earthq Eng Struct Dynam 2014;41:1361–80.

[18] Wu X, Wei G, Hu P, et al. Seismic performance evaluation of building-damper system under near-fault earthquake. Shock Vib 2020;1:2763709.

[19] Chopra AK. Dynamics of structures. Englewood Cliffs, NJ: Prentice Hall; 1995.

[20] Chen MH. Research on real-time hybrid simulation test methods and software development. Nanjing: Southeast University; 2018 [In Chinese].

[21] Gu Q, Zhang DY, Guo W, et al. An efficient computation method for real-time hybrid testing of the vehicle-track-bridge coupling system of high-speed railway. J S China Univ Technol 2021;49(3):123–30 [In Chinese].

[22] Sun BY, Gu Q, Zhang PZ, et al. Elastoplastic numerical substructure method of frame structure considering *P-Δ* effects. Eng Mech 2018;35(2):153–9 [In Chinese].

[23] Sun BY, Gu Q, Zhang PZ, et al. Numerical substructure method for structural nonlinear analysis. Chin J Comput Mech 2015;32(4):465–72 [In Chinese].

[24] Darby AP, Williams MS, Blakeborough A. Stability and delay compensation for real-time substructure testing. J Eng Mech 2002;128:1276–84.

[25] Dong XH, Tang ZY, Li ZB, et al. Research on real-time hybrid simulation by Python-GPU computing. J Vib Eng 2023;36(2):517–25 [In Chinese].

[26] Horiuchi T, Inoue M, Konno T, et al. Real-time hybrid experimental system with actuator delay compensation and its application to a piping system with energy absorber. Earthq Eng Struct Dynam 1999;28(10):1121–41.

[27] Wallace MI, Wagg DJ, Neild SA. An adaptive polynomial based forward prediction algorithm for multi-actuator real-time dynamic substructuring. Proc Royal Soc A-Mathemat Phys Eng Sci 2005;461(2064):3807–26.

[28] Wang Z, Ning X, Xu G, et al. High performance compensation using an adaptive strategy for real-time hybrid simulation. Mech Syst Signal Process 2019;133:106262.
[29] Bonnet PA, Lim CN, Williams MS, et al. Real-time hybrid experiments with Newmark integration, MCSmd outer-loop control and multi-tasking strategies. Earthq Eng Struct Dynam 2007;36(1):119–41.
[30] Zhou H, Xu D, Shao X, et al. A robust linear-quadratic-Gaussian controller for the real-time hybrid simulation on a benchmark problem. Mech Syst Signal Process 2019;133:106260.
[31] Chae Y, Kazemibidokhti K, Ricles JM. Adaptive time series compensator for delay compensation of servo-hydraulic actuator systems for real-time hybrid simulation. Earthq Eng Struct Dynam 2013;42(11):1697–715.
[32] Stoten DP, Gómez EG. Adaptive control of shaking tables using the minimal control synthesis algorithm. Philos Trans R Soc London, Ser A: Math Phys Eng Sci 2001;359 (1786):1697–723.
[33] Shao P, Guo W, Lei Q, et al. Adaptive compound control for the real-time hybrid simulation of high-speed railway train-bridge coupling vibration. Struct Control Health Monit 2021;28(11):e2816.
[34] Phillips BM, Wierschem NE, Spencer Jr BF. Model-based multi-metric control of uniaxial shake tables. Earthq Eng Struct Dynam 2014;43(5):681–99.
[35] Simon D. Optimal state estimation: Kalman, H infinity, and nonlinear approaches. John Wiley & Sons; 2006.
[36] Dorf RC, Bishop RH. Modern control systems. Pearson; 2011.
[37] Wallace MI. Real-time dynamic substructuring for mechanical and aerospace applications: control techniques and experimental methods. University of Bristol; 2005.
[38] Niku SB. Introduction to robotics: analysis, control, applications. John Wiley & Sons; 2020.

CHAPTER 5

Seismic Catastrophe Mechanisms of High-Speed Railway Train–Track–Bridge Systems

Contents

High-Speed Railway Train–Track–Bridge Systems
ISBN 978-0-443-49310-2
https://doi.org/10.1016/B978-0-443-49310-2.00005-3

5.1 Overview

In the previous chapters, we discussed some simulation platforms and experimental methods for train–track–bridge systems under earthquakes. Based on the above discussion, this chapter delves into the seismic damage mechanisms of high-speed railway train–track–bridge systems by analyzing the results from simulation experiments. In the research, a mechanical model for high-speed railway train–track–bridge systems is established, and a brief dynamic characteristic analysis is conducted; then, the force-displacement curves of various components in track-bridge systems are introduced to characterize their mechanical properties. The research then focuses on simulation and experimental analysis for high-speed railway bridge piers, examining the failure modes and damage mechanisms under seismic excitations. Subsequently, Finite Element (FE) simulations are performed under two distinct seismic conditions, namely, far-field earthquakes and near-fault earthquakes, so as to examine the damage mechanisms and failure characteristics of track–bridge systems, and a nonlinear spring model based on the failure mechanisms of track structures is proposed. Finally, the dynamic responses of train–track–-bridge systems are analyzed under both seismic and post-earthquake conditions to examine the influencing factors of operational safety under and after earthquakes.

5.2 Mechanical Properties of Key Components in the Track-Bridge System

5.2.1 Piers

Round-ended solid piers are used, with the dimensions of the 8m and 16m piers listed in Table 5.1. The longitudinal reinforcement ratios of the 8m and 16m piers are 0.15% and 0.4%, respectively, with axial compression ratios of 5% and 10%. Through low-cycle reversed static loading tests, Jiang

Table 5.1 Sectional dimension parameters of circular-ended bridge piers.

Dimensions	Component Parameters	Dimensions	Component Parameters
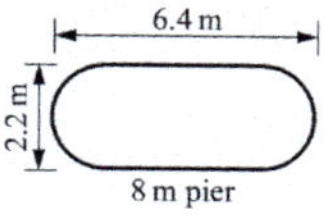 8m pier	Reinforced concrete structure: C35 Design compressive strength: 16.7 MPa	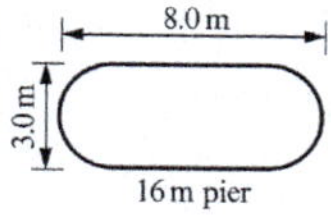 16m pier	Reinforced concrete structure: C35 Design compressive strength: 14.3 MPa

et al. [1] obtained the force-displacement curves of the high-speed railway bridge piers, with the ultimate displacements of the 8m and 16m piers being 266.4 and 409.6 mm, respectively.

To accurately simulate the mechanical properties of bridge piers, this study employs a twoNodeLink element to fit the force-displacement curves of 8 m and 16 m high-speed railway bridge piers, as shown in Fig. 5.1. The numerical curves exhibit good agreement with the experimental curves, demonstrating the accuracy of the pier model.

5.2.2 Main Girders and Friction Plates

The 32m double-track simply supported box girder bridge adopts the standard design. The cross-sectional dimensions of the main girder and friction plate are shown in Table 5.2. Due to its high cross-sectional stiffness, it generally remains in an elastic state under seismic conditions. The box girder is constructed with C50 concrete, featuring a standard span length of 30.5 m (spacing between two bearings), with 0.55 m from the girder end to the bearing and 0.1 m clearance between adjacent spans. In

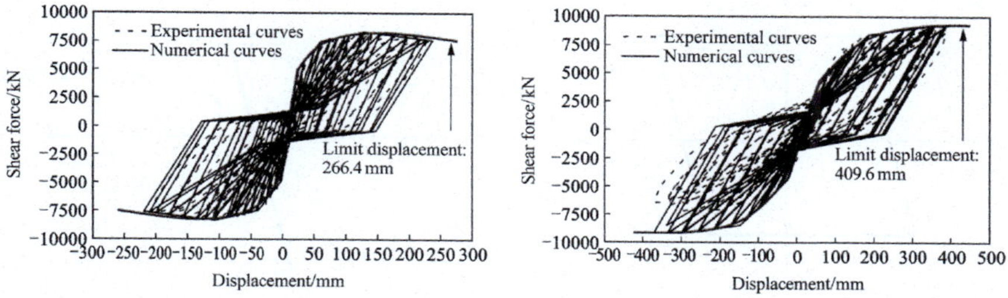

Figure 5.1 Force-displacement relationship curves of high-speed railway bridge piers.

Table 5.2 Cross-sectional dimension parameters of main girder and friction plate.

Dimensions	Component Parameters
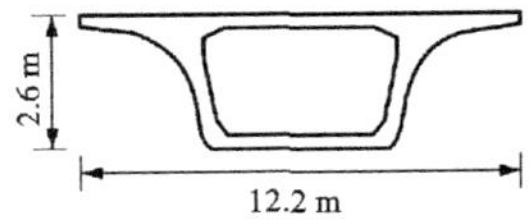 Main girder	Reinforced concrete structure: C50 Design compressive strength: 23.1 MPa Area: $A = 8.34 \times 10^5$ mm^2 Elastic modulus: $E = 4.40 \times 10^4$ N/mm^4 Cross-sectional moment of inertia: $I_z = 9.00 \times 10^{12}$ mm^4
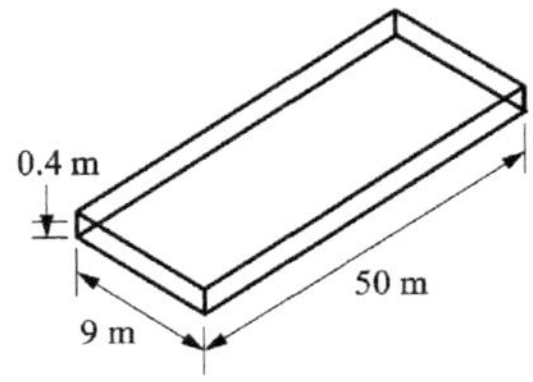 Friction plate	Reinforced concrete structure: C30 Design compressive strength: 14.3 MPa Area: $A = 5.10 \times 10^5$ mm^2 Elastic modulus: $E = 3.60 \times 10^4$ N/mm^2 Cross-sectional moment of inertia: $I_z = 1.70 \times 10^{12}$ mm^4

OpenSees FE modeling, beam elements are simulated using the elasticBeamColumn element.

5.2.3 Bearings

The force-displacement relationship of pot rubber sliding bearings can be defined by a friction model [2], as shown in Fig. 5.2(a). The elements are simulated using the zeroLength Element, with the maximum longitudinal sliding displacement of the sliding bearing not exceeding 60 mm. Assuming the fixed bearing remains undamaged under seismic action, its

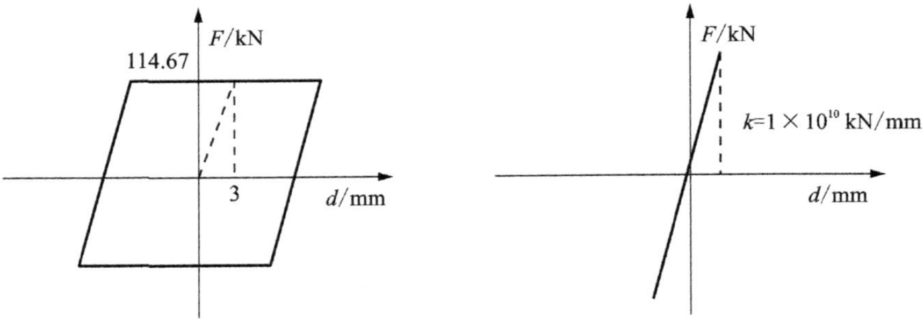

(a) Sliding bearing (b) Fixed bearing

Figure 5.2 Force-displacement curves of bearings.

longitudinal stiffness k is considered very high, at 1×10^{10} kN/mm, as shown in Fig. 5.2(b).

5.2.4 Sliding Layer and CA Mortar Layer

The sliding layer is a component connecting the main girder and the base plate, whose force-displacement relationship can be defined by a friction model. Research by Zhu et al. [3] establishes the longitudinal resistance of the sliding layer as the vertical load per unit length of the track superstructure multiplied by the coefficient of friction. Without train loads, the track superstructure load corresponds to the dead weight of the rail, track slab, mortar layer, and base plate per unit length. With train loads, they are included in the total load. The friction coefficient of the sliding layer typically ranges between 0.2 and 0.3. In this section, the case without train load is considered, where the unit weight of the upper track structure components is 18 KN/mm, with a friction coefficient of 0.2, resulting in a sliding displacement of 0.5 mm, as illustrated in Fig. 5.3(a).

The CA mortar layer, which connects the base plate and track slab, exhibits deformation characteristics similar to the sliding layer under longitudinal forces. Its longitudinal resistance increases with the increase of displacement until the displacement reaches a limit, beyond which the longitudinal resistance will no longer increase even if the displacement continues to grow. Therefore, the force-displacement relationship of the CA mortar layer is described by an elastoplastic model. Based on field test data from Zhu et al. [3], the maximum shear force at failure of the CA mortar layer is 41 kN, with a displacement of approximately 0.5 mm. Consequently, the force-displacement relationship per unit length of the CA mortar layer is illustrated in Fig. 5.3(b).

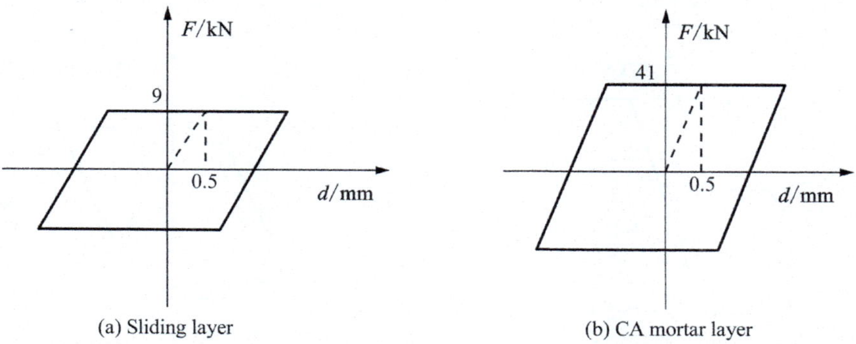

Figure 5.3 Force-displacement relationship curves of the sliding layer and mortar layer.

5.2.5 Shear Grooves and Shear Reinforcement

Located above the fixed bearing of the bridge, shear grooves transfer longitudinal forces such as braking forces to the piers. Considering the slip effect of reinforcement, the longitudinal yield displacement of the shear studs is 1.9 mm, and failure occurs when the maximum displacement exceeds 11.76 mm [4]. Therefore, shear grooves are simulated using elements with the MinMax material model, and its force-displacement relationship is illustrated in Fig. 5.4(a). Zhao et al. [5] conducted quasi-static tests on shear reinforcement in ballastless track structures, revealing that the shear reinforcement exhibits distinct elastoplastic behavior, with a yield force and yield displacement of 92.4 kN and 0.6 mm, respectively. Therefore, the study adopts an elastoplastic model to simulate the force-displacement relationship of the shear reinforcement, with the relationship curves shown in Fig. 5.4(b).

5.2.6 Fasteners

An elastoplastic mechanical model is used as the longitudinal mechanical model of the WJ-8C fastener system, excluding the effect of train loads [6]. Its slip force and maximum sliding displacement are 15 kN and 2 mm, respectively, as illustrated in Fig. 5.5(a). The vertical load-bearing system of the fastener includes iron baseplates, under-rail elastic pads, fastening bolts, and elastic clips. The vertical stiffness of the fastener primarily depends on the load-bearing state of the fastener, as well as the vertical stiffness of the baseplates and elastic clips.

The vertical load-bearing state of the fastener can be divided into three primary stages. The first stage: when the fastener system is subjected to a significant downward load, the compression of the elastic pad exceeds the effective elastic range of the elastic clip, causing the clip to disengage. At this point, the stiffness of the fastener system is solely determined by the

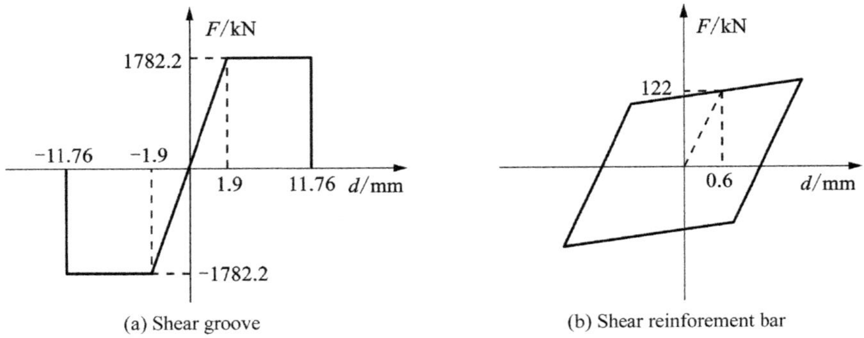

Figure 5.4 Force-displacement relationship curves of the shear grooves and shear reinforcement bar.

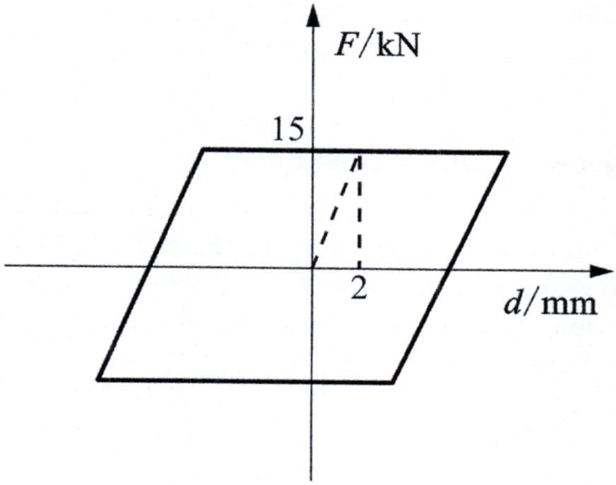

(a) Longitudinal force-displacement relationship of fastener

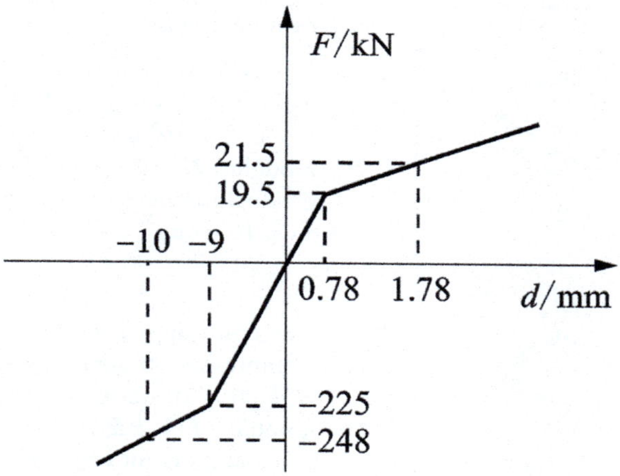

(b) Vertical force-displacement relationship of fastener

Figure 5.5 Force-displacement relationship curves of the fastener system.

stiffness of the elastic pad. The second stage: when the elastic clip and elastic pad work together, the elastic pad is in an unloaded state, while the clamping force of the clip gradually increases. The third stage: when the fastener system is subjected to a significant upward uplift force, causing the rail and elastic pad to separate, the elastic pad disengages from the load-bearing mechanism. At this point, the joint stiffness of the fastener system is determined by the stiffness of the elastic clip. Therefore, the vertical force on the fastener system can be described by the force-displacement relationship shown in Fig. 5.5B.

5.2.7 Base Plates, Track Slabs, and Rails

The base plates, track slabs, and rails are all longitudinal continuous components. They should remain undamaged under seismic action, with their dimensions and parameters shown in Table 5.3. In FE modeling in OpenSees, all longitudinal continuous components are simulated using elastic beam-column elements [7].

Table 5.3 Cross-section dimensions and parameters of longitudinal continuous components in the slab track system.

Dimensions	Component Parameters
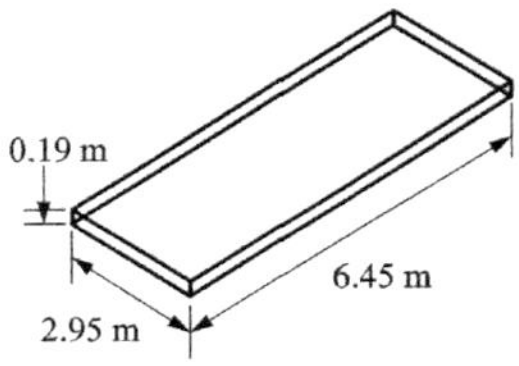 Base plate	Reinforced concrete structure: C50 Design compressive strength: 23.1 MPa Area: $A = 3.60 \times 10^6$ mm^2 Elastic modulus: $E = 3.0 \times 10^4$ N/mm^2 Cross-sectional moment of inertia: $I_z = 4.80 \times 10^{12}$ mm^4
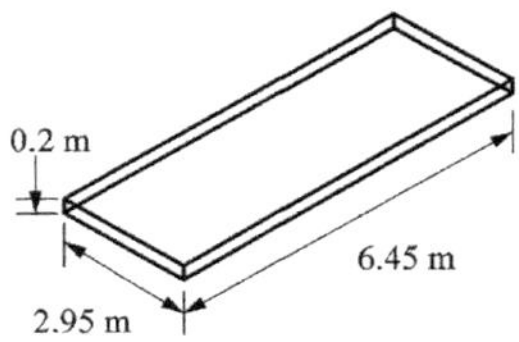 Track slab	Reinforced concrete structure: C30 Design compressive strength: 14.3 MPa area: $A = 5.60 \times 10^5$ mm^2 Elastic modulus: $E = 3.25 \times 10^4$ N/mm^2 Cross-sectional moment of inertia: $I_z = 1.68 \times 10^{12}$ mm^4
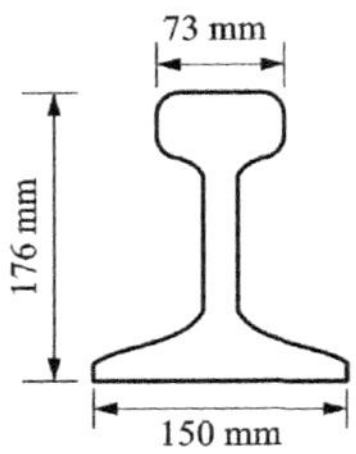 Rail	CHN60 rail Yield strength of rail: 500 MPa Area: $A = 7.75 \times 10^3$ mm^2 Elastic modulus: $E = 2.06 \times 10^5$ N/mm^2 Cross-sectional moment of inertia: $I_z = 5.24 \times 10^6$ mm^4 Mass: $m = 60.64$ kg/m

5.3 Flexural-Shear Strength Degradation Model for High-Speed Railway Bridge Piers under Seismic Damage

5.3.1 Shear Capacity Theoretical Formula

Research by Fan Chao [8] demonstrates that under seismic action, reinforced concrete bridge piers mainly fail in three modes: flexural failure, flexural-shear failure, and shear failure. Flexural failure exhibits ductile behavior, whereas flexural-shear failure and shear failure demonstrate brittle characteristics. Brittle failure typically occurs abruptly, potentially leading to catastrophic consequences such as superstructure collapse. In seismic design, therefore, such failure should be avoided as much as possible.

Under seismic action, at low displacement ductility, when the shear demand corresponding to the flexural capacity of the pier exceeds its shear strength, brittle failure occurs in the pier. Extensive studies have demonstrated that the shear strength of pier columns degrades as displacement ductility increases. The shear strength degradation mechanism effectively accounts for the influence of diagonal crack propagation in concrete, reinforcement yielding, and excessive compressive stress in the concrete shear-compression zone on shear capacity.

Fig. 5.6 illustrates the interaction between the shear capacity envelope and the shear demand of the bridge pier. Where V_n denotes the shear capacity, which includes the initial segment V_i, the descending segment V_d, and the residual segment V_r; V_m denotes the shear demand, which varies as displacement ductility changes. If the shear demand curve of the pier V_m intersects with its shear capacity curve V_n at low ductility, brittle shear failure will occur in the pier. If the shear demand curve never intersects with the shear capacity curve, ductile flexural failure will occur.

The shear capacity expressions of round-ended solid piers for high-speed railways under longitudinal seismic excitation are given by Eqs. (5.1) and (5.2). The model accounts for shear strength degradation due to increasing displacement ductility by introducing a displacement ductility correlation coefficient K.

$$V_n = K(V_c + V_s) \tag{5.1}$$

$$V_n = K\left(\frac{1}{\lambda}0.395f_c^{0.55}\sqrt{1+\frac{P}{0.395f_c^{0.55}}}0.8A_g + 0.445\frac{A_u f_y h_0}{s}\right) \tag{5.2}$$

where K is the influence coefficient of displacement ductility with respect to the shear capacity of high-speed railway piers (conservative and economical

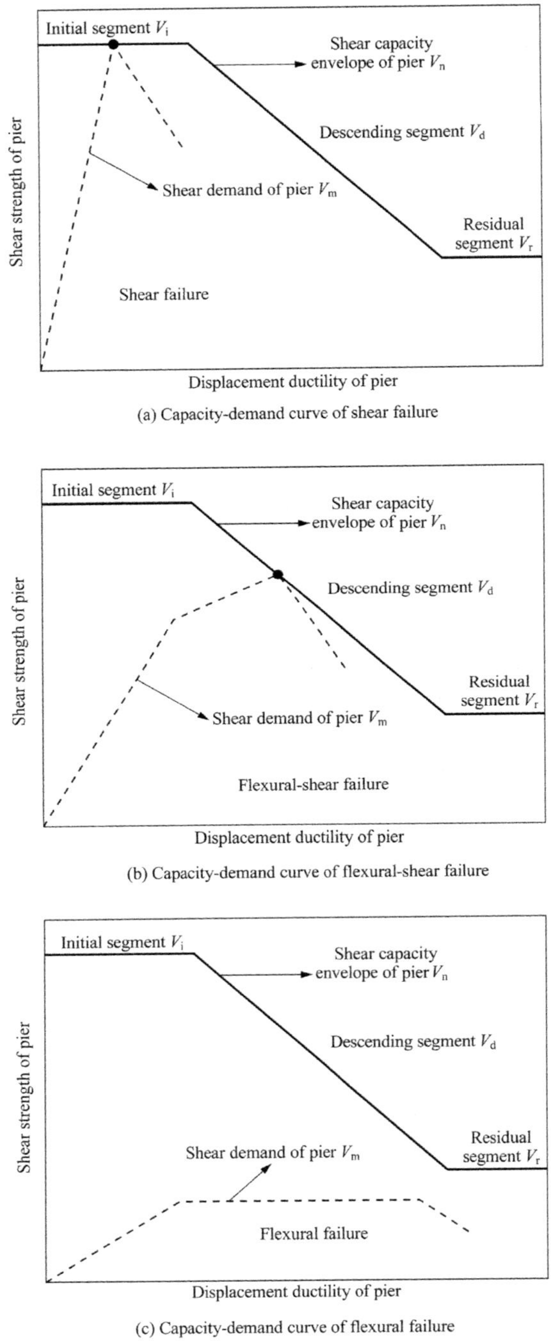

(a) Capacity-demand curve of shear failure

(b) Capacity-demand curve of flexural-shear failure

(c) Capacity-demand curve of flexural failure

Figure 5.6 Shear capacity-demand curve model for a bridge pier [9].

types); V_c denotes the concrete shear contribution; V_s represents the lateral reinforcement shear contribution; λ is the shear span-to-effective depth ratio of the pier; f_c denotes the concrete compressive strength; P denotes the axial pressure at the pier top under seismic action; A_g represents the cross-sectional area of the pier; $0.8A_g$ represents the effective shear area of high-speed railway piers; A_u is the shear area of lateral reinforcement within the pier section; f_y represents the yield strength of stirrups configured in the pier section; h_0 denotes the effective height of the pier section; and s indicates the spacing of shear reinforcement in the pier.

The correction of the displacement ductility correlation coefficient K enhances the accuracy of the proposed shear strength model and provides a certain safety margin, as illustrated in Fig. 5.7 [8]. The quantitative expression of the corrected displacement ductility correlation coefficient K is given by Eq. (5.3). The computational results of the corrected model more accurately reflect the shear strength of the structure, enabling the verification of whether high-speed railway piers will undergo brittle failure under seismic action.

$$K = \begin{cases} 1.0 & (0 < \mu \leqslant 1.33) \\ -0.18\mu + 1.24 & (1.33 < \mu \leqslant 4.93) \\ 0.35 & (\mu > 4.93) \end{cases} \tag{5.3}$$

5.3.2 Trilinear Shear Strength Model

Flexural-shear failure and shear failure of high-speed railway piers under seismic action exhibit brittle characteristics and therefore, should be prevented in seismic design. Flexural failure of bridge piers is typically numerically simulated using fiber models. Flexural-shear failure exhibits a more complex mechanism than pure flexural failure. Therefore, fiber models alone often yield unsatisfactory results due to their inability to accurately capture the shear response of piers during earthquakes. Currently, flexural-shear failure is simulated at three levels: section level, material level, and element level. Section-level simulation superimposes nonlinear shear stiffness in the stiffness matrix, but does not account for the coupling among bending moment, shear force, and axial force. Represented by the modified compression field theory, material-level simulation is theoretically complex, computationally intensive, and cumbersome in practical applications. Currently, element-level flexural-shear failure simulation is more widely adopted, where the total response of the pier

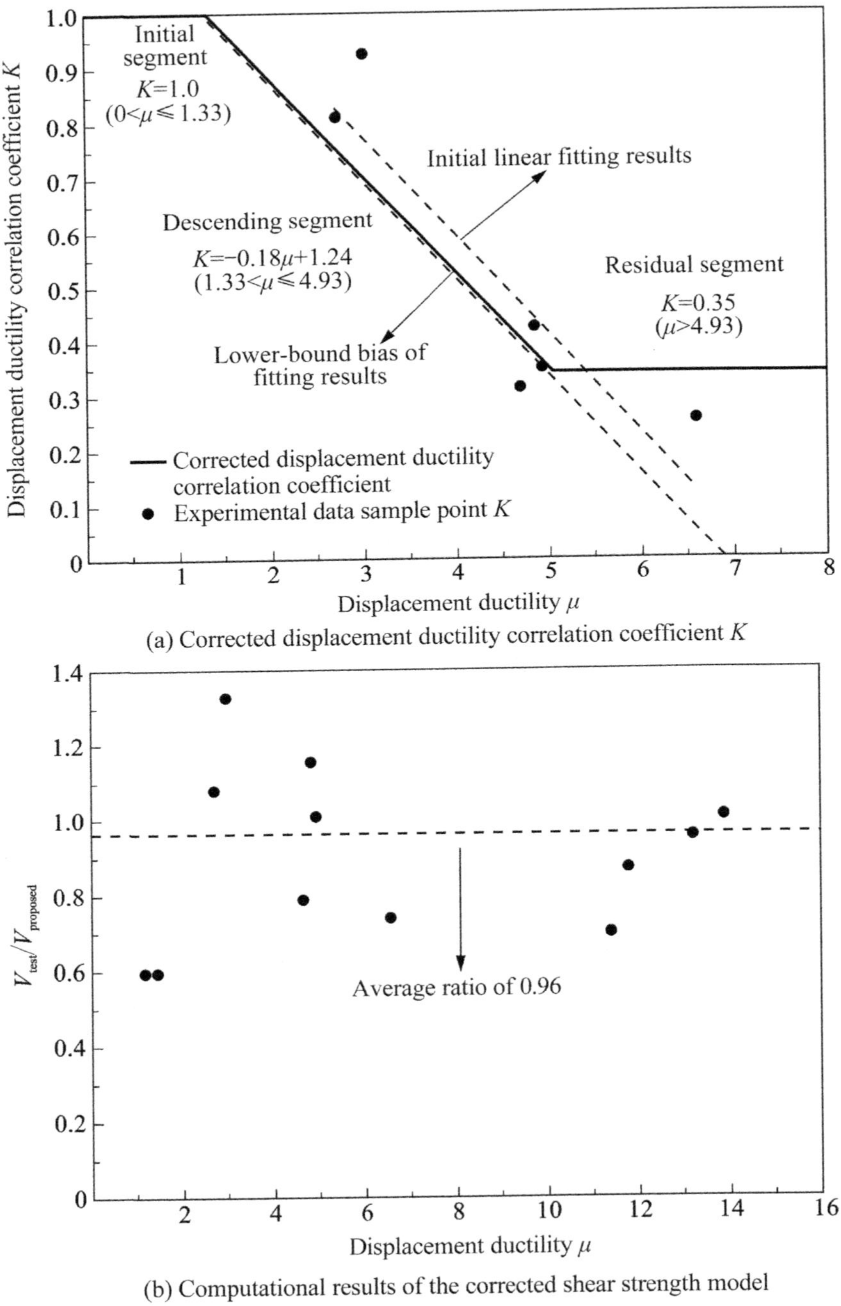

Figure 5.7 Corrected shear strength model.

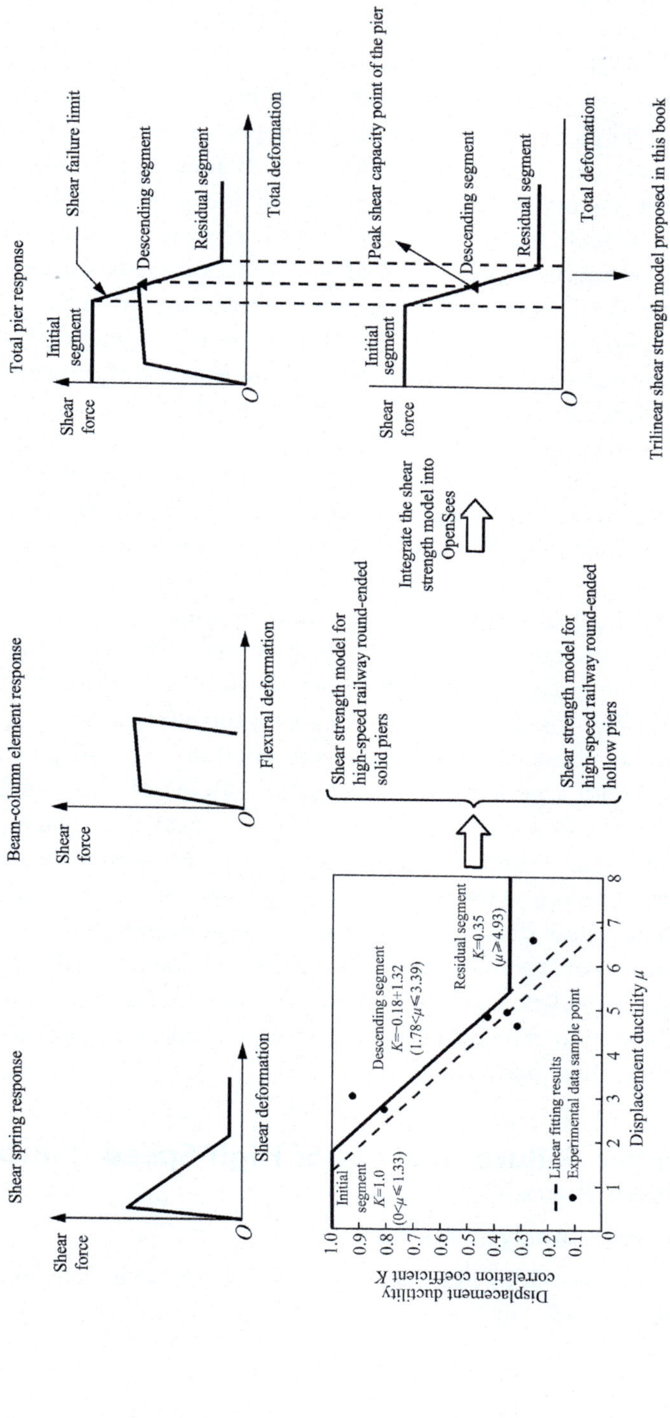

Figure 5.8 Schematic diagram of the trilinear shear strength model.

(shear force–total horizontal displacement) is decomposed into bending response (shear force–bending displacement) and shear response (shear force–shear displacement), coupled through a shear failure criterion.

The low-cycle reciprocating load tests of bridge piers demonstrate that under different pier failure modes, the development patterns of flexural and shear responses exhibit significant differences. Before the peak shear force is reached, the flexural and shear responses of the piers develop synchronously. Once the hysteretic backbone curve enters the degradation phase, a flexural-shear failure in the pier leads to shear-dominated behavior. The shear spring element can effectively simulate the transition process from flexural-dominated to shear-dominated mechanisms in bridge piers under brittle failure models. The crack initiation positions and propagation paths in bridge piers can be inferred from the strain characteristics of the pier materials. As the displacement at the top of the pier increases, the number of cracks grows, and their positions progressively shift upward, forming an overall cross-pattern. Subsequently, the cracks gradually exhibit a downward inclination trend, ultimately developing into critical inclined shear cracks.

Fan Chao [8] evaluated the shear resistance of bridge piers using a trilinear shear strength model, with the three lines representing the initial stage, descending stage, and residual stage of shear strength of the piers, as illustrated in Fig. 5.8. This model accounts for the displacement-dependent degradation of shear capacity. Under the condition of 8-degree seismic fortification with a peak ground acceleration (PGA) of 1.5 g, high-speed railway round-ended solid piers will reach their peak shear capacity, followed by strength degradation (descending segment), ultimately leading to flexural-shear failure. However, such conditions are extremely rare in practice. Tests show that high-speed railway piers exhibit excellent seismic performance. For round-ended solid piers experiencing flexural-shear failure, the degradation of shear strength with increasing displacement must be considered. The trilinear shear strength model can more accurately evaluate their shear performance.

5.4 Seismic Failure Analysis of High-Speed Railway Bridge Piers

5.4.1 Pushover Analysis Method

Based on the research by Hu Yao et al. [10], the pushover analysis method is a static inelastic analysis method that relies on two fundamental assumptions: ① The seismic force distribution pattern of bridge piers is dominated by the first mode vibration; ② the load distribution pattern of bridge piers remains unchanged during the pushover analysis. Based on the first mode vibration, the seismic load distribution pattern can be determined by Eq. (5.4):

$$F_{ij} = \alpha\beta_j\gamma_j x_{ij} m_i \tag{5.4}$$

where F_{ij} represents the horizontal seismic force at node i of the structure in the j-th vibration mode (in kN); α denotes the basic horizontal seismic acceleration; β_j indicates the dynamic amplification factor of the j-th vibration mode; γ_j is the modal participation coefficient of the j-th vibration mode; x_{ij} is the modal coordinate of the mass center at node i of the j-th vibration mode; m_i is the mass at node i.

According to Eq. (5.4), the base shear-to-top displacement curves of bridge piers with different heights can be obtained. To evaluate the seismic capacity of bridge piers, the base shear-to-top displacement curves of bridge piers are converted into capacity spectrum curves. First, the base shear-to-top displacement curves are converted into a bilinear model based on the energy equivalence principle, as shown in Fig. 5.9. Then, the

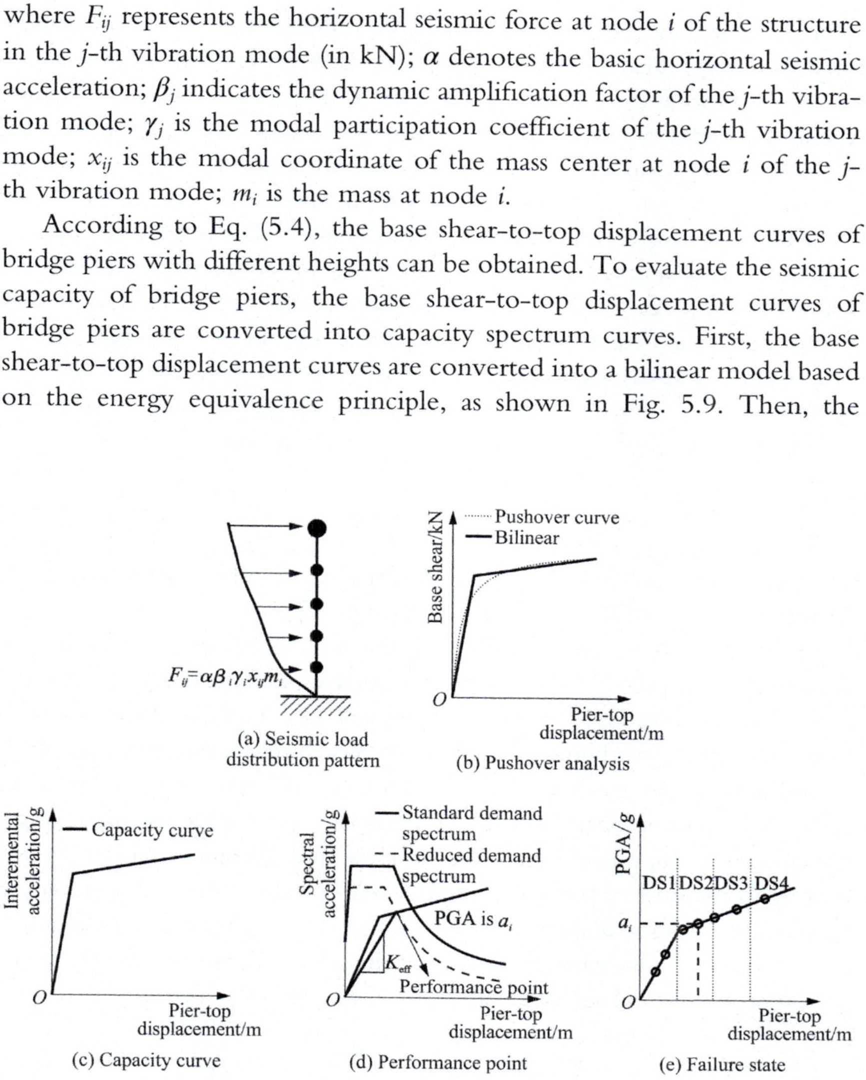

Figure 5.9 Steps of the pushover analysis method.

bilinear model is transformed into a capacity spectrum curve according to Eq. (5.5).

$$S_a = \frac{V_b}{M_j}, S_d = \frac{D_b}{\Gamma_j \varphi_{nj}}, M_j = \frac{\left(\sum_{i=1}^{n} m_i \varphi_{ij}\right)^2}{\sum_{i=1}^{n} m_i \varphi_{ij}^2}, \Gamma_j = \frac{\sum_{i=1}^{n} m_i \varphi_{ij}}{\sum_{i=1}^{n} m_i \varphi_{ij}^2} \tag{5.5}$$

where, S_a denotes pseudo-acceleration; S_d represents pseudo-displacement spectrum; V_b indicates the total base shear; D_b represents the roof displacement; M_j denotes the effective mass of the j-th vibration mode; φ_{ij} denotes the modal displacement at node i in the j-th vibration mode; Γ_j denotes the modal participation factor of the j-th vibration mode; φ_{nj} denotes the modal displacement at the top of the pier in the j-th vibration mode.

The Pushover analysis method, combined with the seismic demand spectrum, can approximately determine the structural response under a specific earthquake. However, to obtain the seismic responses of individual bridge piers, the elastic acceleration spectrum should be converted into an acceleration–displacement demand spectrum, which represents the demand spectrum of structural responses under a specific earthquake, as shown in Fig. 5.9C. The displacement spectrum S_d is expressed as:

$$S_d = \frac{T_n^2}{4\pi^2} S_a g \tag{5.6}$$

where T_n represents the fundamental period of the elastic structure, and g denotes the gravitational acceleration (9.8 m/s^2).

The seismic response of the linear elastic structure can be determined by the intersection point of the structural capacity spectrum curve and the demand spectrum curve. However, bridge piers may undergo plastic deformation under strong earthquakes. Elastoplastic structures tend to undergo plastic deformation under earthquakes. Therefore, the demand spectrum must be reduced to account for the structural nonlinearity. The reduction of the elastic spectrum is achieved through the stiffness and damping related to the maximum response of the structure. The secant stiffness method is employed to account for structural stiffness reduction [11]. Since structural stiffness is related to the natural vibration period of the structure, the equivalent natural period of the elastoplastic structure T_{eq} can be expressed as:

$$T_{eq} = T_n \sqrt{\frac{\mu}{1 + \alpha_s \mu - \alpha_s}} \tag{5.7}$$

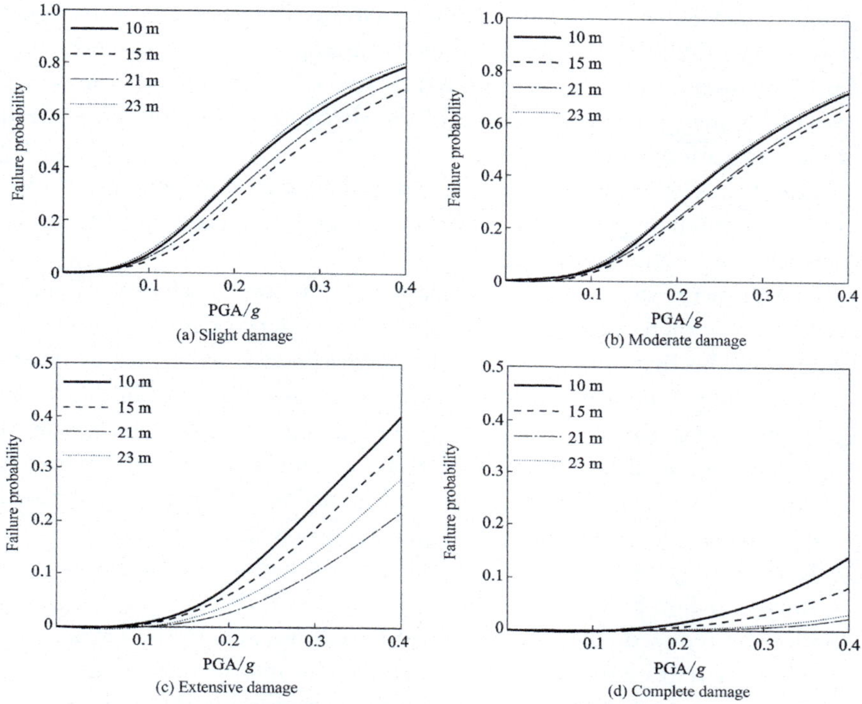

Figure 5.10 Fragility curves of different piers on Type II sites.

where T_n denotes the natural vibration period of the elastic system; α_s represents the ratio of postyield stiffness to preyield stiffness; μ is the ductility coefficient, defined as the ratio of the maximum displacement of the structural system to its yield displacement.

Chopra [12] proposed the concept of the equivalent viscous damping ratio in 1995, defining it as the ratio of energy dissipated per cycle of vibration between an elastoplastic body and an elastic body. Lin et al. proposed reducing the elastic demand spectrum based on the equivalent damping ratio, accounting for the potential pinching in the actual hysteresis curve of the structure. The corrected equation for calculating the equivalent damping ratio [13,14] is expressed as:

$$\xi_{\text{eff}} = \xi_0 + \frac{2}{\pi}\eta\frac{(1-\alpha_s)(1-1/\mu)}{1-\alpha_s+\mu\alpha_s} \tag{5.8}$$

where ξ_0 represents the viscous damping ratio of the structure in the linear elastic stage, taken as 0.05 for reinforced concrete structures; η denotes the damping modification factor, defined as the effective ratio of the actual hysteretic energy dissipation curve of the structure to that of an idealized bilinear vibration model.

Based on the above equation, a reduced spectrum suitable for elasto-plastic structural analysis can be obtained, thereby determining the structural response under seismic action.

In summary, the steps for calculating the seismic responses of high-speed railway piers using the Pushover analysis method are as follows:

① Obtain the first-order mode of the high-speed railway pier to determine the seismic force distribution pattern, as shown in Fig. 5.9A.

② Conduct Pushover analysis on the pier using the first-order mode of seismic force distribution pattern to obtain the base shear-to-top displacement curve, which is then idealized as a bilinear model based on the energy equivalence principle, as shown in Fig. 5.9B.

③ Based on Eq. (5.5), convert the bilinear curve into a capacity curve, as shown in Fig. 5.9C.

④ Plot the reduced elastoplastic demand spectrum within the same coordinate system of the capacity spectrum. The intersection point of the capacity spectrum and reduced spectrum represents the performance point of the pier, which signifies its approximate seismic response, as illustrated in Fig. 5.9D.

⑤ By adjusting the level of PGA, the seismic responses of high-speed railway piers under varying earthquake intensities can be obtained, thereby assessing the failure states of the piers.

5.4.2 Pier Damage Evaluation Metrics

To analyze the seismic performance of high-speed railway bridge piers, the damage states corresponding to pier damage must first be defined. Due to the scarcity of research on failure metrics for high-speed railway bridge piers in existing literature, this section defines the damage states of high-speed railway piers using the displacement ductility ratio, a failure metric proposed by Hwang et al. [15] for conventional bridge columns. Its expression is given as:

$$\mu = \frac{d}{\Delta_{\mathrm{cyl}}} \tag{5.9}$$

Table 5.4 Damage states of bridge piers defined by displacement ductility ratios.

Damage State	Description of Damage State	Criterion
No damage	No reinforcement yielding, only fine cracks in concrete	$0 < \mu < \mu_{cy1}$
Slight damage	Theoretical yielding of the first steel bar, distinct cracks in concrete	$\mu_{cy1} \leq \mu < \mu_{cy}$
Moderate damage	Local plastic hinge, nonlinear deformation, concrete cover spalling, and visible crack propagation	$\mu_{cy} \leq \mu < \mu_{c4}$
Extensive damage	Full formation of plastic hinge, with significant crack width and complete concrete spalling in the plastic hinge zone	$\mu_{c4} \leq \mu < \mu_{cmax}$
Local failure or complete damage	Strength degradation, yielding of the main reinforcement, yielding of stirrups, and crushing of core concrete	$\mu_{cmax} \leq \mu$

where d denotes the maximum pier-top displacement of the pier relative to ground level; Δ_{cy1} represents the displacement of the pier relative to ground at first yield of longitudinal reinforcement in the pier base.

By HAZUS99 (1999) [16] definition, seismic-induced damage in bridge piers can be characterized by displacement ductility ratios, with five damage states: no damage, slight damage, moderate damage, extensive damage, and local failure or complete damage, as detailed in Table 5.4.

In Table 5.4, μ_{cy1} is the displacement ductility ratio corresponding to the initial yielding of longitudinal reinforcement in concrete piers, indicating the threshold for minor pier damage; μ_{cy} denotes the displacement ductility ratio corresponding to equivalent section yielding, representing the limit for moderate pier damage; μ_{c4} is the displacement ductility ratio when the concrete strain at the edge of the column section reaches 0.004, representing the limit for extensive pier damage; μ_{cmax} is the maximum displacement ductility ratio [15], defining the critical value for complete damage as follows:

$$\mu_{cmax} = \mu_{c4} + 3 \tag{5.10}$$

5.4.3 Evaluation Methods for Bridge Pier Fragility

The fragility analysis of bridge piers indicates the probability of damage severity of the piers under specific seismic intensity. Based on four damage levels (slight damage, moderate damage, extensive damage, and complete damage), the fragility curves corresponding to the performance states of bridge piers with varying heights are calculated using the Pushover method. Drawing on research by Kappos et al. [17], the fragility of piers is assumed to follow a log-normal distribution. The fragility is calculated by:

$$P(\mathrm{DS} \geqslant \mathrm{PL}|\mathrm{PGA}) = \varphi\left(\frac{\ln(\mathrm{PGA}) - \ln(\mathrm{PGA_{PL}})}{\beta_\mathrm{C}}\right) \tag{5.11}$$

where P denotes the probability that a bridge pier exceeds a specific performance level (PL) under a given PGA; $\mathrm{PGA_{PL}}$ represents the PGA corresponding to a 50% probability of reaching a certain damage state in the pier, as calculated by the Pushover analysis; β_C indicates uncertainty factors related to structural and seismic characteristics, taken as 0.6.

5.4.4 Fragility Analysis of Bridge Piers of Varying Heights

Based on Eq. (5.11), the performance fragility curves of bridge piers under different site conditions can be obtained using the Pushover analysis method. Due to limited space, the Type II site is taken as an example. Fig. 5.10 presents the fragility curves corresponding to various damage states for bridge piers with heights of 10 m, 15 m, 21 m, and 23 m [18].

Figs. 5.10A and B indicate minimal differences in the probabilities of slight damage and moderate damage under seismic action among bridge piers of varying heights. Under the same seismic intensity, 15 m and 21 m bridge piers exhibit lower probabilities of damage compared to other pier heights, while 10 m and 23 m piers are more susceptible to slight damage or moderate damage. Figs. 5.10C and D demonstrate that 10 m and 15 m piers exhibit higher probabilities of severe damage compared to 21 m and 23 m piers. The probability of extensive damage decreases significantly as pier height increases. However, the probability of extensive damage or complete damage for all piers is significantly lower compared to that of slight or moderate damage. Analysis of Fig. 5.10 reveals that high piers on Type II sites exhibit superior seismic performance, whereas low piers demonstrate relatively poor antiseismic capacity. As the pier height increases, the probability of the pier reaching any given damage state decreases, indicating enhanced seismic resistance. This outcome arises because high-speed railway piers are designed based on stiffness control, where their fragility evaluation metrics are derived from ductility

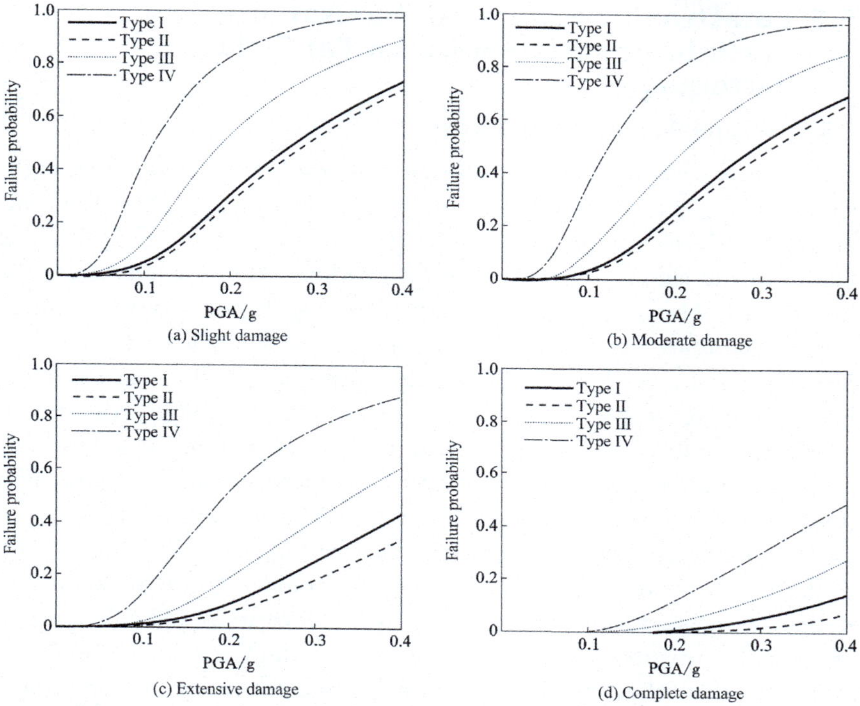

Figure 5.11 Fragility curves of 15m-high bridge piers on four types of sites.

performance metrics. As the pier height gradually increases, the displacement demand of the pier does not rise sharply, while the displacement capacity grows significantly. Research by Zhang Juhui and Guan Zhongguo [19] further demonstrates that the fragility resistance capacity of high-speed railway piers increases as the height increases.

5.4.5 Fragility Analysis of Bridge Piers under Different Site Conditions

For piers of identical height, fragility analysis is conducted under varying site conditions. Taking a 15 m pier height as an example, Fig. 5.11 presents the performance fragility curves under different site conditions, revealing that different site conditions have a similar impact on the seismic performance of 15 m piers. Under the same seismic intensity, the 15 m pier exhibits the worst seismic performance on Type IV sites (soft soil), better performance on Type III sites (medium-soft soil), the best performance on Type II sites (hard soil), and good performance on Type I sites (medium-hard soil).

5.5 Catastrophic Failure Mechanism of a Track-Bridge System Under Far-Field Earthquakes

5.5.1 Seismic Failure Mechanism

Research reveals that failure in simply supported girder bridges for high-speed railway predominantly occurs at pier locations, sliding layer interfaces, and shear reinforcement positions of track structures [20]. The failure of the sliding layer involves longitudinal sliding failure along the bridge axis. However, this longitudinal sliding failure exerts a limited impact on train running safety on bridges, as neither bridge stiffness degradation nor additional track irregularity occurs. In this case, the post-earthquake failure of high-speed railway track-bridge structures is categorized into two types: ① Bridge component failure, including longitudinal residual displacement of piers, excessive opening of expansion joints between girders, and degradation of pier lateral stiffness; ② Uneven settlement of piers, primarily caused by post-earthquake foundation failure.

Under strong seismic action, bridge piers would enter the plastic phase, potentially generating horizontal residual deformation. The horizontal residual deformation would induce a vertical displacement component, while the vertical residual deformation component may cause additional track irregularities (similar to pier settlement), potentially compromising train operational safety. Therefore, it is imperative to research the magnitude of horizontal residual deformation and its impact on track irregularities. Research by Wang Futian [21] indicates that under the Imperial Valley 06 seismic excitation, the horizontal displacement time history at the tops of piers P1–P6 is shown in Fig. 5.12. Although the 8 m and 16 m piers possess different horizontal stiffness, their displacement responses under seismic excitation are nearly identical. This is because the track-bridge structure exhibits excellent integrity under the constraint of a continuous rail structure. The structural integrity induces horizontal residual forces in the piers, preventing them from returning to their original positions. Consequently, the residual displacements of the piers are similar and show little correlation with the pier height. For instance, post-earthquake residual deformations measure 21.51 mm for the 8 m Pier P6 and 21.75 mm for the 16 m Pier P3, demonstrating comparable magnitudes between them. Fig. 5.12 illustrates the post-earthquake longitudinal residual displacements of different piers. As shown in Fig. 5.12, the residual deformations of piers remain around 20 mm under varying seismic excitations. The vertical components generated by pier deformation induce additional track irregularities, which superimpose with short-wavelength

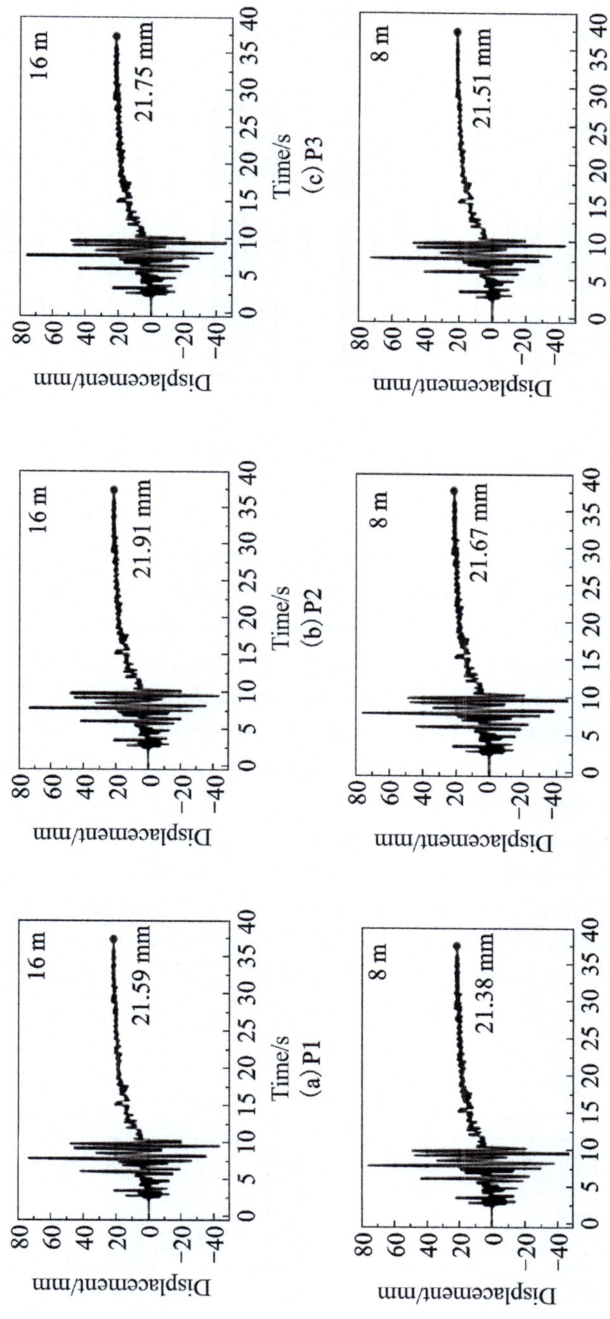

Figure 5.12 Horizontal displacement time history at pier tops under seismic excitation (Imperial Valley 06).

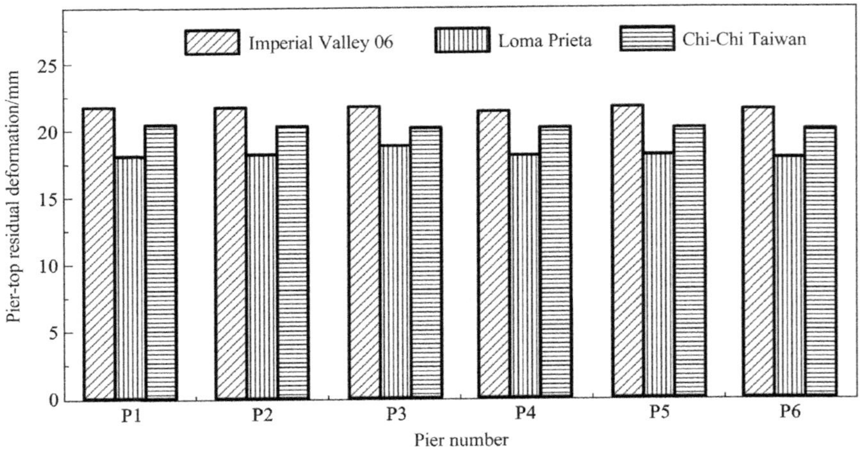

Figure 5.13 Post-earthquake longitudinal residual deformation of piers (PGA = 0.57 g).

track irregularities to form the ultimate track irregularities. However, given the comparable horizontal residual displacements across piers, their vertical components are nearly identical. Moreover, the difference in vertical components between piers and abutments is also minimal. For instance, the differential between Pier P6 and the abutment is approximately 0.03 mm, far below the millimeter-level short-wave track irregularity. Consequently, the horizontal residual deformations of piers and abutments have a negligible impact on train operational safety.

Fig. 5.14 illustrates how the expansion joint spacing expands due to the horizontal residual displacements of piers. After the earthquake, the main girder connected to the pier through fixed bearings undergoes longitudinal displacement induced by the horizontal residual displacement of the pier. However, the main girder connected through expansion bearings is not constrained by piers or abutments. Take the bridge model in Fig. 5.14 as an

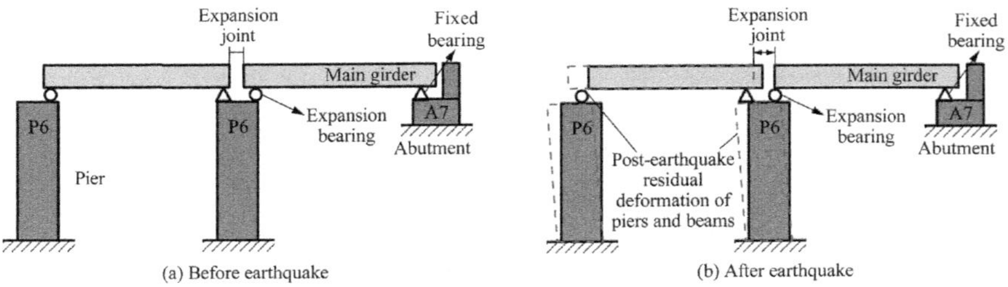

Figure 5.14 Expansion of expansion joints caused by pier-top residual displacement at Pier P6.

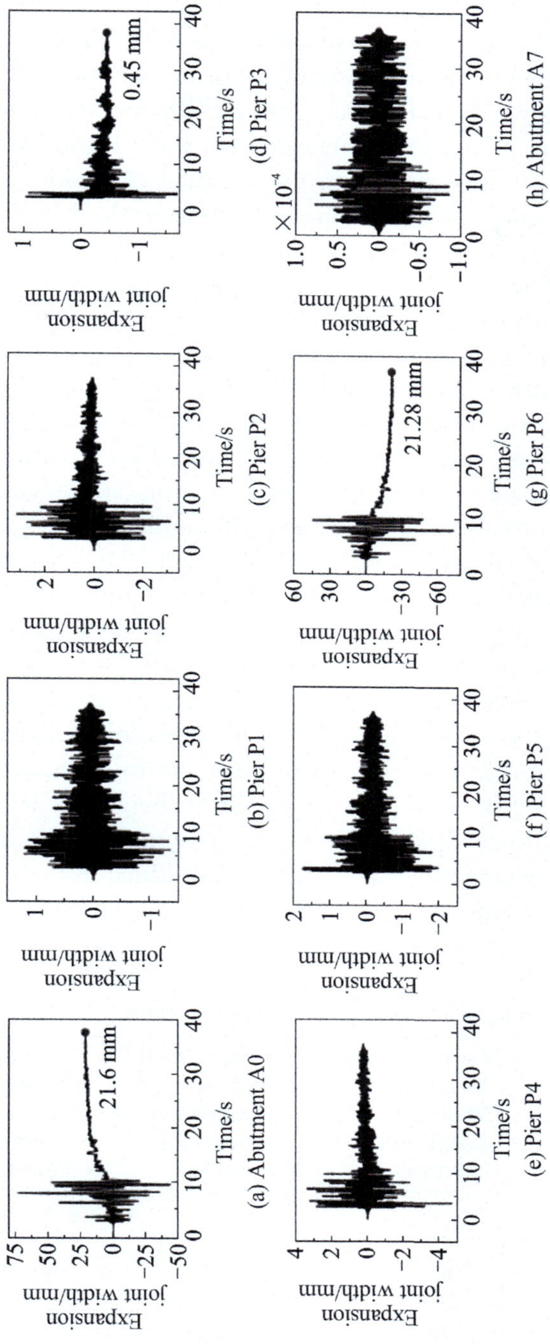

Figure 5.15 Post-earthquake expansion of expansion joints at piers and abutments (Imperial Valley 06 record).

example. When Pier P6 undergoes residual displacement, the main girder connected to Pier P6 moves horizontally as well, while the girder connected to Pier P6 via expansion bearings remains in its original position due to the tensile force from Pier A7. This leads to the expansion of bridge expansion joints. Given that high-speed trains are highly sensitive to the stiffness of track-bridge structures, the widened expansion joints weaken the supporting effect of track structures between adjacent girders, potentially compromising train running safety.

Fig. 5.15 illustrates the time-dependent variation of expansion joints under Imperial Valley 06 seismic excitation, with residual deformation observed post-earthquake. The maximum joint expansion is observed at Pier P6. Meanwhile, except for the residual expansion joint at Pier A7 Table. 5.5 shows a reduction, the residual variations in expansion joints at other locations are minimal, all below 1 mm. The maximum expansion joint variations under different seismic excitations are remarkably consistent, as shown in Table 5.5. Under ground motion records 1, 2, and 3, the expansion values are 21.28 mm, 17.85 mm, and 19.92 mm, respectively.

Under seismic excitation, all bridge piers enter nonlinear states, as illustrated by the hysteretic curves in Fig. 5.16. The maximum displacement responses of piers P2 and P3 are 74.85 and 77.08 mm, respectively, below their failure displacement limits of 266.4 and 409.6 mm. As shown in Fig. 5.16, the hysteresis curves of all 8-m piers are nearly identical, while the 16-m piers exhibit consistent hysteretic behavior. Therefore, for piers of the same height, it is sufficient to calculate the stiffness degradation of only one pier. In this section, piers P2 and P3 are selected to calculate the longitudinal stiffness degradation percentages corresponding to their heights, as shown in Fig. 5.17.

Table 5.5 Information on selected ground motion records.

No.	Residual Peak Value of Widened Expansion Joint/mm	Longitudinal Stiffness Degradation of 8-m Bridge Piers/%	Longitudinal Stiffness Degradation of 16-m Bridge Piers/%
Ground motion record 1	21.28	31.57	14.01
Ground motion record 2	17.85	25.88	11.27
Ground motion record 3	19.92	27.51	12.93

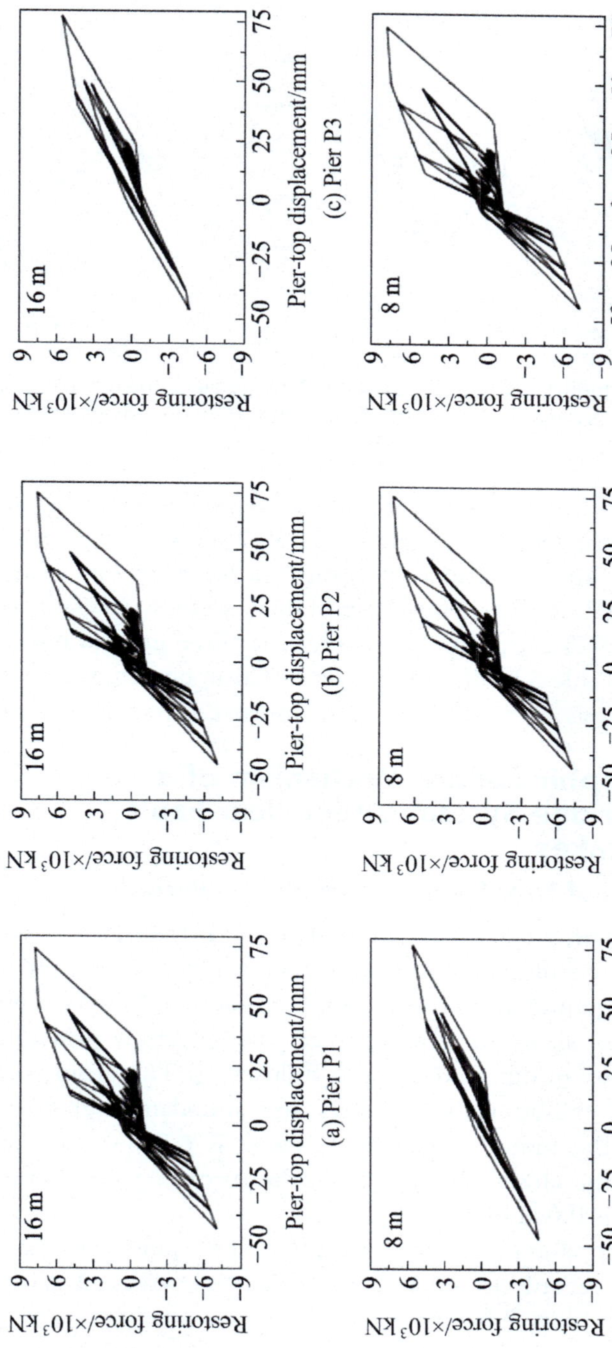

Figure 5.16 Horizontal hysteretic curves of bridge piers under seismic excitation (Imperial Valley 06).

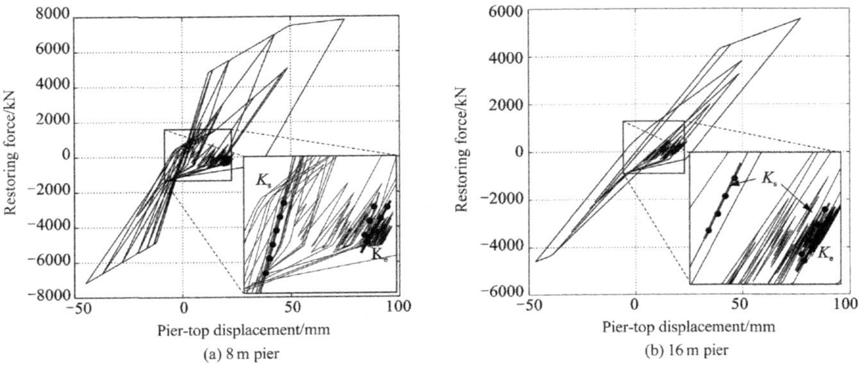

Figure 5.17 Longitudinal stiffness degradation of bridge piers. The 8-m pier exhibits a longitudinal stiffness degradation of 31.57%, while the 16-m pier shows a degradation of 14.01%.

Where K_s represents the initial stiffness of the piers, while K_e denotes the post-earthquake residual stiffness. Fig. 5.17 shows that the 8-m piers exhibit greater stiffness reduction compared to the 16-m piers, with piers P2 and P3 experiencing longitudinal stiffness degradations of 31.57% and 14.01%, respectively. Under the three ground motion records listed in Table 5.5, the stiffness degradation percentages extracted from the hysteresis curves of piers are summarized in the same table.

5.6 Catastrophic Failure Mechanism of a Track-Bridge System Under Near-Fault Earthquakes

5.6.1 Selection of Near-Fault Ground Motion Records

The velocity pulse effects and abundant high-frequency vibrations recorded in near-fault ground motion would play a significant role in structural damage under seismic action. Considering these two primary characteristics of near-fault earthquakes, the selection principles for seismic excitation in this section are as follows: ① The ground motion records should exhibit velocity pulses and abundant high-frequency vibrations; ② the moment magnitude range is 6.5–8.0; ③ the fault distance R_{rup} (the closest distance from the recording site to the fault rupture plane) is 0.0–20.0 km.

Based on the aforementioned principles, eight pulse-type near-fault ground motion records are selected from the NGA-West 2 database of the PEER, as listed in Table 5.6.

To study the impact of near-fault velocity pulse effects and high-frequency vibrations on high-speed railway track-bridge systems, the

Table 5.6 Near-fault ground motion records.

Ground Motion Selected	Earthquake Name	Year	Station Name	R_{rup}/km	Rupture Distance
W1	Imperial Valley-06	1979	EC County Center FF	7.31	6.53
W2	Imperial Valley-06	1979	El Centro Array #6	7.31	6.53
W3	Imperial Valley-06	1979	El Centro Array #7	0.56	6.53
W4	Imperial Valley-06	1979	El Centro Differential Array	5.09	6.53
W5	Cape Mendocino	1992	Petrolia	8.18	7.01
W6	Chi-Chi, Taiwan, China	1999	TCU046	8.18	7.62
W7	Chi-Chi, Taiwan, China	1999	TCU102	1.49	7.62
W8	Chi-Chi, Taiwan, China	1999	TCU128	13.13	7.62

Butterworth filter is used to separate the low-frequency pulse component (PT) and high-frequency component (BG) from the original near-fault ground motion record ONR. The ONR, BG, and PT are then input as excitations into the high-speed railway track-bridge system model. Table 5.7 presents the acceleration time–history curves and corresponding PGAs of the ONR, BG, and PT for the eight selected near-fault ground motions.

5.6.2 Longitudinal Failure Characteristics and Cause Analysis

The low-frequency pulse component (PT) and high-frequency component (BG) obtained by scaling the original seismic excitation ONR to 0.3 *g*, are input into the high-speed railway track-bridge system model [23–24]. The seismic response analysis of the track-bridge system shows that the failure distribution of the track-bridge system under different near-fault earthquakes is similar, as shown in Fig. 5.18. In the figure, S_i (where i ranges from 1 to 7) represents the span. The sliding layer displacement at S7 is significant, and shear grooves fail. The phenomenon arises from inconsistent motion between components: Due to the restraining effect of the track structure, the piers and superstructure at spans S1-S6 always vibrate in-phase. However, due to the constraint of fixed bearings, the box girder and track structure at S7 cannot move synchronously with other spans, resulting in small relative displacements. This leads to large sliding layer displacements at S7 and shear groove failure.

Research found that low-frequency pulses have a minimal impact on the seismic response of the structure, whereas high-frequency vibrations are the primary factor driving the response and failure of structural components. To investigate the causes of such failure characteristics in the track-bridge system, the dynamic coefficient is used to compare the effect of low-frequency pulses and high-frequency vibrations on high-speed railway track-bridge systems. The dynamic coefficient is defined as the ratio of the peak acceleration response of a structure to the PGA. By comparing the dynamic coefficient spectra of the structure under different excitations, the amplification effect of the structural response to seismic excitation can be observed. Fig. 5.19 presents the dynamic coefficient spectra of some ground motions used in the study.

As shown in Fig. 5.19, the dynamic coefficient spectra of BG and PT reach their peak values in the short-period and long-period ranges, respectively. High-frequency vibrations may play a significant role in the response of structures with greater stiffness, while low-frequency pulses dominate the response of long-period structures. In this section, the

Table 5.7 Acceleration time–history curves of ONR, BG, and PT for eight selected near-fault ground motions.

	ONR	BG	PT
W1	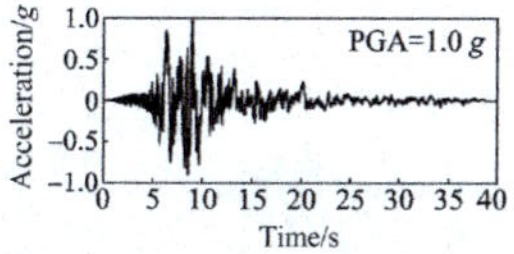	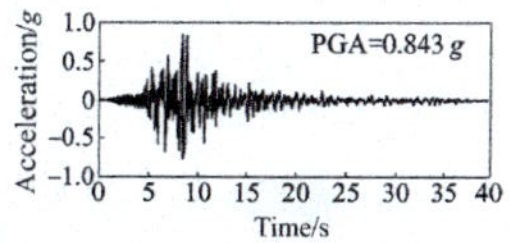	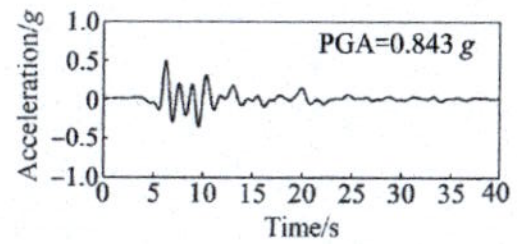
W2	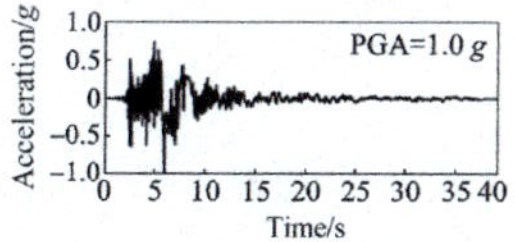	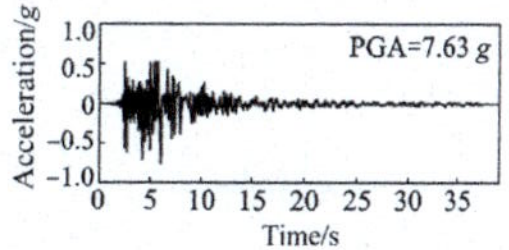	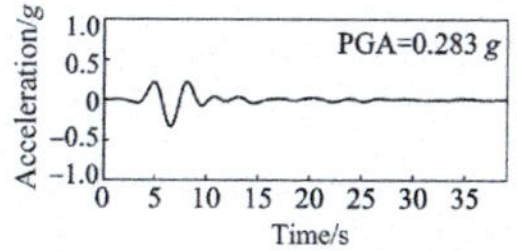
W3	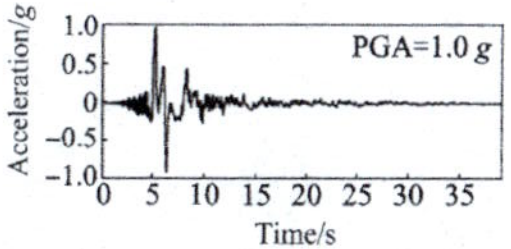	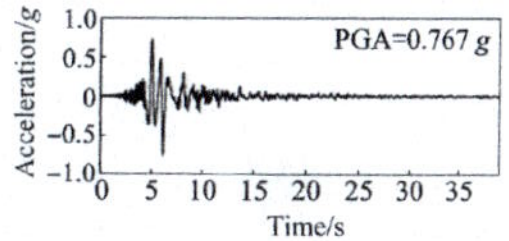	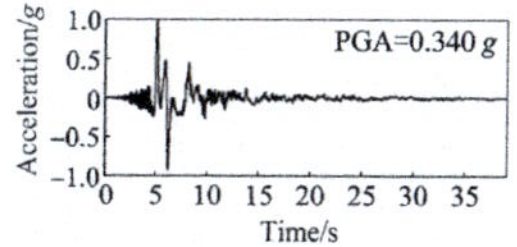
W4	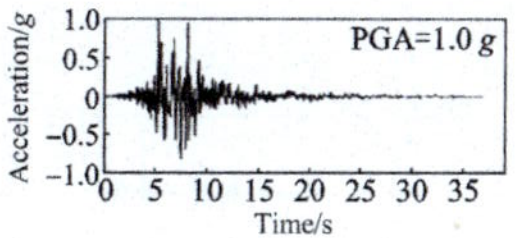	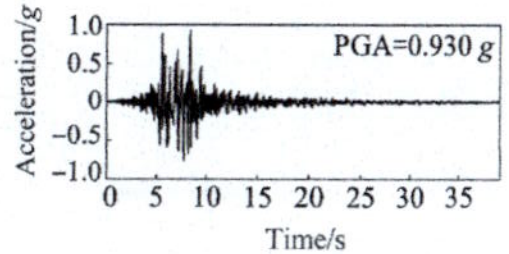	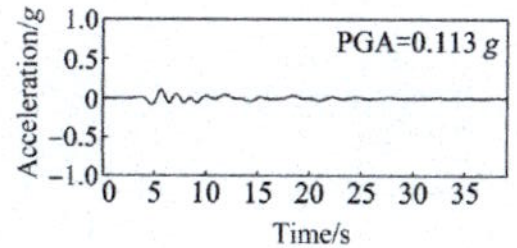

Continued

Table 5.7 Acceleration time–history curves of ONR, BG, and PT for eight selected near-fault ground motions.—cont'd

	ONR	BG	PT
W5	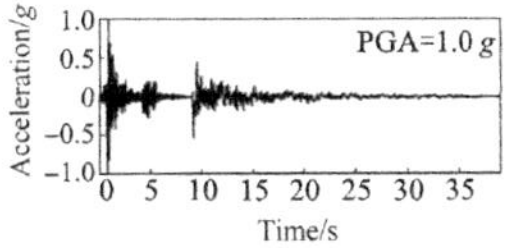	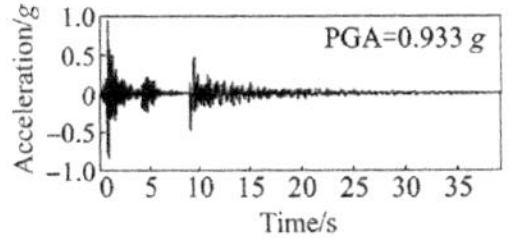	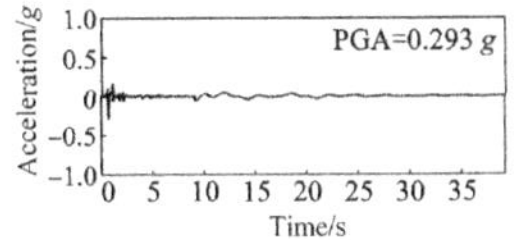
W6	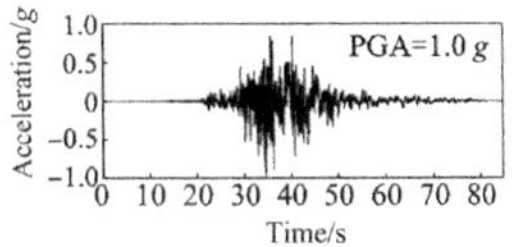	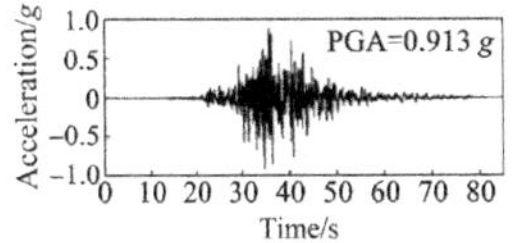	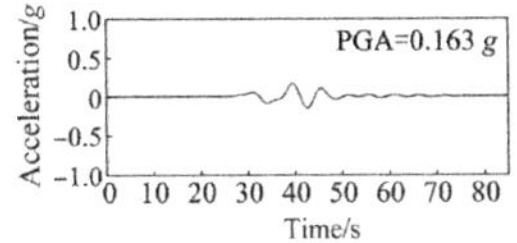
W7	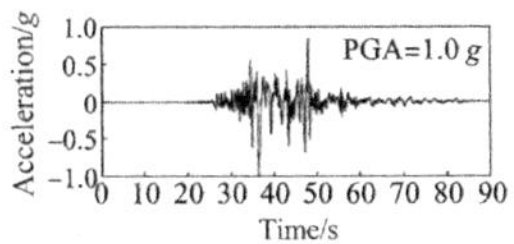	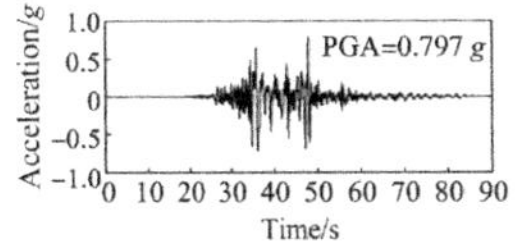	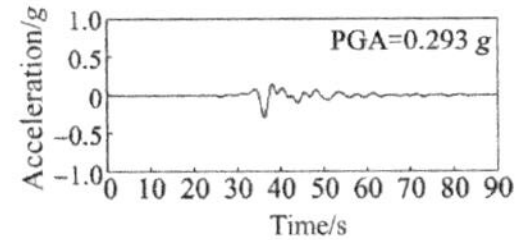
W8	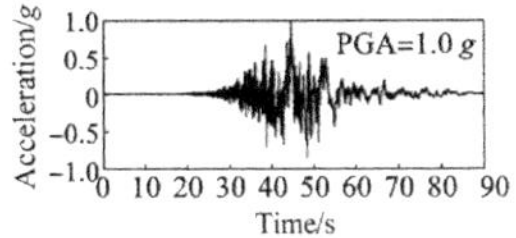	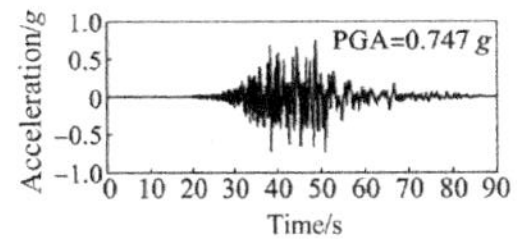	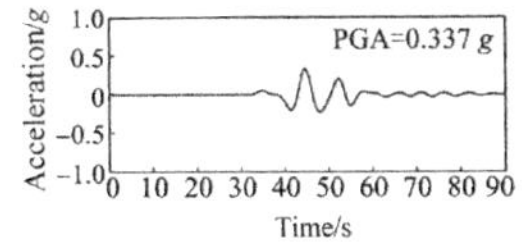

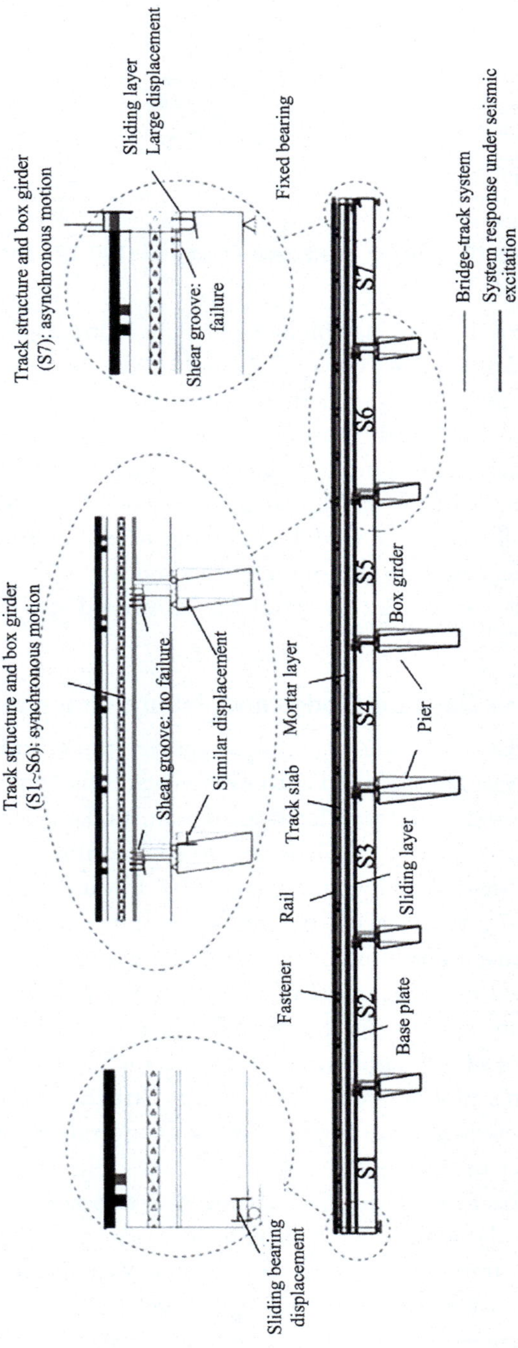

Figure 5.18 Schematic diagram of the response of track-bridge systems under near-fault earthquake excitations.

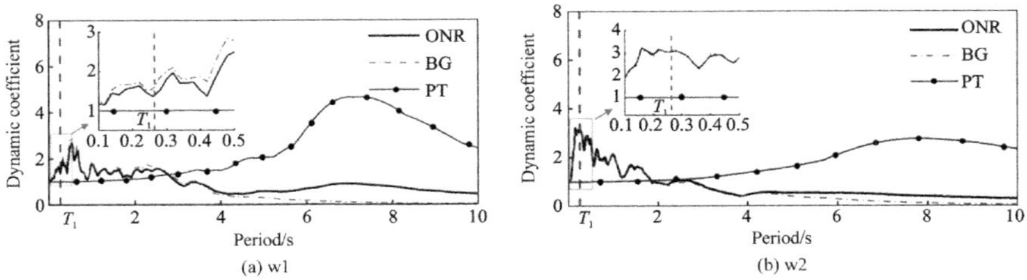

Figure 5.19 Dynamic coefficient spectra of ONR, BG, and PT.

fundamental period T_1 of the high-speed railway track-bridge system is 0.264 s. The dynamic coefficients of BG and ONR are close to each other and both are significantly higher than that of PT. According to the definition of the dynamic coefficient, under BG excitation, the amplification factor of the structural response to the input is comparable to that under ONR excitation, while at the fundamental period T_1 of the track-bridge system, the amplification under PT excitation is relatively small. This indicates that the seismic response of high-speed railway track-bridge systems under near-fault earthquakes is primarily governed by high-frequency vibrations.

5.6.3 Transverse Damage Modes and Failure Characteristics

The transverse damage modes of simply supported track-bridge systems for high-speed railway are classified into two types: One type assumes that both sliding bearings and fixed bearings remain undamaged in the transverse direction, meaning the laterally fixed bearings maintain elastic behavior under seismic action; the other type accounts for the failure of longitudinal sliding bearings and fixed bearings in the transverse direction. The bearing damage constitutive relationship mentioned earlier indicates that laterally fixed bearings initially exhibit linear elastic behavior. When the bearing load reaches its horizontal ultimate capacity, the lateral displacement exceeds the ultimate shear displacement, causing the shear-off of the restraining blocks within the bearing structure. This leads to a transition into a sliding friction phase. If the friction displacement reaches 450 mm, a girder unseating will occur.

When the bearings remain intact in the transverse direction, the damage mode of the high-speed railway track-bridge system is illustrated in Fig. 5.20. Under transverse seismic action, since high-speed railway bridge piers are designed for stiffness with large cross-sectional dimensions and high rigidity (particularly in the transverse direction), they are less prone to failure. Additionally, the constraint effect of the track structure in the high-

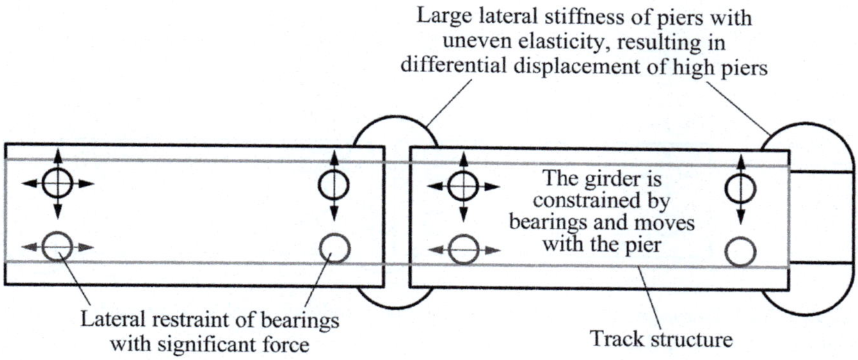

Figure 5.20 Damage mode of a high-speed railway track-bridge system without transverse bearing damage.

speed railway track-bridge system may cause adjacent piers of equal height to exhibit synchronized movement, while height differences between piers could lead to slight deformation discrepancies due to stiffness variations. Under the assumption that the transverse bearing remains intact, longitudinal sliding bearings and fixed bearings will be subjected to significant seismic forces. The transverse constraint imposed by these bearings may prevent excessive displacement across the entire bridge, causing the girder to vibrate laterally with the piers. Meanwhile, the friction plates at the road–bridge transition section will move with the ground motion due to the high stiffness of the embankment. Based on this, the longitudinal continuity of the track structure enhances its overall integrity. Due to the transverse restraint provided by shear grooves, the track may move synchronously with the girder, resulting in minimal relative displacement between the track and the girder. The sliding layer, with its low friction coefficient, operates in a sliding state, while the deformation of the track structure is primarily influenced by inertial effects.

The damage mode of high-speed railway track–bridge systems under transverse bearing damage is shown in Fig. 5.21. Under transverse seismic action, longitudinal sliding bearings and fixed bearings may experience transverse failure and sliding, leading to a reduction in the overall stiffness of the bridge–track system. Due to the high lateral stiffness of piers, they are less prone to damage and can maintain elastic vibration. When adjacent piers have the same height, they exhibit synchronized motion; when pier heights differ, their displacement responses vary significantly. For bearings, the displacement difference caused by adjacent piers of varying heights can lead to excessive bearing forces and potential damage. Under such

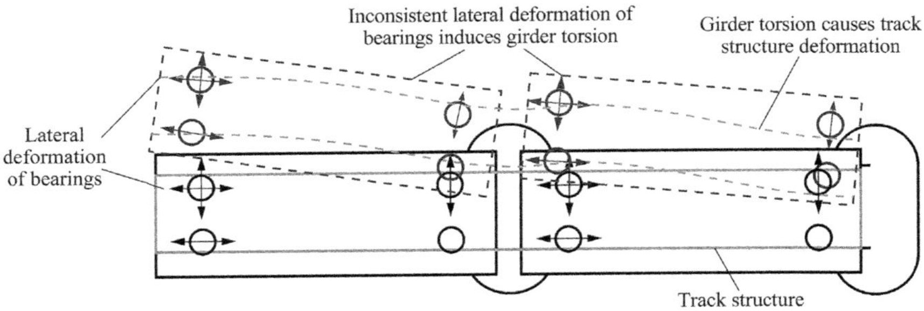

Figure 5.21 Damage mode of the high-speed railway track-bridge system under transverse bearing damage.

conditions, longitudinal sliding bearings and fixed bearings will exit the elastic working stage and transition into the friction-sliding stage, while lateral sliding bearings and bi-directional sliding bearings will maintain their sliding friction state. The differences in their constitutive relationships result in variations in force magnitude and displacement response. Inconsistent displacement at the bearings of both ends of the girder can induce torsional deformation in the girder. Since the upper track structure maintains good overall continuity, the torsion of the girder may lead to deformation in the track structure.

Fig. 5.22 compares the lateral force on the pier and the peak stress of the base plate under the intact bearing condition and bearing failure condition. Here, the analysis focuses on the response of an 8 m high pier with adjacent piers of varying heights. As shown in Fig. 5.22(a), when considering transverse bearing damage, the seismic force on the pier is smaller than that on the pier without transverse bearing damage. Fig. 5.22(b) indicates that transverse bearing damage reduces the stress on the base plate, as the lateral bearing displacement is significantly smaller than the

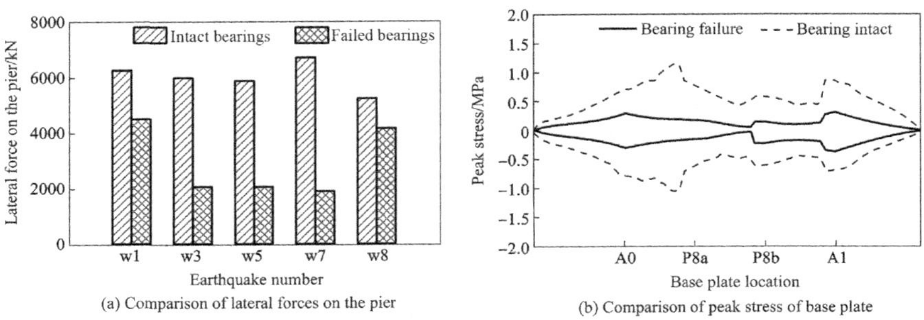

Figure 5.22 Comparison of component responses between bearing failure condition and intact bearing condition.

displacement that would cause girder unseating. The friction energy dissipation of the bearing provides seismic damping benefits, which is advantageous for the elastic components of the track structure.

5.7 Catastrophic Failure Mechanism of Train–Track–Bridge Systems Under Earthquakes

5.7.1 Train–Track–Bridge System Model

As shown in Fig. 5.23, the three-dimensional dynamic model of the train–track–bridge structure [24] under earthquakes is composed of the train subsystem, bridge subsystem, and wheel-rail contact. The train travels at a speed of v, with seismic excitation applied transversely. The equation of motion of the train–bridge subsystem is given by Eq. (5.12), where the subscripts v and b denote the train and bridge, respectively.

$$\begin{bmatrix} \boldsymbol{M}_{\mathrm{vv}} & \\ & \boldsymbol{M}_{\mathrm{bb}} \end{bmatrix} \begin{pmatrix} \ddot{\boldsymbol{X}}_{\mathrm{v}} \\ \ddot{\boldsymbol{X}}_{\mathrm{b}} \end{pmatrix} + \begin{bmatrix} \boldsymbol{C}_{\mathrm{vv}} & \boldsymbol{C}_{\mathrm{vb}} \\ \boldsymbol{C}_{\mathrm{bv}} & \boldsymbol{C}_{\mathrm{bb}} \end{bmatrix} \begin{pmatrix} \dot{\boldsymbol{X}}_{\mathrm{v}} \\ \dot{\boldsymbol{X}}_{\mathrm{b}} \end{pmatrix} + \begin{bmatrix} \boldsymbol{K}_{\mathrm{vv}} & \boldsymbol{K}_{\mathrm{vb}} \\ \boldsymbol{K}_{\mathrm{bv}} & \boldsymbol{K}_{\mathrm{bb}} \end{bmatrix} \begin{pmatrix} \boldsymbol{X}_{\mathrm{v}} \\ \boldsymbol{X}_{\mathrm{b}} \end{pmatrix} = \begin{pmatrix} \boldsymbol{F}_{\mathrm{v}} + \boldsymbol{F}_{\mathrm{v}}^{*} \\ \boldsymbol{F}_{\mathrm{b}} + \boldsymbol{F}_{\mathrm{b}}^{*} \end{pmatrix} \tag{5.12}$$

where $\boldsymbol{M}$, $\boldsymbol{C}$, and $\boldsymbol{K}$ denote the mass, damping, and stiffness matrices; $\boldsymbol{X}$ represents the displacement vector; $\boldsymbol{F}_{\mathrm{v}}$ indicates the load vector of the bridge acting on the train, $\boldsymbol{F}_{\mathrm{v}}^{*}$ is the train load vector caused by seismic forces; $\boldsymbol{F}_{\mathrm{b}}$ is the load vector acting on the bridge from the train, and $\boldsymbol{F}_{\mathrm{b}}^{*}$ is the load vector of the bridge caused by seismic forces.

5.7.2 Dynamic Response of a Train–Track–Bridge System under Earthquakes

Train derailment is categorized into climbing derailment, jumping derailment, and overturning derailment depending on the derailment process [25]. A climbing derailment occurs when the wheel flange gradually rides up onto the top of the railhead, while a jumping derailment occurs when a transient impact causes the wheel to jump onto the railhead. The derailment process of a train is categorized into the normal state, critical state, and derailment state, as shown in Fig. 5.24. When the train enters a critical state, as shown in Fig. 5.24(b), if the vertical wheel-rail force P_1 is sufficiently large, the wheelset may slide downward and return to a stable state; if the vertical wheel-rail force P_1 is inadequate, the excessive lateral wheel-rail force Q_1 may lead to derailment.

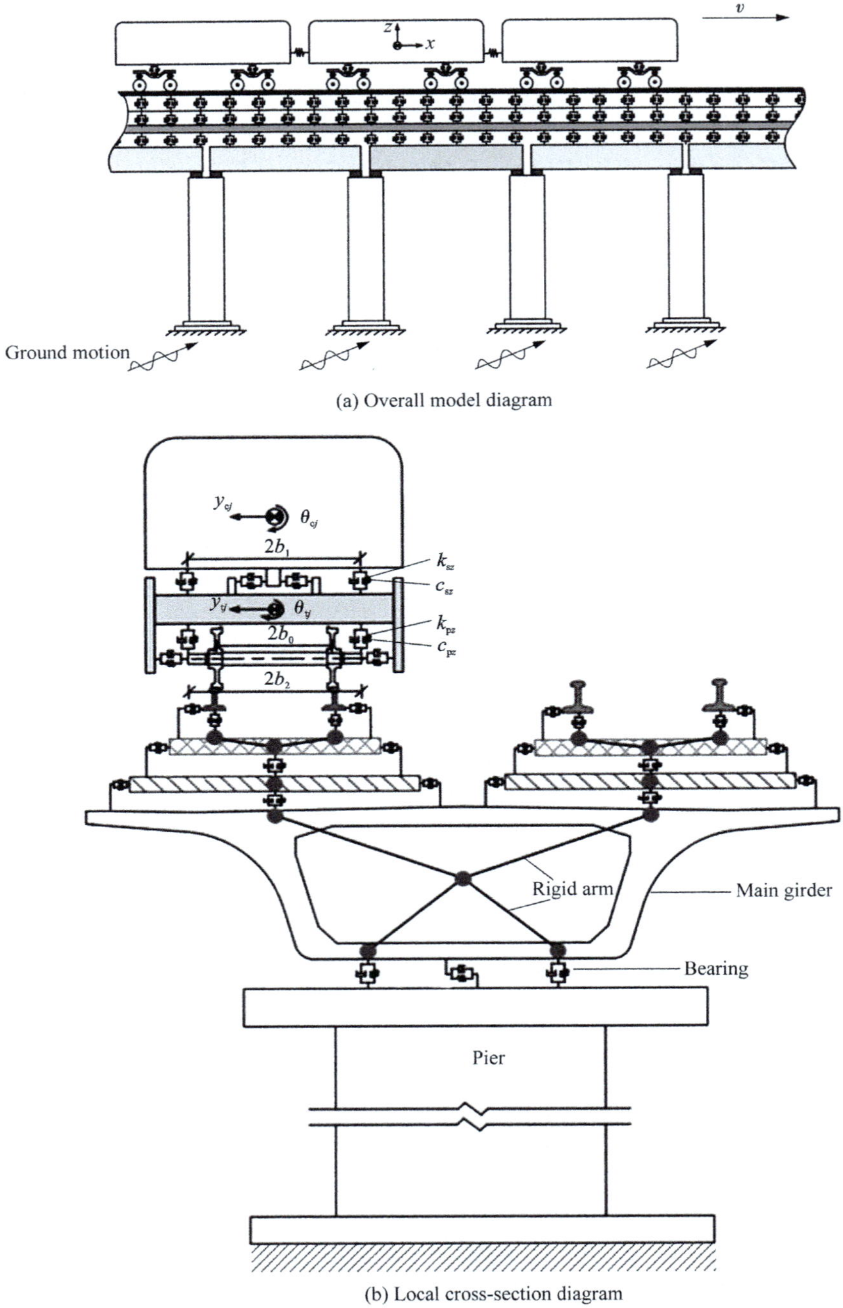

Figure 5.23 Three-dimensional dynamic model of train–track–bridge structure.

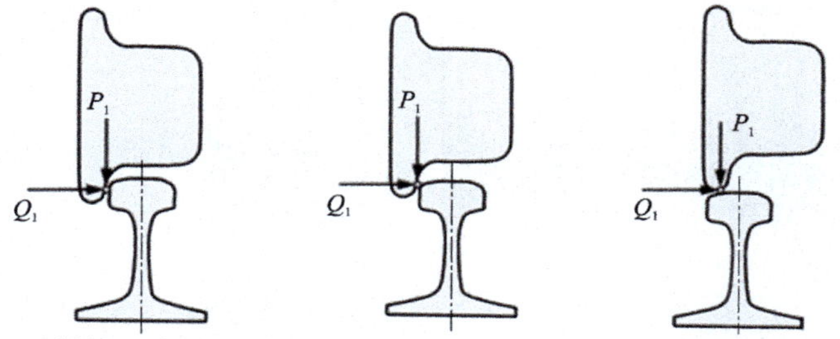

Figure 5.24 Train derailment process.

In China's high-speed railway standards, the derailment coefficient, wheel load reduction rate, and wheel-rail (axle) load reduction rate are adopted to evaluate train operational safety, which are also commonly used criteria in engineering and research. The derailment coefficient (Q/P) is the ratio of the lateral force Q to the vertical force P acting on a single wheel at a given moment. Wheelset derailment is primarily associated with the vertical force P and lateral force Q acting on the wheelset. However, even if the lateral wheel-rail force Q is not significant, severe vertical force reduction (i.e., small vertical force) may still lead to derailment. The wheel load reduction rate is defined as the ratio of the reduced load on the measured wheelset to its static wheel load. On the one hand, excessive lateral wheel-rail forces may damage the connection between the track and the girder, leading to instability in the track structure. On the other hand, excessive wheel-rail forces may alter the track gauge, causing track buckling or derailment. Meanwhile, the wheelset may experience jumping derailment due to excessive impact loads, and the higher the lateral force, the greater the likelihood of derailment.

Using Liu Hanyun's MATLAB-based train–track–bridge dynamic simulation program [26], the dynamic equations are solved via the Newmark-β method, with wheel-rail contact relationships based on Hertz elastic contact theory and Kalker's creepage assumptions. The response of the train–track–bridge system under transverse seismic excitation is analyzed. For comparison with the subsequent velocity spectrum index evaluation results, the train speed is uniformly set to 100 km/h. Under different intensity levels of transverse seismic excitation (I-W1), the response results of the train and bridge are shown in Fig. 5.25. In this figure, the time histories of lateral wheel-rail forces, lateral acceleration of the vehicle body, vertical displacement at mid-span of the girder, and lateral acceleration of the girder are compared. Fig. 5.25 indicates that as

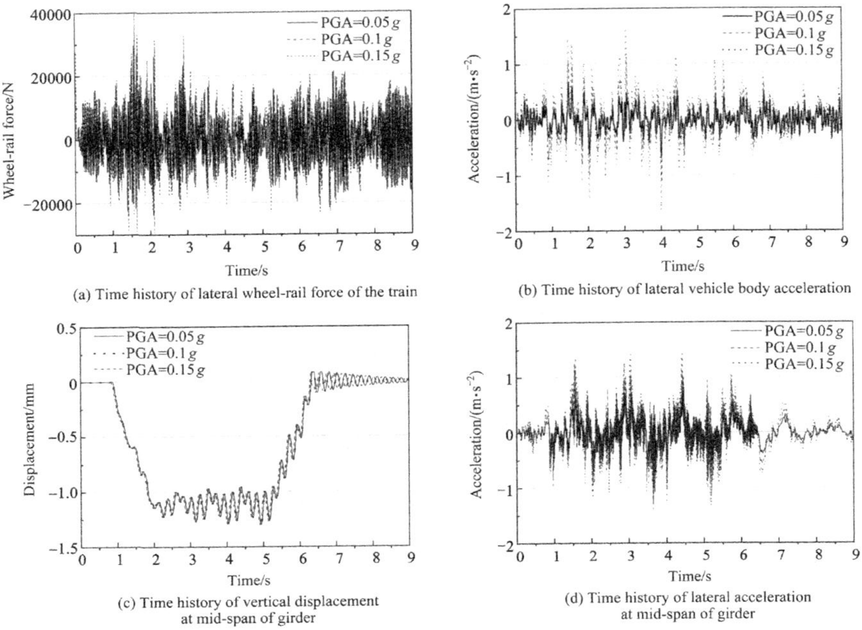

Figure 5.25 Response time history of the train and bridge.

the PGA increases, the wheel-rail forces and the lateral acceleration of the vehicle body exhibit an upward trend, and the lateral acceleration exceeds the specified limit at 0.1 *g* seismic motion intensity. When a train passes over a bridge during an earthquake, the vertical displacement at the mid-span of the bridge exhibits minimal correlation with seismic ground motion because vertical seismic effects are not considered. The peak displacement remains around 1.3 mm, which complies with the standards. The lateral acceleration at the mid-span of the bridge increases along with the increase in the PGA of the earthquake.

Fig. 5.26 compares the variations in train stability and safety performance metrics under seismic actions of different types of sites, as a function of seismic PGA ranging from 0.05 to 0.20 *g*. It can be observed that the peak derailment coefficient, peak wheel-rail unloading rate, and lateral acceleration of the vehicle body all increase with the PGA, exhibiting an approximately linear relationship, which aligns with Xu Xiang et al.'s research findings [27]. The peak lateral acceleration of the vehicle body first exceeds the stability limit (1 m/s^2). Then, the derailment coefficient exceeds its safety limit (0.8). Under the seismic effects considered in this section, the wheel-rail unloading rate of the train does not exceed its safety limit (0.6) before 0.2 *g*.

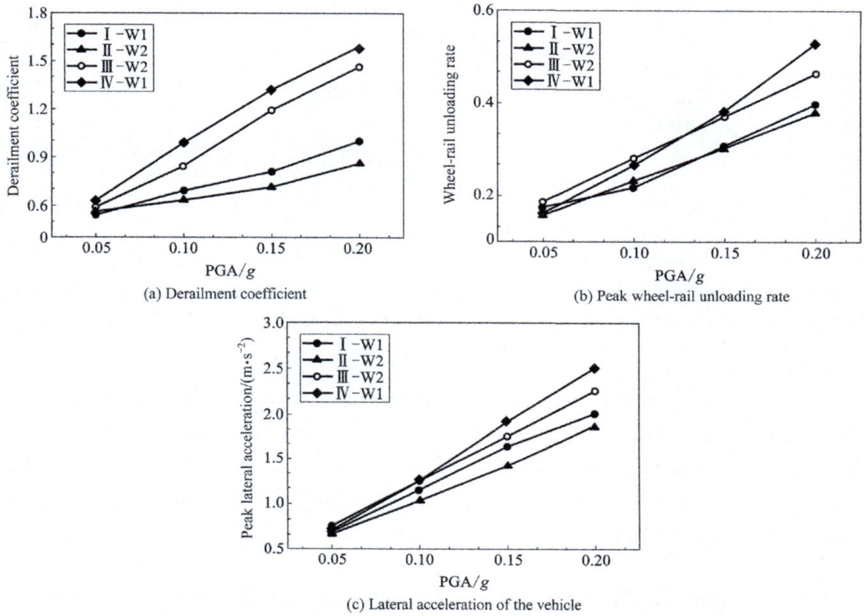

Figure 5.26 Train performance metrics under seismic excitation.

5.7.3 Impact of Bridge Damage on Post-Earthquake Train Operational Safety

5.7.3.1 Impact of Expansion of Expansion Joint Spacing on Train Operational Safety

In the FE model of the track-bridge structure, the vertical force between the track structure and the girder is transmitted through a sliding layer. The lower nodes of the sliding layer element are connected to the girder, while the upper node is connected to the base plate. The expansion of expansion joints is achieved by eliminating the sliding layer elements near the expansion joints to increase the bearing distance of the beam body to the track structure. Based on the previous nonlinear seismic analysis of the track-bridge structure, the expansion magnitudes of the expansion joints are set at 0 cm, 10 cm, 30 cm, and 50 cm, where 0 cm indicates no joint expansion. The expansion joint is located at Pier P1. In the train–bridge coupling simulation, the operating speed of the train is 300 km/h. Short-wavelength track irregularities are excluded from the train–bridge coupling vibration to highlight the influence of the expansion of expansion joints. The impact of expansion of joint expansion on vehicle vertical

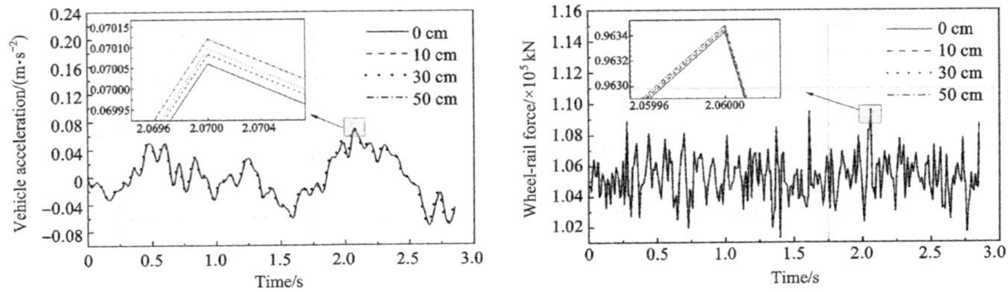

Figure 5.27 Impact of expansion of joint expansion on vehicle vertical acceleration, wheel-rail forces and other ride safety evaluation metrics.

acceleration, wheel-rail forces, and other ride safety evaluation metrics is illustrated in Fig. 5.27. It can be observed that the peak vertical acceleration of the vehicle body increases as the joint expands, but the growth rate is minimal, indicating that the dynamic performance of the train is not highly sensitive to expansion joint.

5.7.3.2 Impact of Pier Stiffness Degradation on Train Operational Safety

Nonlinear seismic time-history analysis of the track-bridge structure indicates that the post-earthquake stiffness reduction of bridge piers ranges from 10% to 30% [24]. Accordingly, the horizontal stiffness of the piers in the bridge model adopted in the test is modified to simulate the degradation of their horizontal stiffness. As shown in Fig. 5.28, the vehicle acceleration response indicates that, regardless of the presence or absence of short-wavelength track irregularities, the reduction in horizontal pier stiffness in this study has little effect on vehicle acceleration or wheelset displacement (with a slight increase in response after stiffness enhancement).

The impact of the horizontal stiffness reduction of the pier on train operational safety metrics is minimal, as the vertical stiffness of the piers remains unchanged after the earthquake, and this section does not consider vertical stiffness degradation of the piers. Additionally, He et al. [28] demonstrated that after an earthquake, the horizontal stiffness reduction of circular-end piers remains compliant with the requirements set forth in relevant standards. In summary, when only considering longitudinal seismic input, the variation in vehicle acceleration caused by track-bridge structural damage is minimal, and the post-earthquake structural damage considered in this section has a limited impact on train operational safety.

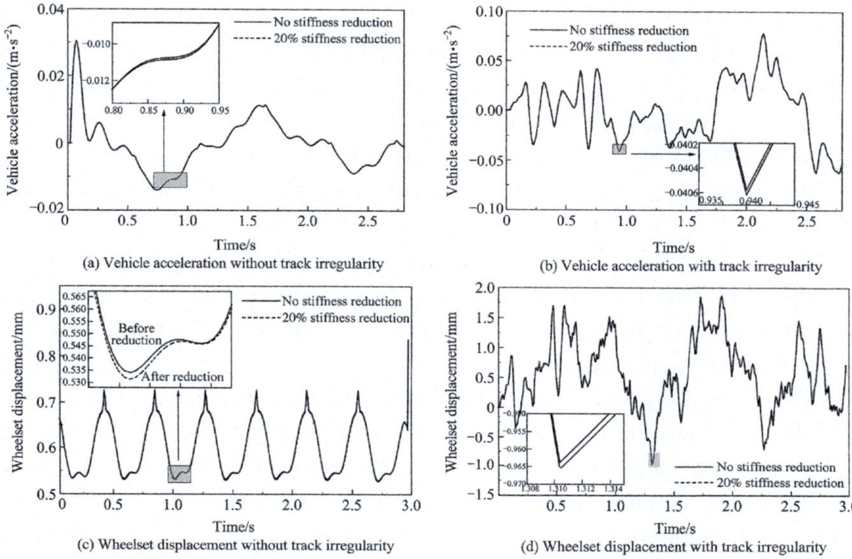

Figure 5.28 Results of the vehicle acceleration response.

5.8 Summary

This chapter first presents the mechanical model of the high-speed railway train–track–bridge system, and introduces the force-displacement curves of the track-bridge system components. Then it studies the failure models and mechanisms of bridge piers. Subsequently, it elaborates on the catastrophic failure mechanisms of the track-bridge system under far-field and near-fault earthquakes. Finally, it analyzes the catastrophic failure mechanisms of the train–track–bridge system under seismic action, with a focus on train operational safety, leading to the following main conclusions.

① The track-bridge system incorporates the CRTS II slab ballastless track system, where the upper track structure provides additional stiffness to the bridge structure and imposes significant constraints. This results in similar displacement responses among bridge piers of varying heights under seismic action. Therefore, the seismic performance analysis of high-speed railway bridges should account for the restraining effect of the upper track structure.

② The proposed shear strength model for high-speed railway round-ended solid piers comprehensively accounts for the influences of factors such as shear span-to-depth ratio, lateral reinforcement ratio, axial compression ratio, and displacement ductility on shear strength, providing an accurate evaluation of the shear strength of the piers. The trilinear shear strength model proposed in this chapter accurately evaluates the shear performance of piers under flexural-shear failure and precisely identifies their failure modes.

③ When the CRTS II slab ballastless track structure is simplified as a nonlinear spring based on stiffness equivalence, its force-displacement curves essentially represent the combined effect of all sliding layers and shear grooves within a single-span bridge. The inertial effect of the track structure has a negligible impact on the seismic response of the simplified bridge system. Therefore, it is recommended to adopt a mass reduction factor of 0.5–0.7.

④ The high-frequency vibrations in near-fault earthquakes significantly affect the seismic response of high-speed railway track-bridge systems. This is due to the relatively large overall stiffness of the system, where its natural vibration period closely matches the dynamic characteristics of high-frequency seismic waves, resulting in a pronounced amplification effect on the response of the high-speed railway track-bridge systems. Therefore, the abundant high-frequency components in near-fault earthquakes must not be overlooked in the study of structures with comparable stiffness to track-bridge systems.

⑤ A comparative analysis of train stability and safety performance metrics under different site conditions reveals that the peak derailment coefficient, peak wheel-rail unloading rate, and lateral vehicle acceleration all increase with the increase of seismic PGA, exhibiting an approximately linear relationship.

⑥ The damage to high-speed railway track-bridge structures caused by longitudinal seismic motion, such as reduced longitudinal stiffness of piers, longitudinal residual displacement of piers, and expanded expansion joints, has a limited impact on train operational safety, indicating that the track-bridge system exhibits excellent seismic performance in the longitudinal direction.

References

[1] Jiang LZ, Shao GQ, Jiang JJ, et al. Experimental study on seismic performance of solid piers with round ended cross-section in high-speed railway. China Civ Eng J 2013;46(3):86–95 [In Chinese].

[2] Code for seismic design of railway engineering (GB 50111—2006). Beijing: China Planning Press; 2009 [In Chinese].

[3] Zhu S, Cai C. Interface damage and its effect on vibrations of slab track under temperature and vehicle dynamic loads. Int J Non Lin Mech 2014;58:222–32.
[4] Huang YC, Wang JW, Wang SJ. Parametric study of mechanical performances of ballastless track on high-speed railway bridge isolated with FPS during earthquake. Railway Stand Design 2016;60(1):27–33 [In Chinese].
[5] Zhao L. Spatial refined analysis method and its application for ballastless track in high-speed railways. Beijing: Beijing Jiaotong University; 2015 [In Chinese].
[6] Code for design of railway continuous welded rail (TB 10015—2012). Beijing: China Railway Publishing House; 2012 [In Chinese].
[7] Guo W, Hu Y, Hou WQ, et al. Seismic damage mechanism of CRTS-II slab ballastless track structure on high-speed railway bridges. Int J Struct Stabil Dynam 2019;20 (1):2050011.
[8] Fan C. Study on shear performance of high-speed railway bridge piers under earthquake action. Changsha: Central South University; 2020 [In Chinese].
[9] Mayes R, Sharpe R. Seismic design guidelines for highway bridges. Berkeley, CA: Berkeley: Final Report Applied Technology Council; 1981.
[10] Guo W, Hu Y, Liu HY, et al. Seismic performance evaluation of typical piers of China's high-speed railway bridge line using pushover analysis. Math Probl Eng 2019;2019(1):9514769.
[11] ATC-40. Seismic evaluation and retrofit of concrete buildings, Volume 1. Redwood City, USA: Applied Technology Council; 1996.
[12] Chopra AK. Dynamics of structures: theory and applications to earthquake engineering. Englewood Cliffs, N J, USA: Prentice Hall; 2000.
[13] Lin YY, Chang KC. Effects of site classes on damping reduction factors. J Struct Eng 2004;130(11):1667–75.
[14] Lin YY, Miranda E, Chang KC. Evaluation of damping reduction factors for estimating elastic response of structures with high damping. Earthq Eng Struct Dynam 2005;34(11):1427–43.
[15] Hwang H, Liu JB, Chiu YH. Seismic fragility analysis of highway bridges. Technical Report, Center for Earthquake Research and Information, The University of Memphis; 2001.
[16] HAZUS99 user's manual. Washington, D.C: Federal Emergency Management Agency; 1999.
[17] Kappos A, Moshonas I, Paraskeva T, et al. A methodology for deviation of seismic fragility curves for bridges with the aid of advanced analysis tools. Proc. 1st ECEES Congress, Geneva; 2006.
[18] Hu Y, Guo W. Seismic response of high-speed railway bridge-track system considering unequal-height pier configurations. Soil Dynam Earthq Eng 2020;137:106250.
[19] Zhang JH, Guan ZG. Seismic vulnerability analysis of regular continuous girder bridges. Vibrat Shock 2014;33(20):140–5 [In Chinese].
[20] Hu Y. Research on seismic performance and failure mechanism of simply supported girder track-bridge system for high-speed railway. Changsha: Central South University; 2019 [In Chinese].
[21] Wang FT. Vehicle system dynamics. Beijing: China Railway Publishing House; 1981 [In Chinese].
[22] Gao X. Study on failure characteristics of simply supported girder track-bridge system for high-speed railway under near-fault earthquake. Changsha: Central South University; 2020 [In Chinese].
[23] Guo W, Gao X, Hu P, et al. Seismic damage features of high-speed railway simply supported bridge-track system under near-fault earthquake. Adv Struct Eng 2020;23 (8):1573–86.

[24] Wang Y. Operational safety evaluation and control of simply-supported girder bridges in high-speed railways under far-field earthquakes. Changsha: Central South University; 2021 [In Chinese].
[25] Guo W, Wang Y, Liu HY, et al. Running safety assessment of trains on bridges under frequent earthquakes based on spectral intensity theory. Int J Struct Stabil Dynam 2021;21(14):2140008.
[26] Liu HY. Dynamic response and operational safety evaluation of high-speed railway long-span continuous steel truss bridges under seismic loads based on the moving element model. Changsha: Central South University; 2019 [In Chinese].
[27] Xu X, Cai CB, He QL, et al. Simulation analysis of dynamic behavior of suspension-type monorail system under earthquake. Railway Stand Design 2020;64(11):1–7 [In Chinese].
[28] He X, Wu T, Zou Y, et al. Recent developments of high-speed railway bridges in China. Struct Infrastruct Eng 2017;13(12):1584–95.

CHAPTER 6

Evaluation Metrics for Train Operation Performance on High-Speed Railway Bridges Under Earthquakes

Contents

High-Speed Railway Train–Track–Bridge Systems
ISBN 978-0-443-49310-2
https://doi.org/10.1016/B978-0-443-49310-2.00006-5

6.1 Overview

Under transverse seismic excitation, a train may experience derailment or overturning. Such derailment can lead to significant casualties and property damage. Therefore, seismic design of high-speed railway bridges must incorporate operational safety evaluation, where accurate and practical evaluation metrics are of critical importance. Research by Lin Yusen and Chen Lingkun indicates that bridges with great stiffness exhibit relatively high operational safety at the same speed under earthquakes [1,2]. *Code for Design of High-Speed Railway* (TB 10621–2014) specifies stiffness limits for piers and girders to ensure operational safety, yet structural limit metrics for evaluating train safety during earthquakes remain absent in bridge design [3]. Xiu et al. proposed quasistatic metrics such as track deformation and girder end rotation, but actual derailment cases demonstrate that trains may derail even without significant track deformation [4], rendering these quasistatic metrics inadequate for assessing train operational safety on the bridge under earthquakes. Miyamoto et al. [5] developed a 58–degrees of freedom (DOF) Shinkansen train model to analyze sinusoidal excitation displacement limits that satisfy operational safety under lateral track excitation. Luo et al. connected sinusoidal excitation limits to random earthquakes using the Spectrum Intensity (*SI*) theory, calculating the *SI* Limit (SI_L) for the operational safety of the Shinkansen train. This method is validated through multiple simulation examples and shake table tests [4,6,7] subsequently being incorporated into Japan's railway seismic design standards.

This chapter first conducts a comparative analysis of the seismic responses of the train-track-bridge system under both seismic and non-seismic conditions, and investigates the dynamic response characteristics and operational safety of the system under varying train speeds and seismic intensities. Subsequently, the operational safety of the CRH380A train [formation: 2× (1T + 6M + 1T), where T denotes the trailer and M the traction] crossing the Zhengzhou Yellow River Grand Bridge on the Zhengzhou-Jinan Railway under seismic conditions is investigated. An OpenSees Finite Element (FE) model for a high-speed train–ballastless track–continuous steel truss bridge system is established. Using the corrected wheel load reduction rate and derailment coefficient limits as evaluation metrics, the study investigates the operational safety of high-speed trains on bridges under varying seismic intensities and running speeds. Finally, based on the safety evaluation methodology and metrics developed by Japan's Railway Technical Research Institute, displacement limits under sinusoidal excitations at different frequencies are calculated for CRH380A trains. These limits are used to derive the SI_L applicable to China's high-speed railways. The proposed corrected operational safety performance metrics are corrected under varying speed conditions, leading to the establishment of a speed-graded SI_L. Based on these metrics, the train operational safety performance is evaluated for bridges with varying pier heights across different site conditions.

6.2 Safety Evaluation Metrics for Operation on Long-Span Railway Bridges under Earthquakes

Because the safety evaluation metrics of trains on high-speed railway simply supported girder bridges under seismic conditions are detailed in Section 1.3.2 of Chapter 1, this chapter focuses solely on the evaluation metrics for long-span railway bridges under seismic action.

Considering the particularity of seismic load effects and the individual differences of bridges, this section appropriately corrects the corresponding train operational safety evaluation metrics for long-span bridges, with the correction approach illustrated in Fig. 6.1.

6.2.1 Correction of Seismic Influencing Factors

Earthquakes typically cause damage to highways and railways or disrupt their operations, complicating rescue efforts. For instance, during the 2008 Wenchuan earthquake, military airdrops were the only viable rescue

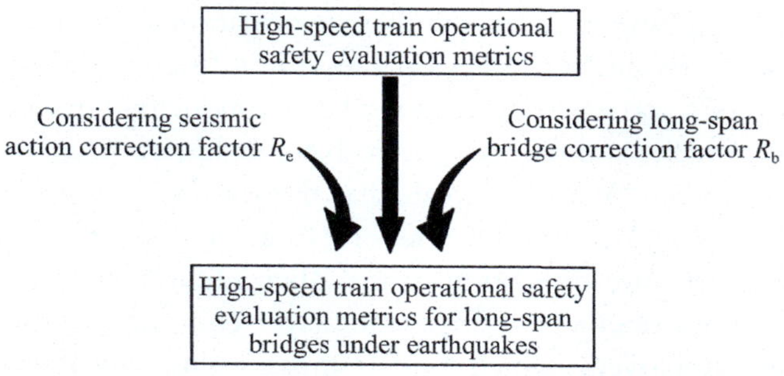

Figure 6.1 Operational safety evaluation metrics of high-speed trains on long-span bridges under earthquakes [8]

method in the early stages [9–11]. High-speed trains are particularly vulnerable to more severe consequences in the event of seismic accidents. Long-span bridges typically span across major rivers, deep canyons, and other geologically hazardous areas. In the event of a high-speed train derailment, there is a high risk of trains falling into water bodies or ravines. Seismic loads exhibit significant variability, with earthquakes occurring at different times on the same site, potentially inducing dynamic responses in bridges and high-speed trains that differ by several orders of magnitude. Therefore, drawing upon seismic safety design concepts from other industries [13,14], such as *Code for Seismic Design of Railway Engineering* (GB 50111–2006) and *Specification of Seismic Design for Highway Engineering* (JTG B02–2013), a seismic disaster consequence severity correction factor R_e is introduced, with a provisional value range of 1.1–1.15.

6.2.2 Correction of Long-Span Bridge Factors

According to research by Yao Junchuan, long-span continuous steel truss bridges exhibit significant dynamic differences from the 32-m simply supported girders widely used in high-speed railways. For instance, the first-order natural frequency of the Zhengzhou Yellow River Grand Bridge on the Zhengzhou-Jinan Railway is 0.92 Hz, with the first 20 natural frequencies within 2.22 Hz. In contrast, the first-order vertical natural frequencies of 24 m and 32 m simply supported girders on 350 km/h ballastless tracks are typically around 9.6 and 6.2 Hz, respectively [14]. The long-span continuous steel truss bridge exhibits lower stiffness compared to simply supported girders, resulting in larger and more complex structural deformations and more pronounced nonlinear behavior. Moreover, long-span continuous steel truss girders adopt ballastless track technology, resulting in more complex deformation of the track structure on the bridge. Extensive studies on the seismic/postearthquake mapping relationship of track irregularities have been conducted both domestically and internationally, and numerous valuable conclusions have been derived from these studies. Analysis reveals that the transmission of substructure deformation to track irregularities involves a certain degree of attenuation. In the track structure simulation of the Zhengzhou Yellow River extra-large continuous steel truss bridge, multilayer equivalent spring-damping models are employed, which underestimates the contribution of the track structure. Therefore, span-dependent correction factor R_b for long-span bridges is introduced.

The corrected operational safety evaluation metrics of high-speed trains on long-span bridges under earthquakes are:

$$\begin{cases} R_e R_b Q/P \leqslant 0.8 \\ R_e R_b \Delta P/P \leqslant 0.6 \end{cases} \tag{6.1}$$

where $R_e \geqslant 1.0$, which represents the seismic disaster consequence severity correction factor and is temporarily taken as $R_e = 1.1$ in this section; $R_b \geqslant 1.0$, which represents the dynamic and nonlinear influence correction factor for long-span bridges and is temporarily taken as $R_b = 1.1$ in this section. In the specific evaluation, if any of these above metrics exceeds the limit, the high-speed train is deemed unsafe under such operating conditions, and the corresponding speed is designated as the restricted operating speed.

6.3 Safety Evaluation of Train Operation on High-Speed Railway Bridges under Earthquakes

Using the previously listed evaluation metrics for train operation performance and train-track-bridge coupling model, this section evaluates the operational safety of high-speed trains on simply supported girder bridges under earthquakes and on the Zhengzhou Yellow River extralarge continuous steel truss bridge.

6.3.1 Train Operation on High-Speed Railway Simply Supported Girder Bridges under Earthquakes

This section takes a single-track high-speed railway line as the research object. A refined computational model of the high-speed train–twin-block ballastless track-bridge coupling system is established using the Simpack and OpenSees co-simulating train-track-bridge (SOTTB) program that integrates multibody system dynamics software Simpack and FE software OpenSees. The system accounts for transverse seismic loading and self-excitation of track irregularities. Based on train operational performance evaluation metrics, the dynamic response laws and operational safety of the train-track-bridge system under different speeds and seismic intensities are analyzed.

6.3.1.1 Computational Analysis Parameters

The CRH2 train is adopted as the live load for high-speed railway operation, with a four-vehicle formation. A 10 span, 32 m standard simply supported girder bridge model is established, with the cross-section

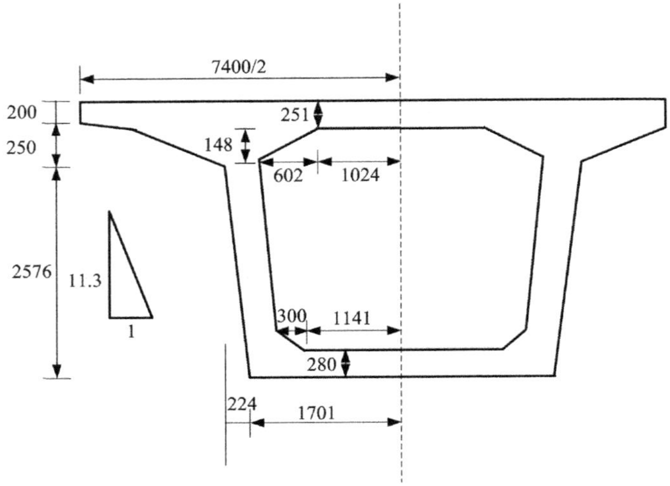

Figure 6.2 Cross-section dimensions of the box girder

dimensions of the box girder shown in Fig. 6.2. A fiber-based pier model is developed, and the piers are directly fixed to the ground, ignoring the influence of soil–pile interaction. The piers are 12.5 m high, with concrete grade C35. The cross-section dimensions of the pier are shown in Fig. 6.3. The plan view of the piers is illustrated in Fig. 6.4. Pot rubber bearings are used for the bearing model, and the damping ratio of the bridge is set to 5%. The bridge is located in a zone with a seismic intensity of 7 degrees. A classic ground motion record, the El Centro earthquake motion, is selected, with its basic characteristics listed in Table 6.1. The corresponding time–history curve is shown in Fig. 6.5. The cosimulation time step is set to $\Delta t = 2.5 \times 10^{-5}$ s. The spatial coupling model of the train-track-bridge system established based on the SOTTB program is illustrated in Fig. 6.6.

6.3.1.2 Analysis of Train Operational Safety on Railway Bridges Under Seismic and Non-seismic Conditions

To analyze the seismic impact on train operational safety, this section compares the operational safety performance metrics of the train and structural responses under seismic and non-seismic conditions. The El Centro earthquake motion is selected as the transverse seismic excitation, with its peak acceleration scaled to 0.035*g*, while the influence of track irregularities is also considered. The train operating speeds are selected as representative values of 250 and 300 km/h. Figs. 6.7–6.11 present the seismic responses of the train-track-bridge system at 250 km/h under

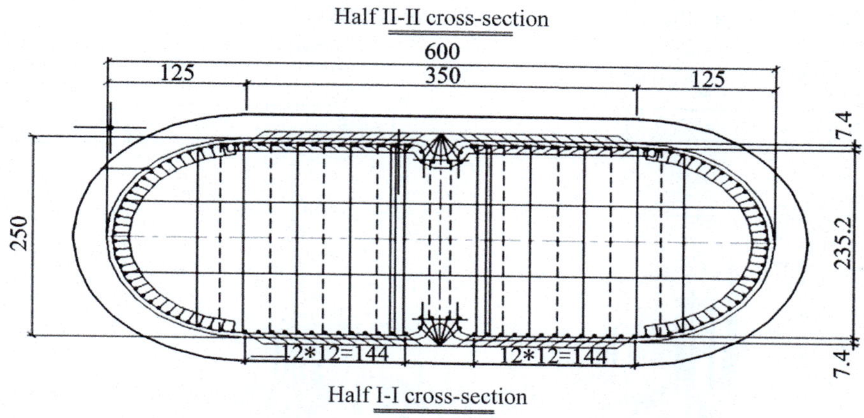

Figure 6.3 Cross-section dimensions of the pier (unit: cm)

seismic and non-seismic conditions. The peak values of passenger comfort metrics and operational safety metrics are listed in Table 6.2, while the response peaks of the multi-span simply supported box girder bridge are presented in Table 6.3.

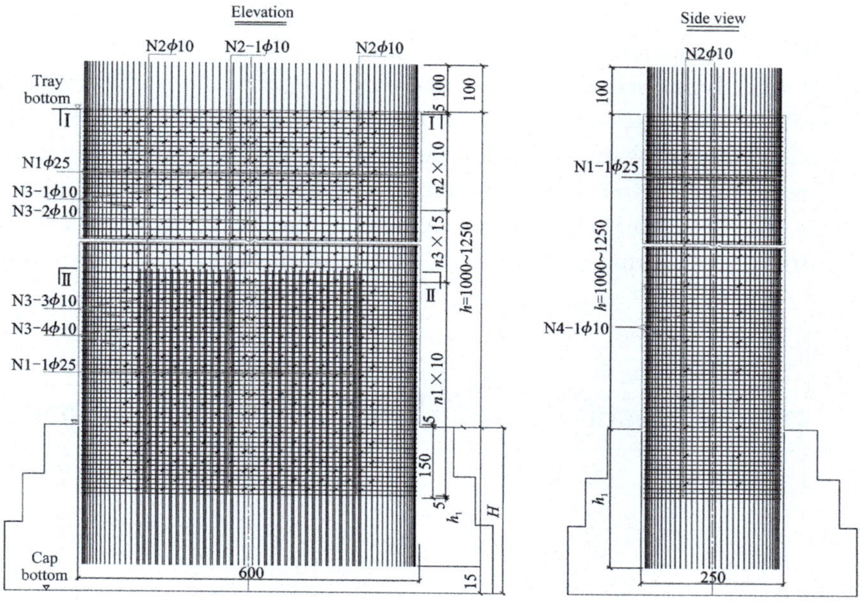

Figure 6.4 Plan view of piers (unit: cm)

Table 6.1 Basic characteristics of ground motion record.

Earthquake Name	Year	Magnitude	Fault Distance/km	Peak Ground Acceleration (PGA)/g	Peak Ground Velocity (PGV)/($cm \cdot s^{-1}$)	Peak Ground Displacement (PGD)/cm
El Centro	1940	6.95	6.09	0.281	31.309	24.150

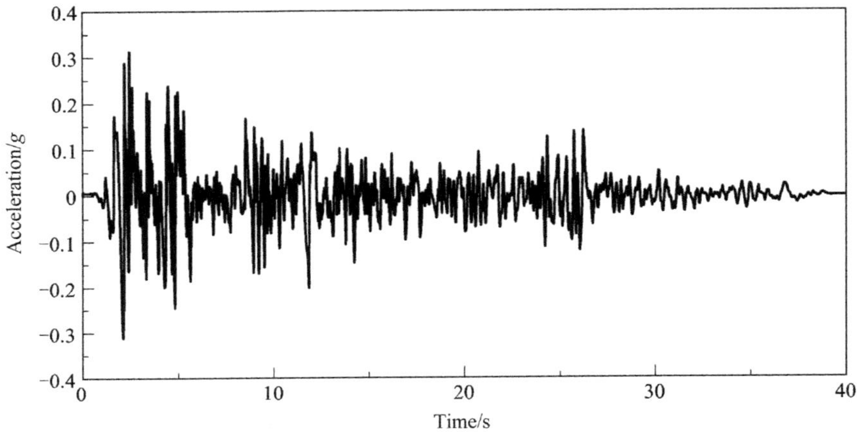

Figure 6.5 Time-history curve of the El Centro ground motion record

The computational results indicate that:

(1) Under lateral seismic load and track irregularities, the lateral response of the train-track-bridge system under seismic irregularities is significantly greater than that without seismic irregularities, while the vertical response shows a minimal difference compared to scenarios without irregularities. This is because the system is subjected to lateral excitation from both seismic loading and track irregularities, whereas the vertical

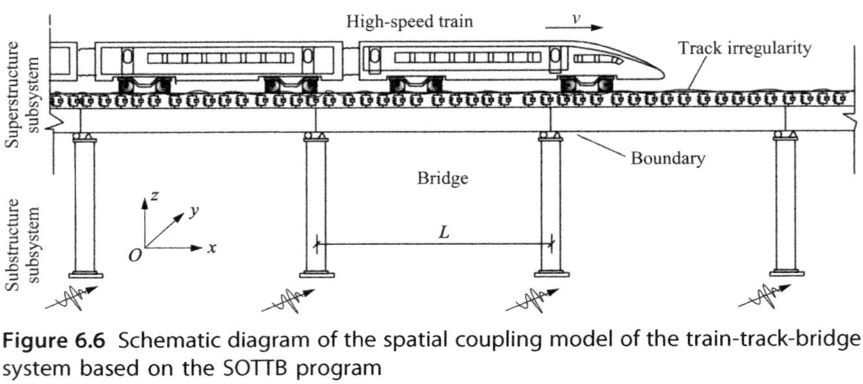

Figure 6.6 Schematic diagram of the spatial coupling model of the train-track-bridge system based on the SOTTB program

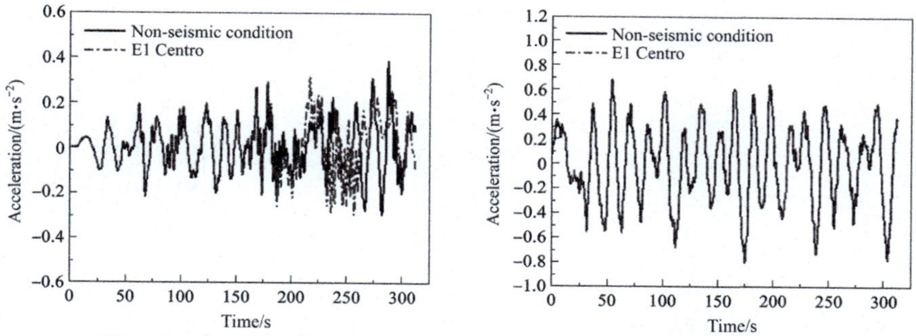

(a) Lateral acceleration time-history curve of the first vehicle section

(b) Vertical acceleration time-history curve of the first vehicle section

Figure 6.7 Comparison of vehicle acceleration time-history curves under seismic and non-seismic conditions

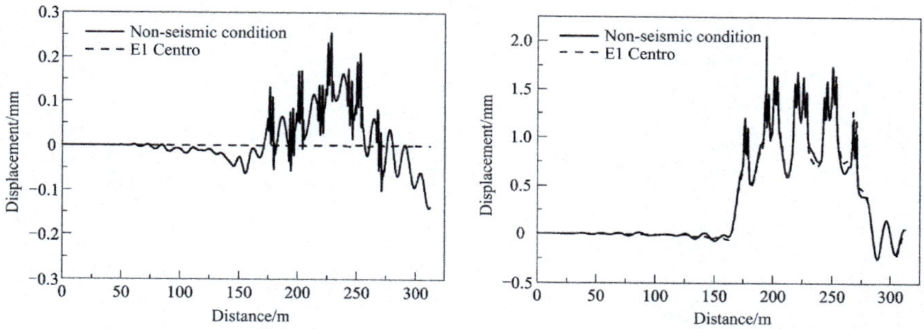

(a) Time-history curve of lateral midspan rail displacement of the 5th span bridge

(b) Time-history curve of vertical midspan rail displacement of the 5th span bridge

Figure 6.8 Comparison of displacement time-history curves for the left rail under seismic and non-seismic conditions

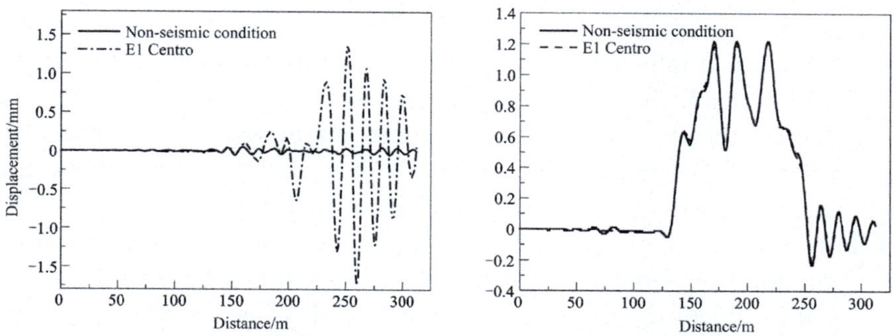

(a) Time-history curve of lateral midspan displacement of the 5th span bridge

(b) Time-history curve of vertical midspan displacement of the 5th span bridge

Figure 6.9 Comparison of midspan displacement time-history curves for the 5th span bridge under seismic and non-seismic conditions

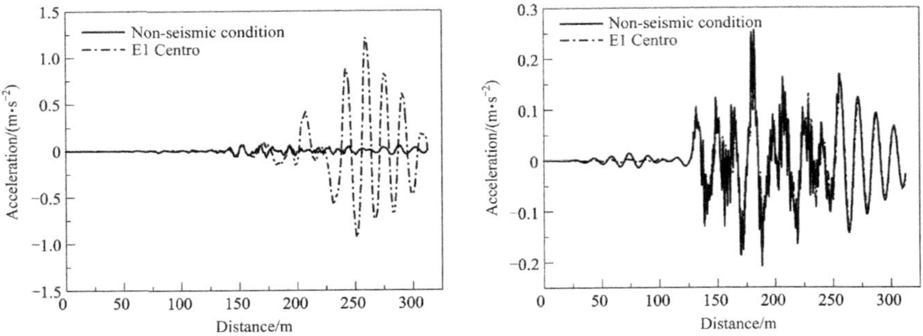

(a) Time-history curve of lateral midspan acceleration of the 5th span bridge

(b) Time-history curve of vertical midspan acceleration of the 5th span bridge

Figure 6.10 Comparison of midspan acceleration time-history curves for the 5th span bridge under seismic and non-seismic conditions

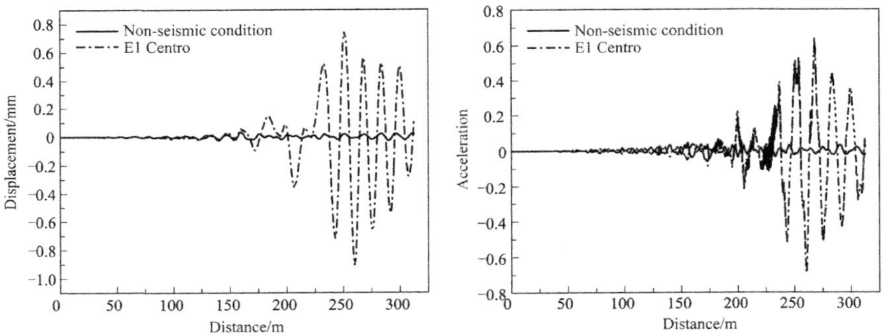

(a) Time-history curve of lateral pier-top displacement of the 5th span bridge

(b) Time-history curve of lateral pier-top acceleration of the 5th span bridge

Figure 6.11 Comparison of midspan lateral response curves for the 5th span bridge under seismic and non-seismic conditions

direction is influenced only by track irregularities. Under the lateral seismic load, the bridge response shows no significant difference at varying speeds. This indicates that this seismic load plays a dominant role in the response of the system, with its impact far exceeding that of train speed and track irregularities on the bridge system.

(2) In terms of train ride comfort and safety, compared to non-seismic conditions, vertical comfort and acceleration show only slight improvement under this lateral seismic load, while the lateral comfort metrics and operational safety metrics exhibit significant increases. This indicates that the seismic load has a pronounced impact on train operational safety, leading to a decline in operational safety.

Table 6.2 Peak values of comfort metrics and operational safety metrics of high-speed trains under seismic and non-seismic conditions.

Excitation Type	Train Speed/ $(km \cdot h^{-1})$	Vehicle Acceleration/$(m \cdot s^{-2})$		Sperling Index		Derailment Coefficient	Wheel Load Reduction Rate	Lateral Force/kN
		Lateral	Vertical	Lateral	Vertical			
Non-seismic condition	250	0.328	0.691	1.815	2.799	0.607	0.265	35.532
	300	0.5	0.795	2.0377	2.964	0.710	0.427	39.25
Seismic condition	250	0.547	0.905	2.227	2.851	0.6607	0.372	43.24
	300	0.663	0.985	2.231	3.003	0.7862	0.5618	56.3

Table 6.3 Peak response values of simply supported girder bridge under seismic and non-seismic conditions.

Excitation Type	Train Speed/(km · h^{-1})	Bridge Midspan Response		Pier-Top Response	
		Lateral Displacement/mm	Lateral Acceleration/(m · s^{-2})	Lateral Displacement/mm	Lateral Acceleration/(m · s^{-2})
Non-seismic condition	250	0.0709	0.0874	0.0407	0.0664
	300	0.127	0.0802	0.219	0.228
Seismic condition	250	1.754	1.139	0.983	0.815
	300	1.75	1.142	0.977	0.782

6.3.1.3 The Impact of Train Speed and Seismic Intensity on Train Operational Safety on High-Speed Railway Bridges

This section further investigates the impact of train speed and seismic intensity on operational safety under transverse seismic excitation and track irregularities. Here, the El Centro ground motion is selected as the transverse seismic excitation. This section calculates the seismic responses of the train-track-bridge system with a pier height of 12.5 m, in which high-speed trains run at 250 and 300 km/h under seismic accelerations of 0.035 *g*, 0.068 *g*, 0.1 *g*, 0.16 *g*, and 0.22 *g*. The peak acceleration responses of the train under different speeds and seismic intensities are compared, as shown in Fig. 6.12. The safety performance metrics of high-speed trains are compared, as shown in Fig. 6.13. The peak dynamic responses of high-speed trains under different speeds and seismic intensities at a pier height of 12.5 m are listed in Table 6.4.

As indicated by the summary of peak seismic responses for the train-track-bridge system under earthquakes:

① The ride comfort metrics of the train increase as the seismic intensity and speed increase. Compared to speed, seismic loading has a more significant impact on the ride comfort of the train. In terms of operational safety, the derailment coefficient and maximum wheel-rail lateral force increase as the seismic intensity and speed increase. The wheel load reduction rate generally increases as the train speed increases, whereas under different seismic intensities, it shows a slight increase, indicating that the wheel load reduction rate is more sensitive to the train speed.

② When the train speed is 250 and 300 km/h, the seismic intensity reaches 0.16 *g* and 0.1 *g*, respectively, causing the stability of the train to fail, with the derailment coefficient exceeding the dynamic safety limit

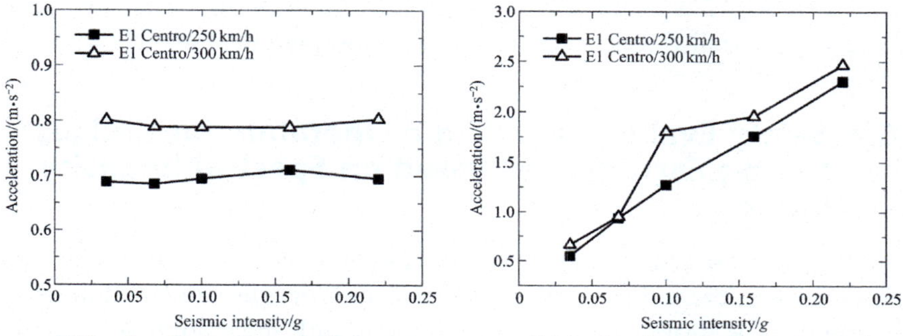

(a) Comparison of peak vertical acceleration responses

(b) Comparison of peak lateral acceleration responses

Figure 6.12 Comparison of peak acceleration responses of the train

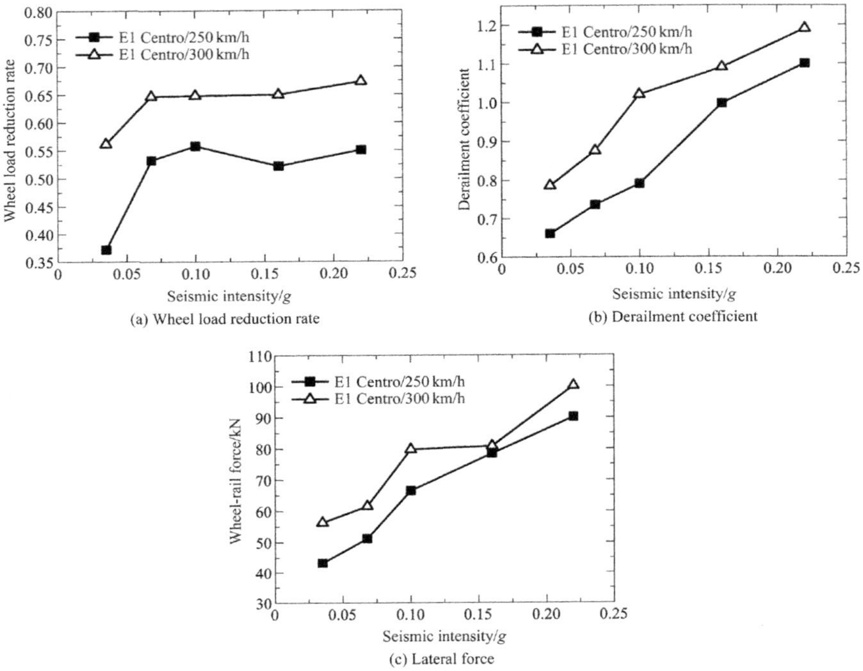

Figure 6.13 Comparison of train operational safety metrics

of 0.8. When the train speed is 300 km/h, the derailment coefficient reaches 1.19, approaching the danger limit of 1.2; the wheel load reduction rate is 0.673, exceeding the permissible standard value of 0.6; the lateral wheel-rail force reaches 100.1 kN. Under these working conditions, the wheel-rail forces of the first wheelset in the forward direction as shown in Fig. 6.14 indicate that at 261.5 m (t = 3.139 s), 281 m (t = 3.372 s), and 297.416 m (t = 3.569 s), the lateral and vertical wheel-rail forces are zero, a sign of jumping derailment.

6.4 Safety Evaluation of Train Operation on Bridges During Earthquakes Based on Spectral Intensity Index

In the seismic design of railway bridges, an essential performance objective is to ensure the operational safety of trains on the bridge during an earthquake. Research indicates that bridges with great stiffness generally exhibit relatively high operational safety at the same speed under

Table 6.4 Peak dynamic responses of high-speed trains during the El Centro earthquake.

Velocity/(km · h^{-1})	Seismic Intensity/g	Vehicle Acceleration/(m · s^{-2})		Sperling Index		Derailment Coefficient	Wheel Load Reduction Rate	Lateral Force/kN
		Lateral	Vertical	Lateral	Vertical			
250	0.035	0.547	0.689	2.227	2.796	0.6607	0.372	43.24
	0.068	0.937	0.685	2.621	2.799	0.7365	0.533	51.22
	0.1	1.273	0.695	2.873	2.799	0.7907	0.558	66.5
	0.16	1.759	0.711	3.178	2.796	0.997	0.522	78.43
	0.22	2.302	0.695	3.517	2.801	1.1	0.551	90.11
300	0.035	0.663	0.801	2.231	2.963	0.786	0.562	56.3
	0.068	0.952	0.789	3.001	2.963	0.876	0.647	61.53
	0.1	1.805	0.788	3.190	3.152	1.020	0.648	79.8
	0.16	1.959	0.788	3.270	3.27	1.090	0.650	80.815
	0.22	2.465	0.803	3.568	3.31	1.190	0.673	100.1

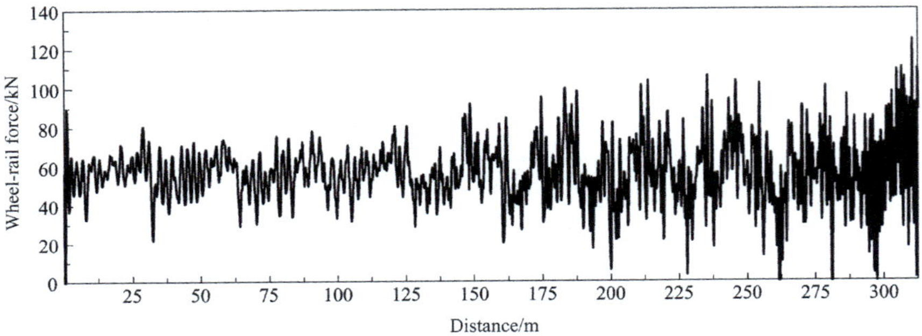

(a) Vertical wheel-rail forces of the first wheelset in the forward direction

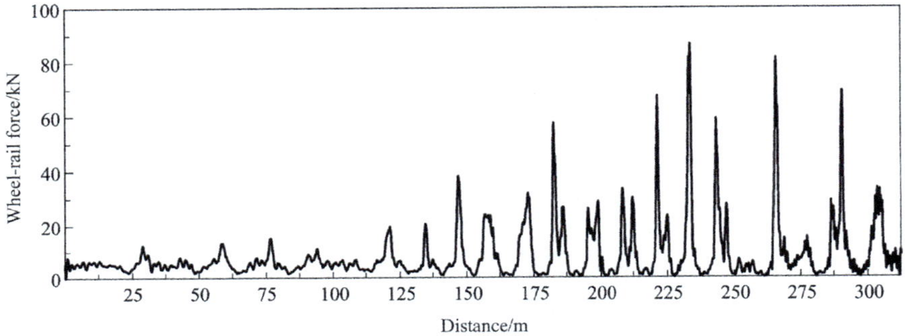

(b) Lateral wheel-rail forces of the first wheelset in the forward direction

Figure 6.14 Wheel-rail forces of the first wheelset in the forward direction

earthquakes [1-2]. Although China's standards for high-speed railways define stiffness limits for piers and girders to ensure operational safety, structural limit metrics for evaluating train safety during earthquakes remain absent in bridge design [3]. Other metrics for evaluating operational safety during earthquakes are quasistatic metrics such as track deformation and girder end rotation, but actual derailment cases demonstrate that trains may derail even without significant track deformation [4], rendering these quasistatic metrics such as track deformation inadequate for evaluating train operational safety on the bridge under earthquakes. Furthermore, although train-track-bridge dynamic simulations under seismic conditions can accurately the responses of trains running on bridges analyze and evaluate operational safety through metrics such as the derailment coefficient and wheel-rail unloading rate, the computer-based numerical simulation method requires the establishment of accurate train models, track-bridge models, and accurate contact relationships, along with the input of extensive train-bridge parameters, rendering it difficult to apply in practical bridge design. Moreover, research on limits for bridge response metrics during seismic-induced train derailments remains relatively limited. To study the safety evaluation of train operation on bridges under earthquakes based on the track-bridge model, map structural responses to operational safety, and provide more precise guidance for bridge design, this study proposes a velocity *SI* index.

6.4.1 Definition of *SI* Index

Because train derailments under earthquakes are highly random, train operational safety should be analyzed from an energy perspective. Research by Zeng Qingyuan et al. demonstrates that when the system resistance work δ_{cr} is less than the external input energy δ_{pr}, the train becomes unstable; when $\delta_{cr} = \delta_{pr}$, the train is in a critical stable state; when $\delta_{cr} < \delta_{pr}$, the train remains stable [15]. By analyzing the relationship among the velocity spectra, the maximum potential energy, and the maximum kinetic energy, the Japan's Railway Technical Research Institute established the *SI* for evaluating train operational safety under earthquakes. Through experimental validation and numerical simulation, it compared the effectiveness of the *SI* index in evaluating train operational safety during earthquakes, confirming its high accuracy. From an energy perspective, this index establishes an energy-based evaluation criterion to quantify the relationship between resistance work and input energy. The method has been incorporated into Japan's railway design standards.

Let us take an SDOF vibration system as an example to illustrate the process of establishing the *SI* index: Let the system mass be M, stiffness be K, and fundamental frequency be $\omega = (K/M)^{0.5}$. The maximum displacement response is x_{max}, the velocity spectrum is S_V, and the displacement spectrum is S_D. The maximum potential energy is then $1/[2(Kx_{2max})]$. Since $x_{max} = S_D$ and $S_V = \omega S_D$, when a train derails and reaches a critically stable state, its potential energy attains the maximum. According to the principle of energy conservation, the maximum kinetic energy at the onset of overturning equals the critical potential energy. The maximum potential energy per unit mass, as shown in Eq. (6.2), demonstrates that the maximum potential energy of the system is directly related to the velocity spectrum. Thus, the critical potential energy of the system can be expressed in terms of the velocity spectrum:

$$\frac{1}{2}(K/M)S_D^2 = \frac{1}{2}S_V^2 \tag{6.2}$$

Ground motion is random, and the predominant periods of different sites vary, whereas the fundamental frequencies of different structures also differ. Consequently, the frequency components of railway structure responses during earthquakes are complex. However, the response periods still fall within a certain range (0.1-2.5 s as specified in Japanese standards). Analytical validation shows that the response frequency ranges of typical Chinese high-speed railway bridges are also generally within this period. To calculate the input energy magnitude (δ_{pr}) of the system, the structural response across all frequency bands must be considered. Therefore, the energy within the target frequency range of interest is summed, and the velocity spectrum is integrated to determine the velocity *SI*. The expression is as:

$$SI(\xi, T) = \int_{0.1}^{2.5} S_{\mathrm{v}}(\xi, T)\mathrm{d}T \tag{6.3}$$

where ξ denotes the structural damping ratio; for high-speed railway bridges (reinforced concrete structures), a value of 0.05 is adopted. T denotes the period range included in the structural response.

6.4.2 *SI* Index Limits

In this section, the CRH380A train is taken as an example to determine the train response under sinusoidal excitation, determine the displacement limit under sinusoidal excitations, and calculate the SI_{L} of the high-speed train based on this displacement limit. The analysis is then extended to multiple speed conditions.

6.4.2.1 Limits of Train Operational Safety Under Sinusoidal Excitation

Based on the dynamic characteristics of the train itself, Xia He [16] classified the instability of the train during seismic excitation into overturning instability, jumping instability, and overturning-jumping instability, corresponding to low-frequency excitation, high-frequency excitation, and excitation close to the natural frequency, respectively. The magnitude of the dynamic response and the type of instability of the train are critically dependent on the frequency of the base excitation. Because seismic ground motions contain both high- and low-frequency components, this study applies lateral sinusoidal waves of varying frequencies to the train to determine its safety limits under different excitation frequencies.

(1) Train-track model under sinusoidal excitation

The analysis model of the train subjected to lateral sinusoidal track load excitation is shown in Fig. 6.15A, where the train and wheel-rail contact parameters remain consistent with the previously established train-track coupling model. The model is simulated using the multi-body system dynamics software Simpack, as illustrated in Fig. 6.15B. According to Japan's railway design standards, the sinusoidal excitation limits are determined with the train remaining stationary, and bench tests have been conducted for validation. The test results show a high degree of agreement with the simulation outcomes. For comparison with Shinkansen train results, vehicle speed v is first set to a relatively low value (20 km/h).

Sinusoidal excitation is applied in the form of displacement. The sinusoidal waveform consists of three segments: the steady-state segment, the transition segment, and the sinusoidal segment, as shown in Fig. 6.16A. To account for initial vibrations during the computational startup phase, a zero-excitation phase is first applied to allow

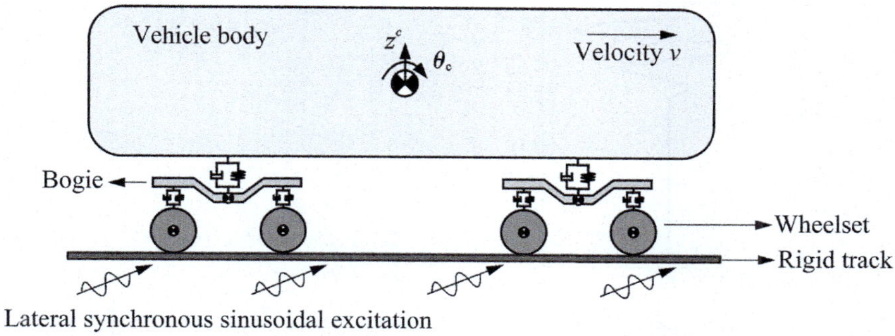

(a) Analysis model of the train under lateral sinusoidal track load excitation

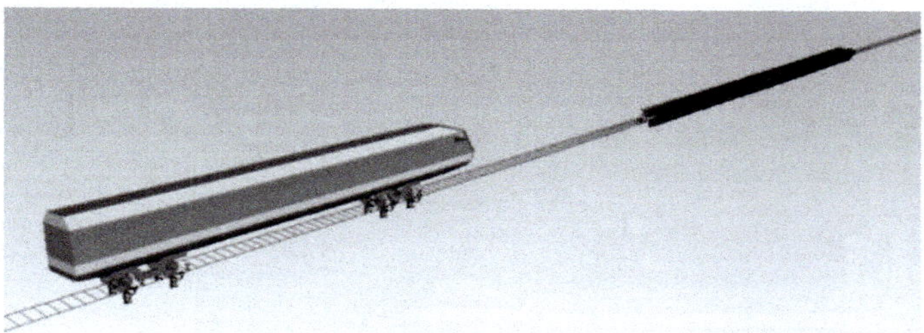

(b) Multibody system dynamics train-track model

Figure 6.15 Dynamic model of the train subject to track excitation

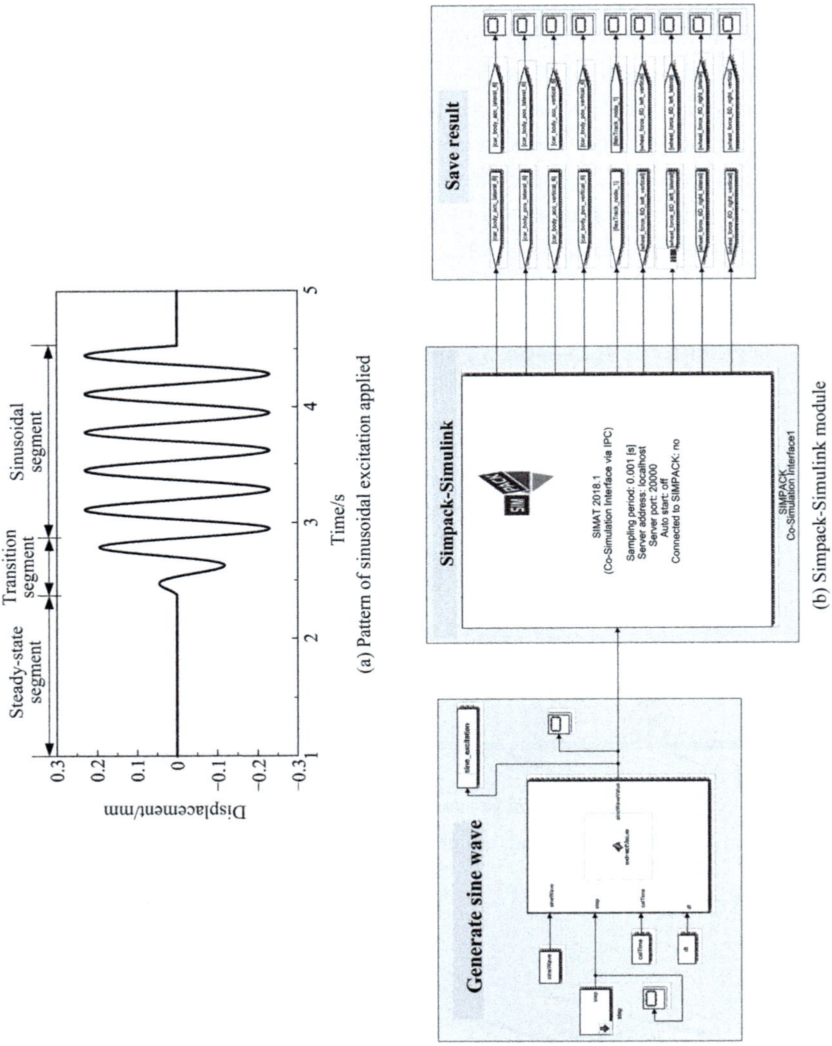

(a) Pattern of sinusoidal excitation applied

(b) Simpack-Simulink module

Figure 6.16 Pattern of sinusoidal excitation applied

the train to reach a stable operating state before sinusoidal track excitation is introduced. A transition segment with a low slope is provided between the steady-state segment and the sinusoidal segment to prevent abrupt load impacts. For sinusoidal waves of different frequencies, the same number of cycles (five cycles in this analysis) should be adopted. The excitation amplitude is gradually increased (5-mm increments) until the train evaluation metric exceeds the limit. The amplitude of the sinusoidal wave at this point is defined as the base displacement limit for the corresponding frequency. To achieve lateral motion of the rail, a portable 6-DOF marker is adopted in Simpack. The input of the 6-DOF marker can be a custom excitation in Simpack. However, to facilitate the input of customized displacement time histories and subsequent processing, SIMAT is employed here to achieve cosimulation between Simulink and Simpack to generate rail displacement excitation, as illustrated in the program flowchart in Fig. 6.16B.

(2) Train response under sinusoidal excitation

1) Vehicle acceleration

Vehicle acceleration is a key metric of train ride comfort. Relevant standards specify the vertical acceleration limit as 1.3 m/s^2 and the lateral acceleration limit as 1.0 m/s^2. Take a sinusoidal load excitation at 3 Hz as an example. The acceleration response of the train is shown in Fig. 6.17. As shown in the figure, the magnitude of the acceleration increases as the excitation amplitude grows. The excitation is laterally applied, so the magnitude of the lateral acceleration is much greater than the vertical acceleration. Under a 0.1 m excitation amplitude, it is 1.141 m/s^2, exceeding the regulatory limit.

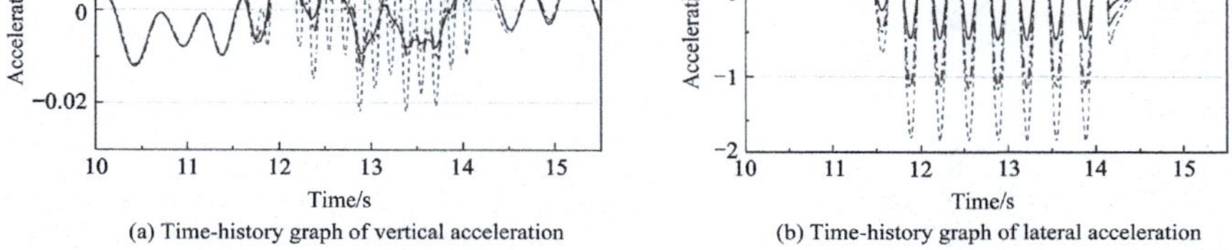

Figure 6.17 Vehicle acceleration of the vehicle under sinusoidal excitation (f = 3 Hz)

2) Derailment coefficients

The peak derailment coefficients of left and right wheelsets during sinusoidal excitation are shown in Fig. 6.18. Fig. 6.18A and B share the same excitation amplitude but differ in excitation frequency. Under identical frequency excitation, the derailment coefficients of left and right wheels on the same wheelsets are comparable. The derailment coefficient magnitude increases as the frequency grows. At an amplitude of 0.15 m, the maximum derailment coefficients at excitation frequencies of 0.75 and 2 Hz are 0.31 and 0.52, respectively, both below the regulatory limit.

① Wheel-rail unloading rate, during sinusoidal excitation, the wheel load reduction rate is shown in Fig. 6.19. The wheel-rail unloading rate increases with the excitation frequency. At an amplitude of 0.15 m, the unloading rates for the left and right wheels are 0.409 and 0.422 at 0.75 Hz, and 0.45 and 0.46 at 2 Hz, respectively. No wheel unloading derailment occurs.

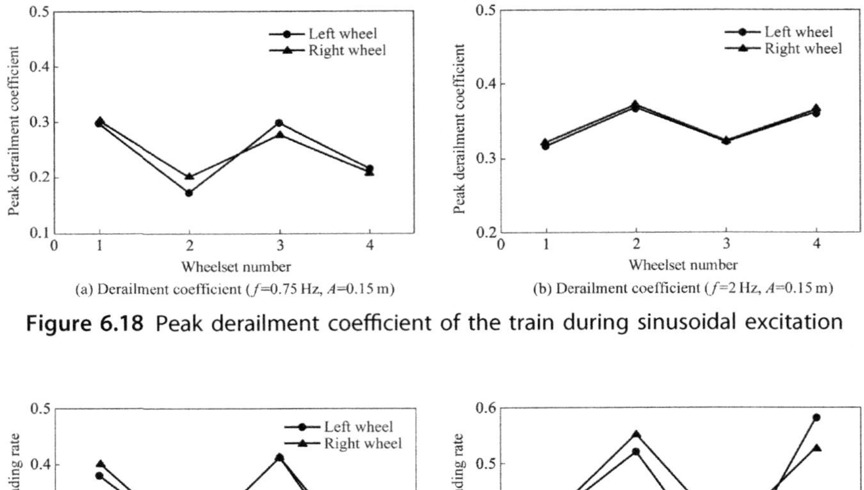

(a) Derailment coefficient (f=0.75 Hz, A=0.15 m) (b) Derailment coefficient (f=2 Hz, A=0.15 m)

Figure 6.18 Peak derailment coefficient of the train during sinusoidal excitation

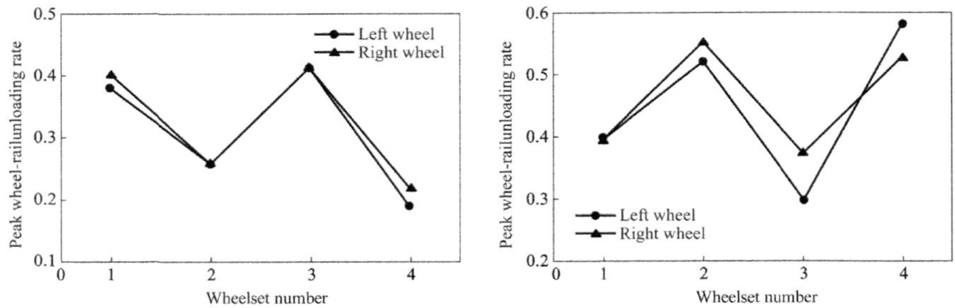

(a) Wheel load reduction rate (f=0.75 Hz, A=0.15 m) (b) Wheel load reduction rate (f=2 Hz, A=0.15 m)

Figure 6.19 Wheel load reduction rate during sinusoidal excitation

② Lateral wheel-rail force. The time history of wheel-rail force response during sinusoidal excitation is shown in Fig. 6.20. Under the excitation of the same frequency, the amplitude of the wheel-rail force increases as the excitation amplitude increases. When the excitation amplitudes are 0.03 m and 0.17 m, the peak wheel-rail forces are 6.335 kN and 43.2 kN, respectively, with the latter exceeding the safety limit. At higher excitation amplitudes, the wheel-rail force response exhibits more high-frequency components. This is because, as the lateral wheel-rail displacement increases, the wheel flange of the wheelset comes into contact with the railhead, causing rapid climbing of the single wheelset. This leads to rapid variations in the wheel-rail contact parameters, which directly affect the calculated wheel-rail force.

③ Limit displacement under sinusoidal excitation. As analyzed in the previous section, the operational safety and ride comfort metrics vary with the excitation amplitude and frequency, but the relationship between these metrics and the amplitude/frequency remains unclear. This section calculates the train operational evaluation metrics under sinusoidal excitations with varying amplitudes and frequencies at the same speed, analyzes their variation patterns with frequency and amplitude, and statistically determines the displacement limits under sinusoidal excitations with different periods that meet operational safety requirements.

The ride comfort metrics under sinusoidal excitation are shown in Fig. 6.21, in which each curve represents a sinusoidal wave with a

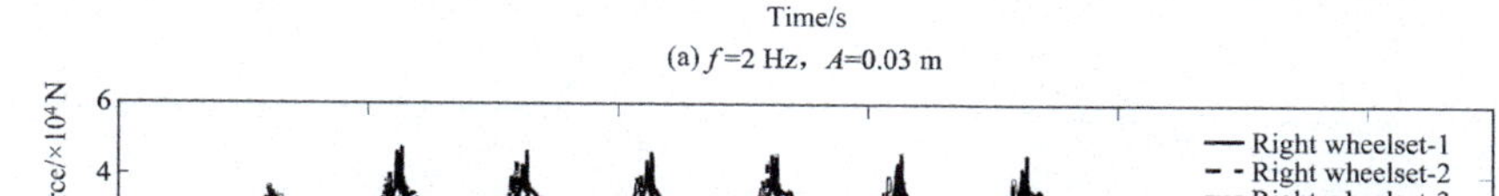

(a) f=2 Hz, A=0.03 m

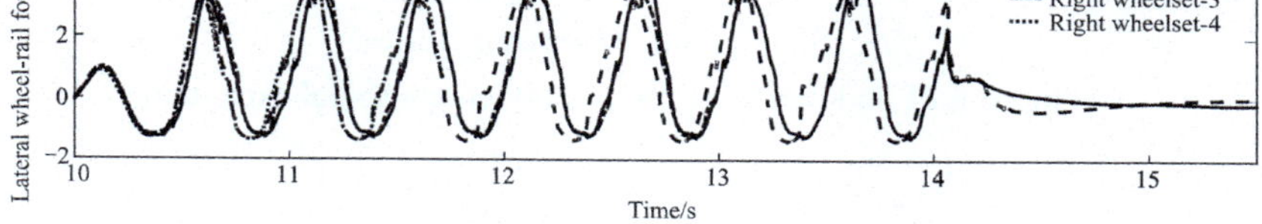

(b) f=2 Hz, A=0.17 m

Figure 6.20 Time history of wheel-rail force response under sinusoidal excitation

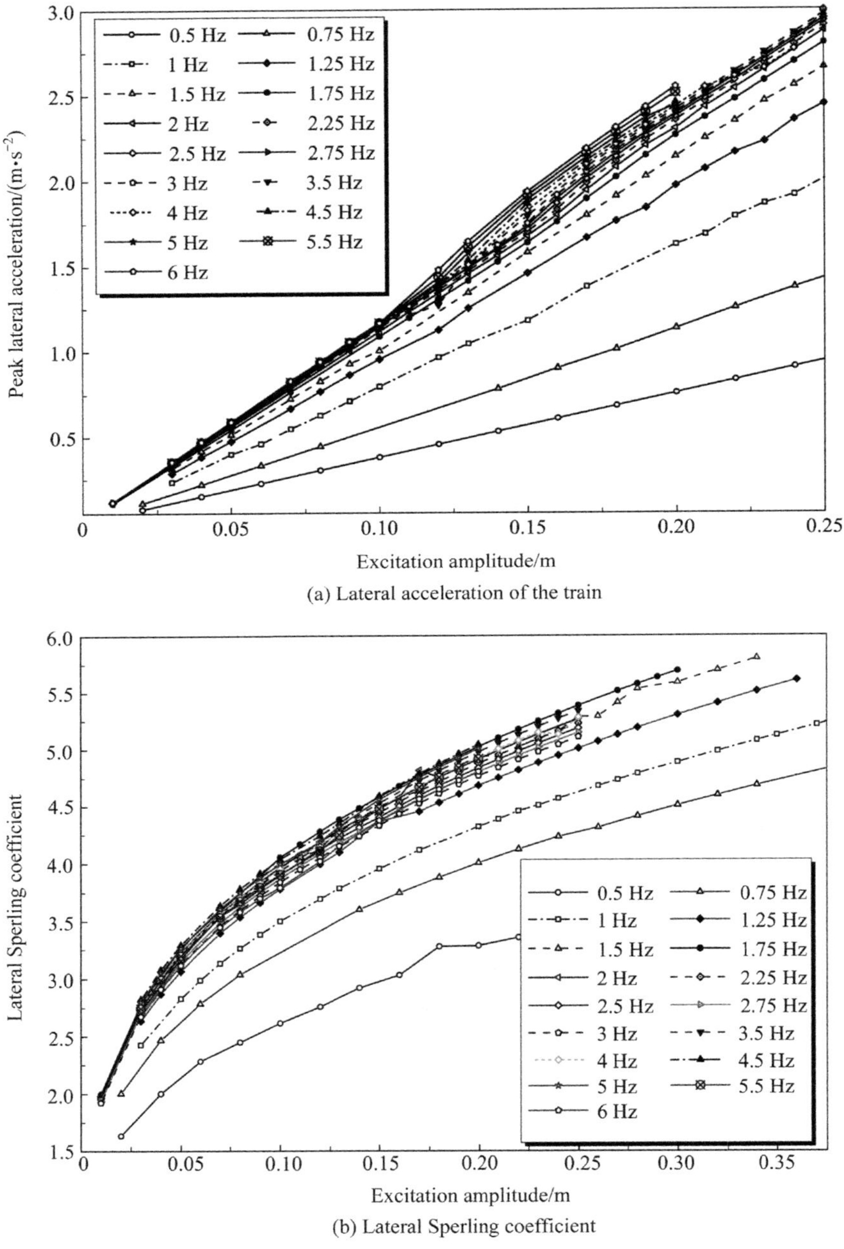

Figure 6.21 Ride comfort metrics under sinusoidal excitation

different frequency. As discussed in the previous section, the primary vibration mode of the train is lateral, so only the lateral acceleration and lateral Sperling coefficient of the train are analyzed here. As shown in the figure, ride comfort metrics increase as the amplitude increases. At a fixed frequency, the peak lateral acceleration of the train exhibits a linear increase as the excitation amplitude increases, while the lateral Sperling coefficient shows a decreasing slope with amplitude growth, approximately following a logarithmic trend. Due to its rapid growth under small-amplitude excitation, the sperling coefficient reaches its limit (3.0) at a smaller excitation amplitude than that of the lateral acceleration of the train (1.0 m/s^2). For instance, under 1 Hz excitation, the Sperling coefficient reaches the limit at 0.060706 m, whereas the lateral acceleration exceeds the limit at 0.124,185 m. Under sinusoidal excitation with the same amplitude, the ride comfort metrics increase with the excitation frequency, but at a gradually lower rate. The ride comfort metrics under different excitation frequencies with an amplitude of 0.1 m are presented in Table 6.5. When the excitation frequency is below 2 Hz, the peak acceleration and Sperling coefficient increase rapidly. As the frequency continues to rise, the growth rate of the ride comfort metrics slows. This is because the vibration frequency of the vehicle body primarily falls below 2 Hz [17], resulting in more significant growth of ride comfort metrics at lower frequencies.

The operational safety metrics under sinusoidal excitation, as illustrated in Fig. 6.22, include the lateral wheel-rail force, derailment coefficient, and wheel load reduction rate. As shown in the figure, all safety metrics increase as the amplitude rises, exhibiting similar growth trends, because both the derailment coefficient and wheel load reduction rate are calculated from the wheel-rail forces. At low excitation amplitudes (below 0.1 cm), all metrics increase in an approximately

Table 6.5 Ride comfort metrics under the same amplitude excitation (0.1 m).

Excitation frequency/Hz	0.75	1	2	3	4	5	6
Peak lateral acceleration/ (m·s^{-2})	0.4445	0.7969	1.117	1.151	1.159	1.164	1.165
Lateral Sperling coefficient	3.039	3.4982	3.7684	3.8421	3.958	3.9949	3.7944

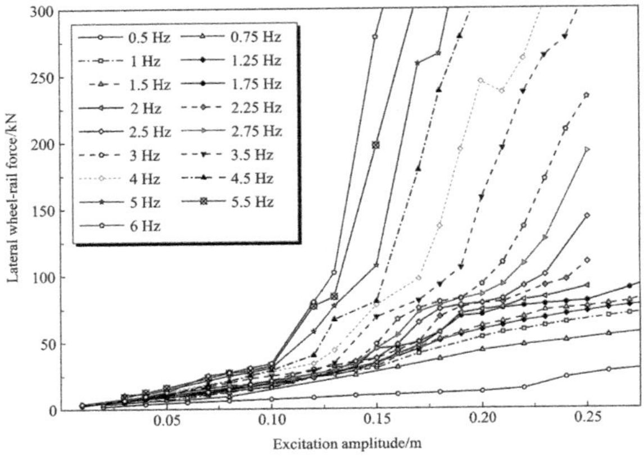

(a) Lateral wheel-rail force

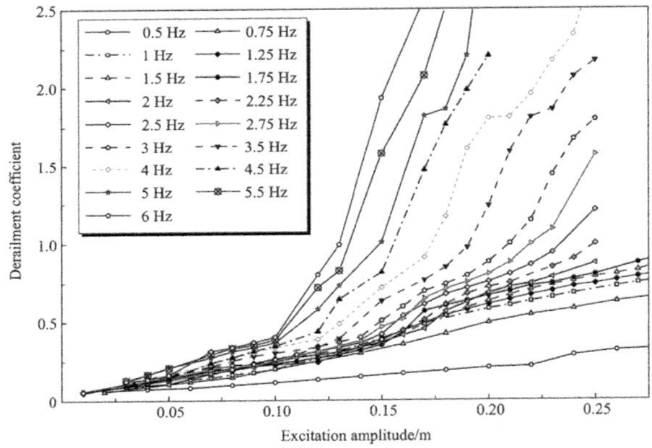

(b) Derailment coefficient

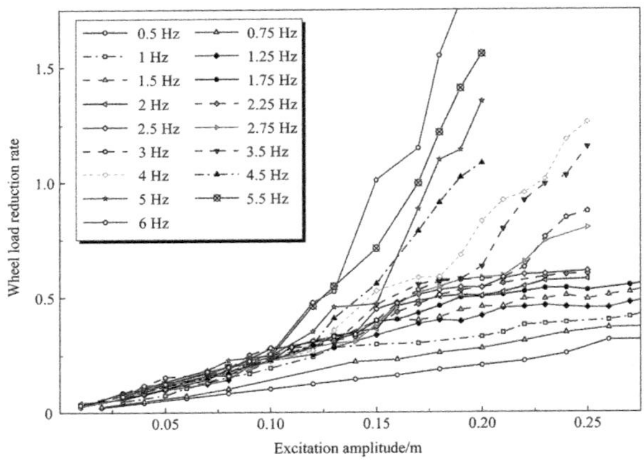

(c) Wheel load reduction rate

Figure 6.22 Operational safety metrics under sinusoidal excitation

Table 6.6 Safety metrics under the same amplitude excitation (0.12 m).

Excitation frequency/Hz	0.75	1	2	3	4	5	6
Lateral wheel-rail force/kN	14.75	16.18	20.16	21.23	28.00	30.95	34.17
Derailment coefficient	0.1232	0.1945	0.2796	0.2854	0.2699	0.3525	0.3785
Wheel load reduction rate	0.1595	0.2029	0.2495	0.2620	0.3361	0.3669	0.4033

linear manner; as the amplitude increases, the metrics grow more rapidly under high-frequency excitation, showing an approximately exponential growth trend. Table 6.6 presents the variation of safety metrics with frequency under the same amplitude excitation, showing that all metrics exhibit a positive correlation with frequency. The lateral wheel-rail force exceeds the limit at 5 Hz, at which point the displacement limit for this frequency can be calculated.

With the same method, the displacement limits that meet various performance requirements can be determined, as shown in Fig. 6.23. The figure indicates that the maximum displacement amplitude decreases with increasing frequency, and the stability-based limit is smaller than the safety-based limit. The ride comfort under seismic conditions is a secondary requirement, so it is not considered here. In the process of evaluating whether a train derailment occurs, it is often observed that one metric exceeds its limit whereas other metrics remain within acceptable ranges. To ensure train operational safety, if any single metric exceeds its prescribed limit, the train is deemed to be at risk of derailment. The most stringent requirements among similar limits from different standards are adopted to account for the worst-case scenarios during operation. By taking the intersection of displacement limits for all metrics, the final displacement limit curve is derived, as illustrated in Fig. 6.24.

6.4.2.2 Establishment of SI (Seismic Impact) Evaluation Metrics for the CRH380A Train

The displacement limit for excitation determined in the previous section is based on a sinusoidal wave, whereas seismic motion is a complex random wave with frequency components that differ significantly from sine wave characteristics. Therefore, the displacement limit derived from sinusoidal excitation cannot be directly applied and requires appropriate conversion.

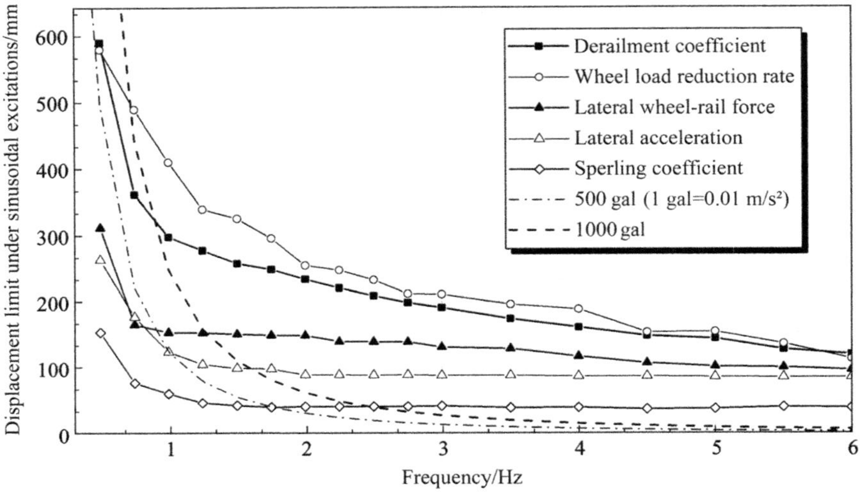

Figure 6.23 Excitation displacement limit

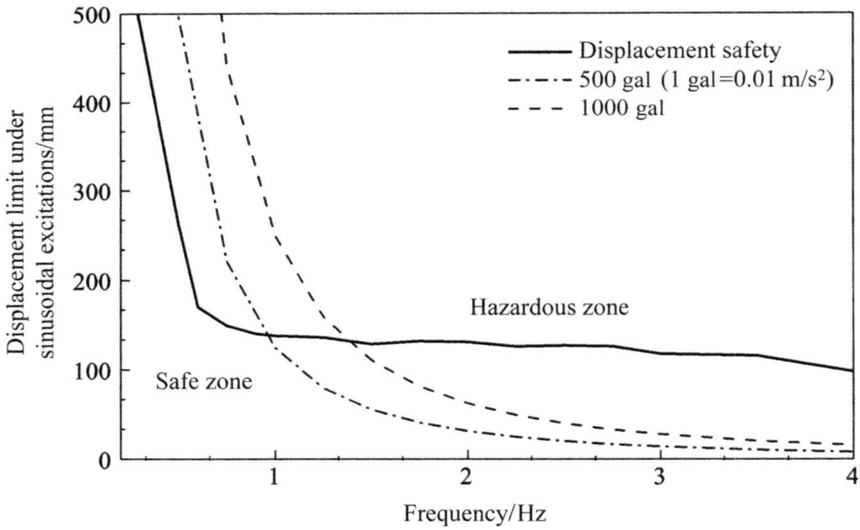

Figure 6.24 Final excitation displacement limit

The flowchart for solving SI_L (Seismic Impact Limit) is illustrated in Fig. 6.25. First, establish the train–track model and determine the velocity range. However, because different train models exhibit distinct dynamic characteristics and wheel-rail parameters vary across high-speed railway systems in different countries, the calculated safety limit magnitudes differ;

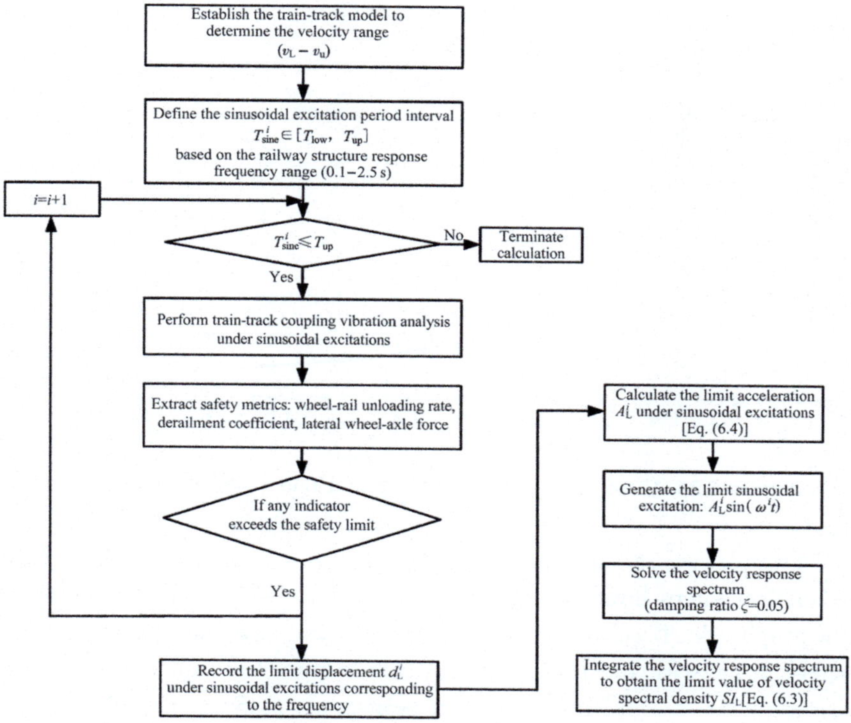

Figure 6.25 Flowchart for solving SI_L

subsequently, compute safety displacement limit d^i_L under sinusoidal excitation with a period of T^i_{sine}, repeat this process until T^i_{sine} falls below the upper limit of the period, thereby obtaining the excitation displacement limits for the frequency range of interest. As shown in Fig. 6.25, the displacement limits decrease continuously with increasing frequency.

Subsequently, acceleration safety limit A^i_L corresponding to the frequency is calculated based on displacement limit d^i_L, as shown in Eq. (6.4), where ω_i denotes the current frequency:

$$A^i_L = \omega_i^2 d^i_L \tag{6.4}$$

The calculated absolute acceleration safety limit under sinusoidal excitation, as shown in Fig. 6.26, indicates that values above the curve represent the hazardous zone, whereas those below are safe. Additionally, the acceleration safety limit gradually decreases with an increasing period of excitation.

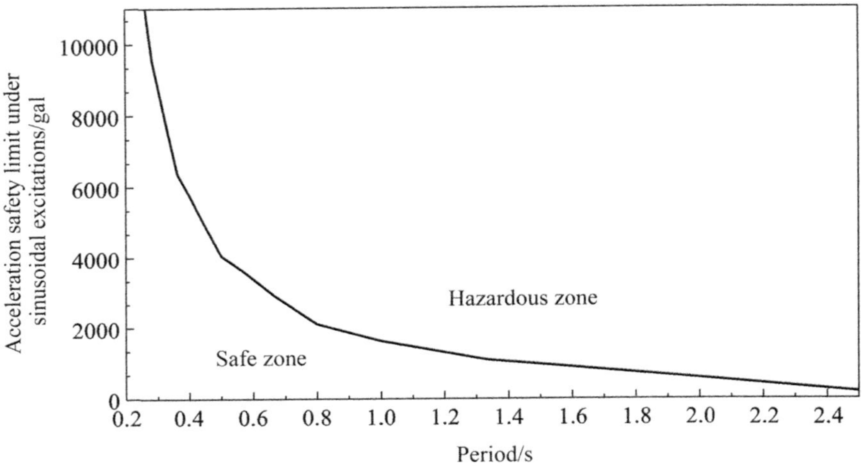

Figure 6.26 Absolute acceleration safety limit under sinusoidal excitation

The process of calculating the SI_L from the sinusoidal excitation acceleration safety limit, as shown in Fig. 6.27, involves first determining the amplitude A_L^i of the limiting sinusoidal wave based on the acceleration safety limit, with the period T_{sine}^i of the limiting sinusoidal wave corresponding to A_L^i, and the excitation periods set to 5 seconds. The subsequent step involves solving the response spectrum of the limiting sinusoidal wave (with a damping ratio of 0.05), in which the amplitude of this limiting sinusoidal excitation corresponds to the displacement limit required for operational safety. The velocity response spectrum of this limiting sinusoidal excitation is shown in Fig. 6.28, indicating that the amplitude of high-frequency excitation is relatively large, but as the frequency gradually decreases, the amplitude no longer shows significant variation.

The SI_L calculated for different frequency excitations in this section is shown in Fig. 6.29. As shown in the figure, the SI_L decreases as the

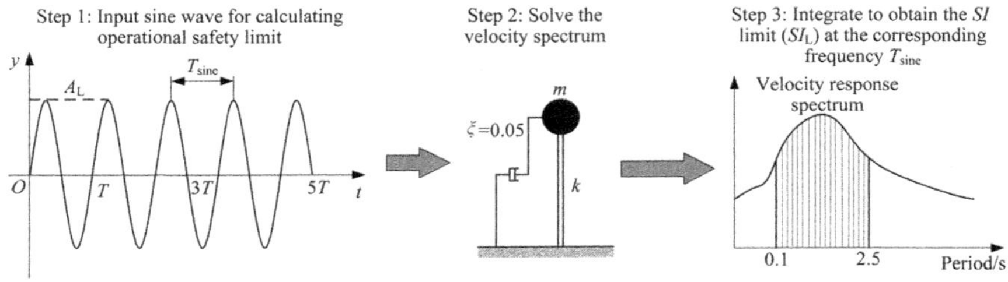

Figure 6.27 Calculation of SI_L from acceleration safety limits

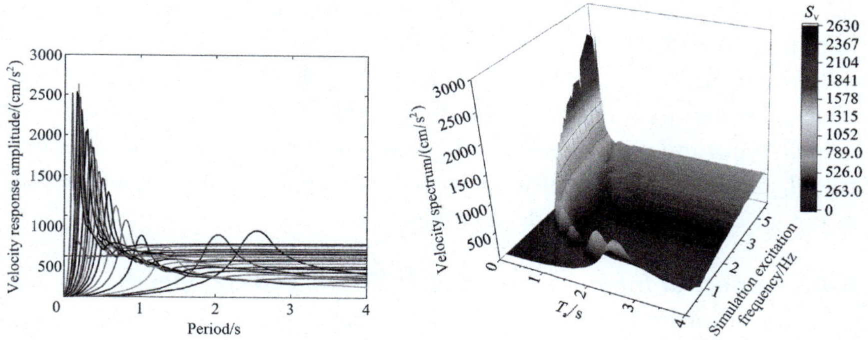

(a) Velocity spectrum (b) Velocity spectrum distribution

Figure 6.28 Velocity response spectrum under limiting sinusoidal excitation (damping ratio $\xi = 0.05$)

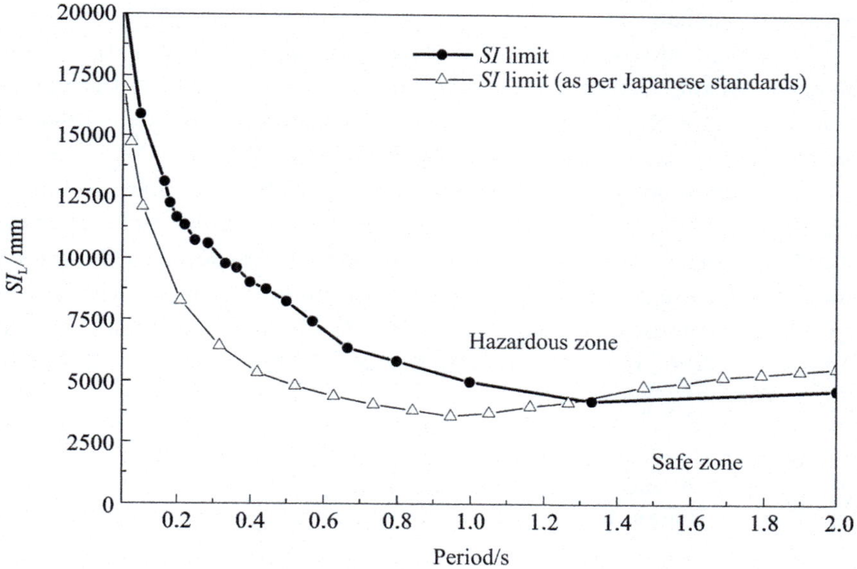

Figure 6.29 SI_L under different frequency excitations

excitation period increases, with a particularly rapid rate of decrease at high frequencies. This indicates that when the input excitation is primarily high-frequency (i.e., when the structural fundamental frequency is high), the safety limit for SI_L is more stringent. On the other hand, as the period increases beyond a certain point, the SI_L reaches a relatively stable state. The SI_L values calculated in this section, when compared with those calculated by the Japanese standards, show generally consistent trends. At high frequencies, the limits obtained in this study are slightly higher than those calculated by the Japanese standards, while the differences are minimal at lower frequencies. This discrepancy arises from differences in vehicle types and their respective dynamic characteristics. Once calculated, the SI_L can be used as a metric to evaluate the operational safety of the train at the given train speed under earthquakes.

6.4.2.3 Spectral Intensity Limits at Different Vehicle Speeds

The operational safety of a train is closely related to its speed. When a train is stationary or running at low speeds, the dynamic excitation effects of track irregularities cannot be accurately captured. In addition, the dynamic response of a train under seismic conditions varies with speed. This study first analyzes the dynamic performance of trains at different speeds under sinusoidal excitation.

When an earthquake occurs, a train generally decelerates due to the function of the earthquake early warning system. Therefore, the operational safety of trains at 300 km/h under earthquakes is not considered here. The lateral wheel-rail force time histories of Wheelset No. 1 on the left side at different speeds (20 km/h, 50 km/h, 100 km/h, and 150 km/h) are shown in Fig. 6.30. The excitation frequency (3 Hz) and amplitude (0.07 m) remain consistent across all four speeds. As shown in the figure, under the same excitation, the lateral wheel-rail force increases with higher speed. The lateral wheel-rail force is 12.73 kN when the train runs at 50 km/h, 40.01 kN at 100 km/h, and 42.02 kN at 150 km/h,

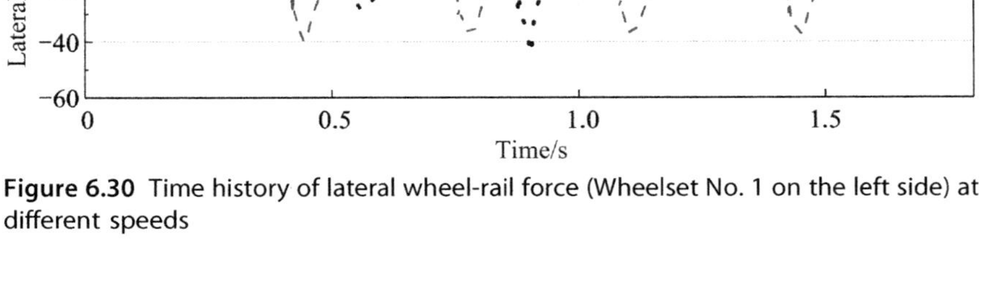

Figure 6.30 Time history of lateral wheel-rail force (Wheelset No. 1 on the left side) at different speeds

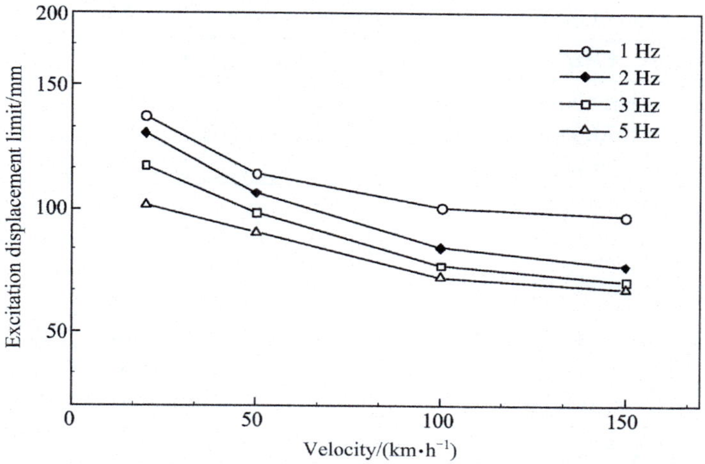

Figure 6.31 Impact of train speed on excitation displacement limits

demonstrating that as the vehicle speed increases, the rate of increase in the peak wheel-rail force shows a slowing trend.

Fig. 6.31 analyzes the excitation displacement limits at different speeds. The four curves correspond to four sinusoidal excitation frequencies. As speed increases, the displacement limits gradually decrease, with the rate of reduction slowing down at higher speeds. This aligns with Nishimura et al.'s [18] experimental results. Moreover, the high-frequency limits are consistently lower than the low-frequency limits at all vehicle speeds.

Based on the calculated displacement limits under sinusoidal excitations for corresponding speeds, the operational safety metrics of trains at low speeds are corrected to obtain the operational safety limits of the train. The ratios of excitation limits at different speeds to the 20 km/h limit are presented in Table 6.7. Since the displacement limit (d_L) is positively correlated with the *SI* limit, the SI_L calculated for 20 km/h in the previous section is adjusted based on the ratio of displacement limits at different speeds. To ensure design safety, the smaller ratio (2 Hz) is adopted. The results are shown in Fig. 6.32. Two evaluation metrics are selected here: the metrics for 20 km/h and 150 km/h, denoted as S1 and S2, under the stationary or low-speed operating conditions and high-speed operating conditions, respectively.

6.4.3 Validation of Spectral Intensity Metrics

This section validates the accuracy of the *SI* evaluation method and SI_L by comparing the *SI* evaluation results with the evaluation results of operational safety under the train–bridge coupling vibration under earthquakes.

Table 6.7 Ratio of excitation limits at different speeds at the 20-km/h limit.

Speed km/h	1 Hz/%	2 Hz/%	3 Hz/%	4 Hz/%	5 Hz/%
50	82.90	81.31	83.65	86.19	89.12
100	73.03	64.51	65.67	68.23	71.20
150	67.24	56.07	57.62	60.41	63.64

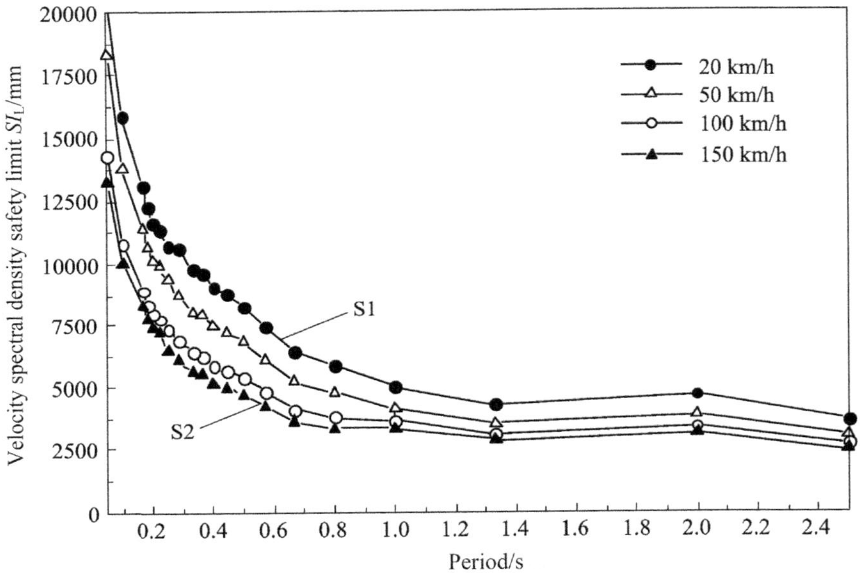

Figure 6.32 Spectral Intensity safety limit of the train at different speeds

6.4.3.1 Selection of Ground Motion

Sites can be categorized into four types based on soil shear wave velocity to account for site effects: type I (hard soil), type II (medium-hard soil), type III (medium-soft soil), and type IV (soft soil). Due to the filtering effect of soft soil on high-frequency seismic signals, the characteristic period (T_g) of the site increases as shear wave velocity decreases, as shown in Table 6.8. The characteristic period zoning of the bridge location should be determined based on the seismic parameter zoning. The railway seismic design standards divide the zoning into three levels. For this bridge, the characteristic period zoning is zone I, with a seismic fortification intensity of 7 degrees (0.10 *g*).

Table 6.8 Classification of site soil type.

Site Type	Type I Site	Type II Site	Type III Site	Type IV Site
Site soil classification	Rock or hard soil	Medium-hard soil	Medium-soft soil	Soft soil
Shear wave velocity $V_{se}/(m \cdot s^{-1})$	$V_{se}>500$	$250 < V_{se} \leq 500$	$150 < V_{se} \leq 250$	$V_{se} \leq 150$
Characteristic period T_g/s	0.25	0.35	0.45	0.65

The dynamic response of the structure is not only related to its dynamic characteristics, but also closely associated with ground motion characteristics, including spectral properties, intensity, and duration. For different site conditions, three ground motion acceleration records corresponding to each type of site are selected from the PEER database, as listed in Table 6.9. A comparison between the selected ground motion acceleration response spectrum and the design acceleration spectrum is shown in Fig. 6.33. The comparison indicates that the deviation between the mean spectrum and the design spectrum remains within 20% across critical periods, confirming that the selected ground motions comply with seismic code requirements. For long-span bridges with widely spaced supports or significantly different site conditions at substructure supports, inconsistent excitation input is required to account for the traveling wave effect and local site effect in ground motion. However, for medium- and short-span bridges, it is permissible to assume uniform ground motion at all supports. For the 32-m simply supported bridge structure under consideration, uniform ground acceleration input is adopted for seismic loading due to its relatively short span.

6.4.3.2 Seismic Response of Bridge–Track Structure

Seismic dynamic time–history analysis of an 8 m-high simply supported girder bridge has been performed, accounting for bridge–track interactions and train-bridge coupling effects, with three ground motions (type II site) input at PGA levels of 0.05, 0.1, and 0.15 g. The midspan rail acceleration responses are extracted as shown in Fig. 6.34. When the PGA is 0.15 g, the peak accelerations of II-W1 to II-W3 were 3.76 m/s^2, 3.832 m/s^2, and 3.814 m/s^2, respectively.

Table 6.9 Ground motion records selected.

Ground Motion Number	Earthquake Name	Seismic Station	Richter Scale	PGA/g	Fault Distance/km	Shear Wave Velocity/($m\cdot s^{-1}$)
I -W1	Iwate, Japan	YMT017	6.9	0.09	36.73	410.57
I -W2	Chi-Chi, Taiwan-03, China	TCU073	6.2	0.034	20.91	473.65
I -W3	L'Aquila, Italy	Lab. Gran Sasso	6.3	0.021	11.15	547
II -W1	Landers	Mission Creek fault	7 28	0.0597	26.96	355.42
II -W2	Christchurch	TPLC (Templeton School seismic station in New Zealand)	6.2	0.1256	16.61	249.28
II -W3	Parkfield	Cholame Shandon Array #12	6.19	0.1262	17.64	408.93
III-W1	Imperial Valley 06	El Centro Array #12	6.53	0.1449	17.94	196.88
III-W2	Chuetsu Oki, Japan	NIG013	6.8	0.1080	29.8	174.55
III-W3	Superstition Hills 01	Imperial Valley Wildlife Liquefaction Array	6.22	0.1308	17.59	179
IV-W1	Tottori, Japan	SMN002	6.61	0.1096	16.61	138.76
IV-W2	Darfield, New Zealand	Christchurch Resthaven	7	0.1795	19.48	141
IV-W3	Imperial Valley 06	Niland Fire Station	6.53	0.2628	36.92	212

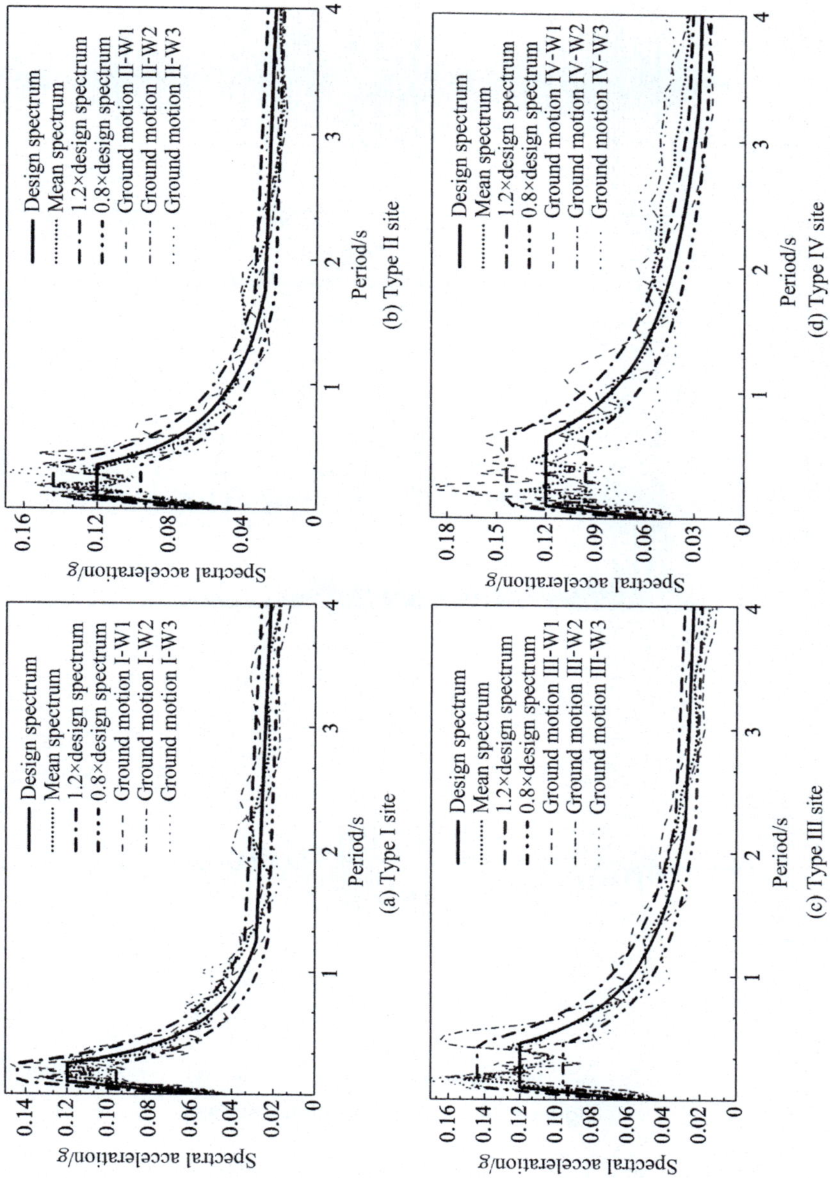

Figure 6.33 Acceleration response spectrum of the selected ground motion

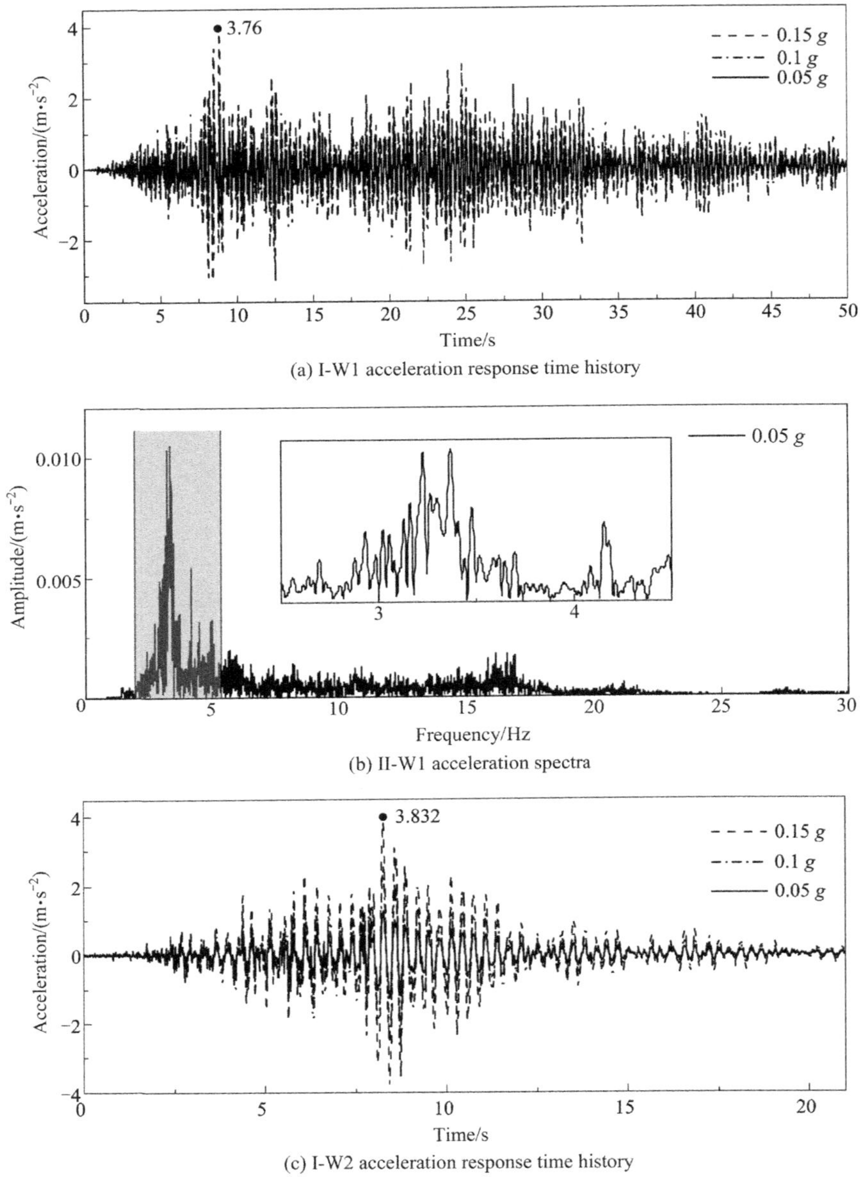

Figure 6.34 Track acceleration response

The frequency components of structural responses under the three ground motions, as shown in Fig. 6.34, are primarily concentrated around 3 Hz. This is consistent with the fundamental frequency of the structure

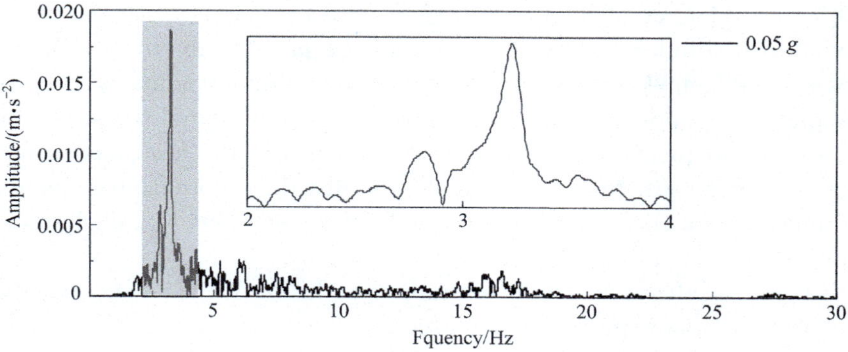

(d) II-W2 acceleration spectra

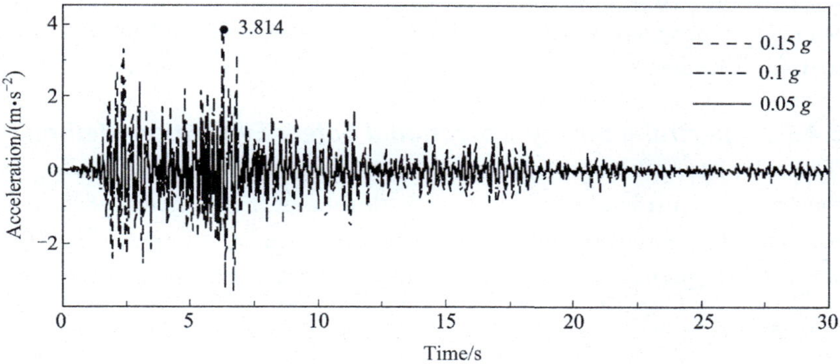

(e) I-W3 acceleration response time history

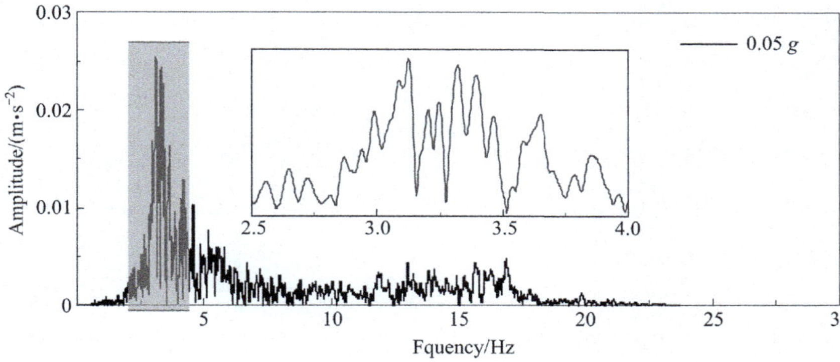

(f) II-W3 acceleration spectra

Figure 6.34 Cont'd

($f_e = 3.06$ Hz) and aligns with Wang Shaolin's conclusion that the lateral synchronous motion frequency range of the pier-girder system falls between 2.485–4.34 Hz [19]. Although the ground motion contains complex frequency components that significantly influence structural responses, the dominant frequencies of the responses remain near the fundamental frequency of the structure. For example, the first three dominant frequency components of II-W3 are 3.125 Hz, 3.319 Hz, and 3.396 Hz, all clustered around 3 Hz.

Based on the acceleration response time history of the structure, the velocity response spectra of the structure can be derived. The velocity response spectra of the three seismic motions under different damping ratios are shown in Fig. 6.35. By integrating the velocity spectra ($\zeta = 0.05$) over the period range of 0.1–2.5 s, the corresponding *SI* indices for the structural responses are 2960 mm, 3940 mm, and 3470 mm, respectively.

6.4.3.3 Spectral Intensity Operational Safety Evaluation Method

The calculated *SI* is compared with SI_L to evaluate the operational safety of the train, in principle at the first k dominant frequencies of structural response. Operational safety is deemed compliant only when $SI < SI_L\ (T_i)$ across all frequencies. The evaluation process, as illustrated in Fig. 6.36a, is

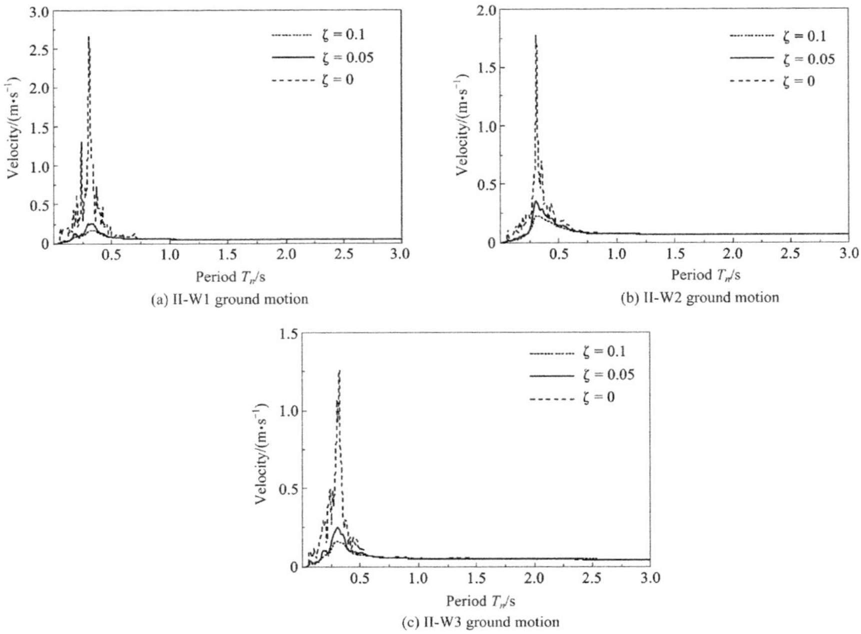

Figure 6.35 Velocity response spectra of the structure

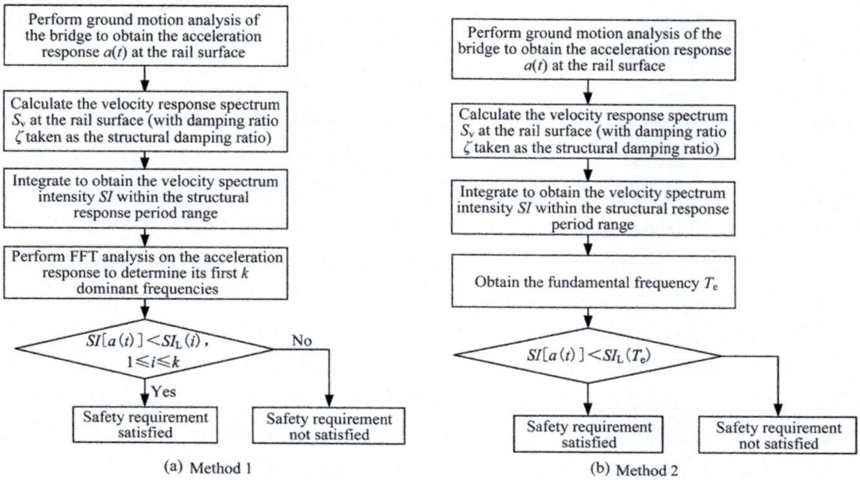

Figure 6.36 Flowchart for assessing operational safety based on spectral intensity indexes

relatively cumbersome because it requires calculating the top k-order frequencies for multiple comparisons. The structural response spectrum and velocity spectrum show that the dominant frequency components of the structural response are concentrated near the fundamental frequency of the structure. Therefore, it is sufficient to compare only the SI with the SI_L at the fundamental frequency. If $SI < SI_L$ (T_e), the safety of the structure is deemed to meet the requirements. The evaluation process is illustrated in Fig. 6.36b. The steps for evaluating the operational safety of the train on bridges during earthquakes using the SI are summarized as follows:

① Establish an FE model for the bridge-track structure, perform ground motion time-history analysis, and extract the acceleration response time-history $a(t)$ at the midspan rail surface of the bridge.

② Based on the acceleration time-history at the midspan rail surface of the bridge, calculate its velocity response spectrum S_v (ζ), with the damping ratio ζ taken as the structural damping ratio of the bridge. Integrate the velocity response spectrum over the frequency range of interest to obtain the SI index of the bridge-track structure under the given ground motion.

③ Perform Fast Fourier Transform (FFT) analysis on the structural response to determine its first k dominant frequencies, or conduct modal analysis to obtain the fundamental frequency T_e. Obtain the velocity spectrum SI_L by referring to Fig. 6.32, compare the SI index with the SI_L to evaluate train operational safety during earthquakes: If

$SI < SI_L$, the operational safety requirement is satisfied. If $SI > SI_L$, the operational safety requirement is not satisfied.

Additionally, grade train operational safety metrics (S1 and S2) to evaluate operational safety at different speeds. The *SI* value calculated from the structural response can be used to evaluate train operational safety without performing any train-bridge coupling vibration analysis under earthquakes. This approach requires only a structural seismic time-history analysis, which is easy for structural designers and holds significant practical value in engineering applications.

Train operational safety is evaluated under earthquakes of different magnitudes on type II sites based on the *SI* index, as shown in Fig. 6.37. When PGA = 0.05 *g*, the *SI* values of all three ground motions are below the S2 limit, indicating that the train running at medium-to-high speeds (⩾150 km/h) and still meets safety requirements. When PGA = 0.1 *g*, the *SI* indices of II-W1 and II-W2 fall between the S1 and S2 limits, indicating that safety requirements of the train are met at low but not at high speed. II-W3 lies on the S2 limit curve, indicating a critical stability state. At PGA = 0.15 *g*, the *SI* values of all three seismic waves fall between the S2

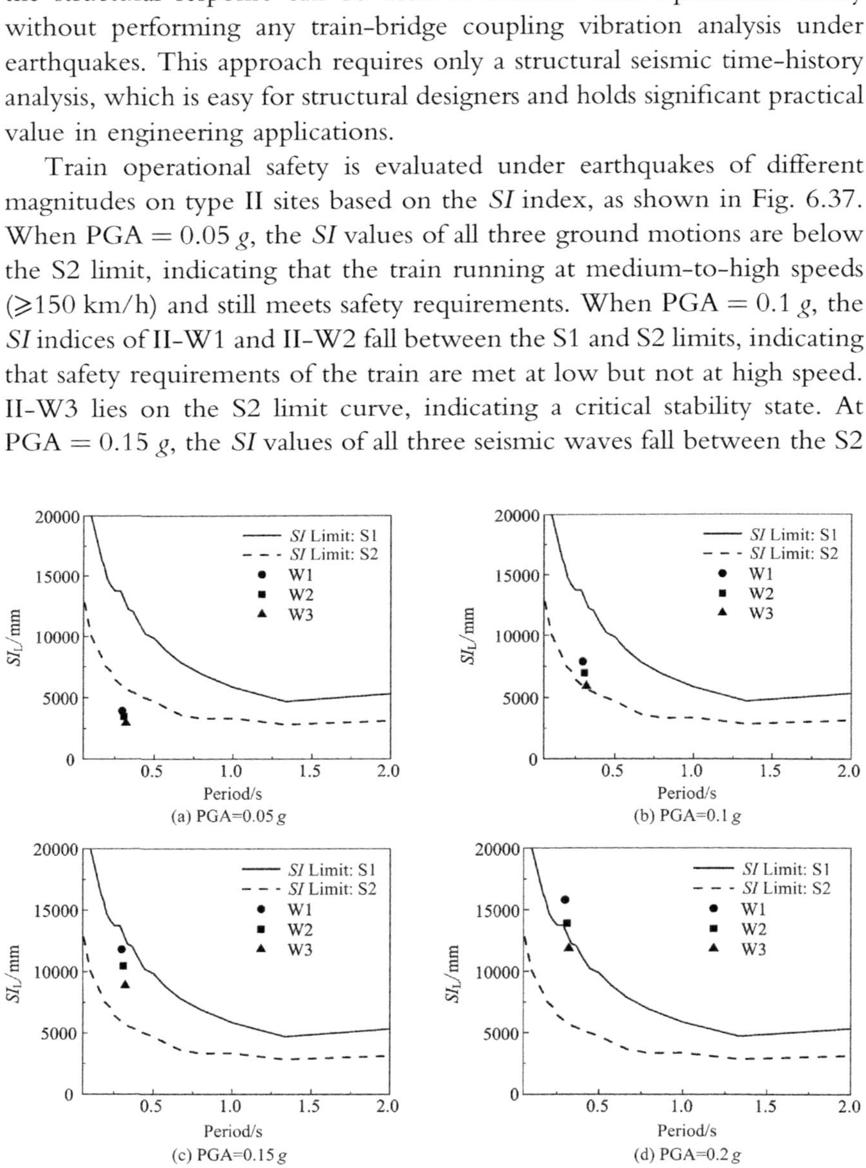

Figure 6.37 Evaluation of train operational safety during earthquakes (type II sites) based on the spectral intensity (*SI*) index

and S1 limits. At PGA = 0.2 *g*, the *SI* indices of II-W1 and II-W2 exceed the S1 limit, making low-speed train operation unsafe on the bridge.

6.4.3.4 Validation of the Spectral Intensity Evaluation Method

The method for evaluating train operational safety using *SI* metrics has been extensively verified by Luo et al. [4,6,7]. They compared *SI* evaluation results with the data from train-track coupling tests under seismic conditions and the numerical simulation results of train-track-bridge coupling, concluding that the experimental verification is the most reliable method. However, because of time constraints, the SI_L metrics and evaluation method are preliminarily validated through comparison with numerical simulation results of the train-track-bridge coupling system under seismic conditions. The derailment coefficients of the train-track-bridge system under earthquakes from Chapter 5 are compared with the evaluation results using the *SI* metrics, with the results shown in Table 6.10. The SI_L corresponding to the current working condition was 6418 mm, as shown in Fig. 6.32.

As shown in Table 6.10, for ground motions I-W1, II-W2, and IV-W1, the *SI* metrics and derailment coefficients yield consistent results in evaluating train operational safety. For ground motion II-W1, the *SI* evaluation method produced relatively conservative results, indicating train derailment at PGA = 0.1 *g*, while the actual derailment coefficient (0.6678) remains below the limit of 0.8. For ground motion I-W2, the evaluation results are consistent with derailment coefficients at PGA = 0.05 *g* and PGA = 0.1 *g*, and the evaluation indicates train derailment at PGA = 0.15 *g*, whereas the derailment coefficient remains below the limit. The results demonstrate that although the *SI* metrics may be conservative under certain working conditions, they generally provide an accurate measure of train operational safety under earthquakes.

Table 6.11 presents the derailment coefficients of high-speed trains on bridges with 8 m piers under the Kobe and EI Centro ground motions

Table 6.10 Comparison between spectral intensity (*SI*) metrics and derailment coefficients during earthquakes, calculated in the previous chapter.

	PGA = 0.05 *g*		PGA = 0.1 *g*		PGA = 0.15 *g*	
Ground Motionpt	**Derailment Coefficients**	***SI*/mm**	**Derailment Coefficients**	***SI*/mm**	**Derailment Coefficients**	***SI*/mm**
I-W1	0.5413	2680	0.6917	5350	0.8034	7998
I-W2	0.5613	2850	0.6317	5710	0.7094	8560
II-W1	0.5181	3940	0.6678	7880	0.8186	11800
II-W2	0.5846	3470	0.8373	6960	1.196	10500
IV-W1	0.6252	3290	0.9856	6570	1.318	9870

Table 6.11 Comparison of spectral intensity metrics in this section with the train derailment coefficients [1] during earthquakes from the literature.

	PGA = 0.05 *g*		PGA = 0.1 *g*		PGA = 0.15 *g*	
Ground Motion	Derailment Coefficients	*SI*/mm	Derailment Coefficients	*SI*/mm	Derailment Coefficients	*SI*/mm
Kobe	0.514	2936	0.980	5874	1.36	8811
EI Centro	0.544	3666	0.985	7341	1.45	11037

obtained by Lin Yusen [1]. It can be observed that at PGA = 0.05 *g*, all derailment coefficients remain below the limit (0.8). At PGA = 0.1 *g*, the derailment coefficients exceed the 0.8 limit, indicating that train operational safety requirements are not satisfied. The *SI* metrics calculated in this section are also listed in Table 6.11. At a speed of 150 km/h, the fundamental frequency of the 8 m pier simply supported girder bridge is determined using Fig. 6.32, yielding an SI_L of 5784.49 mm during high-speed operation. At PGA = 0.05 *g*, the *SI* values for both Kobe and EI Centro seismic waves remain below the limit, indicating that train operational safety requirements are satisfied. At PGA = 0.1 *g*, the *SI* values exceed the limit, indicating that train operational safety requirements are not satisfied. These conclusions are consistent with the train-bridge coupling simulation results. Therefore, by comparing the *SI* evaluation results with the operational safety evaluation results documented in the literature, the accuracy of the *SI* metrics is preliminarily verified. These metrics can then be applied to evaluate train operational safety.

6.5 Summary

This chapter starts with a comparative analysis of the seismic responses of the train-track-bridge system under seismic and non-seismic conditions and their impact on the operational safety of high-speed trains. Subsequently, the operational safety of the CRH380A EMU traversing the Zhengzhou Yellow River Grand Bridge on the Zhengzhou-Jinan Railway during earthquakes is investigated. An OpenSees FE model for a train-ballastless track-continuous steel truss bridge is established. Using the corrected wheel load reduction rate and derailment coefficient limits as evaluation metrics under seismic action, the study examines the operational safety of high-speed trains on bridges under varying seismic intensities and at different running speeds. Finally, the train–track model is utilized to calculate the SI_L for the evaluation of train operational safety on the high-speed railway under seismic conditions at varying speeds. A practical graded evaluation method for train operational safety under earthquakes was proposed and validated. By analyzing the relationships among the velocity spectra, the maximum potential energy, and the maximum kinetic energy, the study established the *SI*, and an energy-based evaluation criterion was established to quantify the relationship between work done by resistance and input energy. The process for solving SI_L includes Step 1: Establish the train–track model and determine the velocity range. Step 2: Compute safety displacement limit d_L^i under sinusoidal excitation with a period of T_{sin}^i. Repeat this process until T_{sine}^i falls below the upper limit of the period. Step 3: Compute acceleration safety limit A_L^i corresponding to the frequency based on displacement limit d_L^i to generate bounded sinusoidal wave excitations. Step 4: Calculate the velocity response spectrum and integrate it to obtain the SI_L. The main conclusions are:

(1) Seismic loads exert a significant impact on train operational safety. Under seismic excitation, the peak responses of the derailment coefficient and wheel load reduction rate of traction are generally lower than those of trailers owing to the additional mass from installed electromechanical equipment in the traction. Both the derailment coefficient and wheel load reduction rate generally increase with higher train speed and greater seismic intensity.

② Without considering the discreteness of earthquakes and the strong nonlinearity of the ballastless track system on long-span bridges, the analyses of three ground motions (Taft, El-Centro, and artificial ground motions) show that operational safety of trains at speeds not exceeding 285 km/h during a frequent earthquake intensity of 0.054 *g* (7 degree) on the Zhengzhou Yellow River extralarge steel truss bridge of the Zhengzhou-Jinan Railway is guaranteed. During design-level earthquake (0.151 *g*) and rare earthquake (0.248 *g*) excitation at seismic fortification intensity VII, the train speed must be reduced to below 180 and 120 km/h, respectively, to maintain operational safety.

③ Train operational safety and ride comfort metrics are positively correlated with the amplitude and frequency of sinusoidal excitation. Ride comfort metrics such as the Sperling coefficient and train acceleration are the first to reach their limits. At low excitation amplitudes, all operational safety metrics exhibit an approximately linear increase.

As the amplitude increases, the metrics grow more rapidly under high-frequency excitation, showing an approximately exponential growth trend. The lateral wheel-rail force is among the first to reach its limit.

④ Under high-frequency excitation, the acceleration limits for sinusoidal excitation and the SI_L gradually decrease with an increase in period. As the excitation frequency gradually decreases, the limits tend to stabilize. In the SI_L-based operational safety evaluation, as the dominant frequencies of the responses remain near the fundamental frequency of the structure, the structural responses can be directly compared with the SI_L at the fundamental frequency, eliminating the need for comparison with SI_L values at all frequencies.

⑤ Sinusoidal excitation analysis for trains at different speeds reveals that as the train speed increases, the displacement limits gradually decrease, with the rate of reduction slowing at higher speeds. Two evaluation metrics (S1 and S2) for low-speed and high-speed conditions are proposed to evaluate train operational safety at different speeds. The accuracy of the *SI* evaluation method and SI_L is validated by comparing the *SI* evaluation results with the evaluation results of operational safety under train–bridge coupling vibration.

References

[1] Lin YS. Study on train operations during an earthquake on a high-speed railway bridge. Chengdu: Southwest Jiaotong University; 2007 [In Chinese].

[2] Chen LK. Research on operation safety and vibration control of dynamic responses for the vehicle-bridge system under earthquake. Changsha: Central South University; 2012 [In Chinese].

[3] National Railway Administration. Code for design of high-speed railway (TB 10621-2014). Beijing: China Railway Publishing House; 2014 [In Chinese].

[4] Luo X. Study on methodology for running safety assessment of trains in seismic design of railway structures. Soil Dynam Earthq Eng 2005;25(2):79–91.

[5] Miyamoto T, Ishida H, Matsuo M. Running safety of railway vehicle as earthquake occurs. Quart Rep RTTI 1997;38(3). 117–122.

[6] Luo X, Miyamoto T. Method for running safety assessment of railway vehicles against structural vibration displacement during earthquakes. Quart Rep RTTI 2007;48(3):129–35.

[7] Luo X, Miyamoto T. Examining the adequacy of the spectral intensity index for running safety assessment of railway vehicles during earthquakes. Beijing: The 14th World Conference on Earthquake Engineering; 2008.

[8] Li JL. High-speed railway bridge operation simulation technology based on client-server architecture and operational safety analysis under earthquake. Changsha: Central South University; 2018 [In Chinese].

[9] Gong B, Kong B, Sun ZH, et al. Research progress on military emergency medical rescue evacuation system in earthquake disaster. Disast Med Rescue (Electr Ed) 2017;6 (1):25–7 [In Chinese].

[10] Xu AH. Research on the classification and grading of China's emergency rescue teams for sudden incidents: cost analysis. Beijing: Institute of Crustal Dynamics (ICD), China Earthquake Administration; 2018 [In Chinese].
[11] Xu JH, Deng D. Comparative analysis and study on rescue organization for major earthquake disaster in China. City Disast Reduct 2019;(2):55–61 [In Chinese].
[12] Code for seismic design of railway engineering (GB 50111—2006). Beijing: China Planning Press; 2009 [In Chinese].
[13] Specification of seismic design for highway engineering (JTG B2—2013). Beijing: China Communications Press; 2013 [In Chinese].
[14] Yao JC. Rating of operation performance for simply supported box beam with common span for high-speed railway. China Railw Sci 2017;38(2):34–41 [In Chinese].
[15] Zeng QY, Xiang J, Lou P, et al. Mechanical mechanism of derailment and theory of derailment prevention. J Railw Sci Eng 2014;1(1):19–31 [In Chinese].
[16] Xia H. Dynamic interaction of vehicle and structure. Beijing: Science Press; 2004 [In Chinese].
[17] Zhai WM. Vehicle-track coupled dynamics. 4th ed. Beijing: Science Press; 2015 [In Chinese].
[18] Nishimura K, Terumichi Y, Morimura T, et al. Experimental study on the vehicle safety by earthquake track excitation with 1/10 scale vehicle and roller rig. J Sys Design Dynam 2010;4(1):226–38.
[19] Wang SL. Coupled vibration of high-speed train-track-bridge under earthquake and its driving safety analysis. Chengdu: Southwest Jiaotong University; 2013 [In Chinese].

CHAPTER 7

Seismic Mitigation and Control Technologies for High-Speed Railway Train–Track–Bridge Systems

Contents

High-Speed Railway Train–Track–Bridge Systems
ISBN 978-0-443-49310-2
https://doi.org/10.1016/B978-0-443-49310-2.00001-6

7.1 Overview

High-speed railway bridges confront severe potential seismic threats in China, a country prone to earthquakes. Under seismic loads, the bridge–track system for high-speed railways inevitably undergoes structural damage, including stiffness degradation and residual deformation. Such critical component damage propagates to rails through interlayer interactions, inducing track geometry irregularity and stiffness irregularity. Such conditions affect the dynamic performance of the train–track–bridge system, potentially endangering operational safety at severe levels [1].

Extensive investigations and research into earthquake-induced bridge failure over recent decades have driven rapid advancements in bridge seismic mitigation technologies. Research by Ken Yeh demonstrates that seismic mitigation technologies primarily utilize vibration control components (e.g., damped braces) or damping devices (e.g., dampers), which undergo plastic deformation or provide supplemental damping during seismic events to dissipate input energy, thereby reducing structural response and controlling structural deformation [2]. Building upon existing research, this study proposes the integration of rotational friction dampers (RFDs), which combine energy-dissipating restraining bearings, and tuned mass dampers (TMDs), into bridge structures as seismic mitigation components, and the seismic mitigation performance of the three different mitigation components is verified through experiments.

Beyond proven seismic mitigation technologies, exploring simple yet effective seismic design methods holds equal significance for ensuring the safety of train–track–bridge systems under earthquake excitations. China's seismic design for bridges has been based on extensive lessons learned from past earthquake disasters. Its shift from early adoption of building-oriented seismic standards to the current performance-based design approach that is integrated into modern standards demonstrates the advancement of seismic design. However, under the current standards, performance-based design is carried out following the philosophy of "design for minor earthquakes and verification for major earthquakes." According to this philosophy, elastic design is adopted for structures under minor earthquakes, while the seismic performance of structures under moderate and major earthquakes only serves as a verification metric, not as a performance-based design metric. Therefore, China's seismic design philosophy for railway bridges remains outdated, and no systematic performance-based design method for high-speed railway bridges has been established. Moreover, in recent years, seismic isolation and energy dissipation technologies for high-speed railway bridges have advanced rapidly. It is imperative to integrate the developed isolation and damping devices with advanced design philosophies to further enhance the seismic performance of high-speed railway bridges. This book establishes an equivalent energy-based design procedure (EEDP) method for high-speed railway bridges based on the energy balance principle and through theoretical derivation, and validates the effectiveness of the EEDP method in practical engineering applications with case studies.

7.2 RFD-Based Seismic Mitigation Technologies for Track–Bridge Systems

Structural vibration control is achieved by rationally installing energy dissipation devices in structures to reduce their dynamic responses under external excitations. Structural vibration control methods are classified into passive control, active control, semiactive control, and hybrid control [3]. First developed in the 1970s, friction dampers are designed for passive control. To accommodate applications in different structures, researchers at home and abroad have developed various types of friction dampers. In 1982, Pall et al. [4] developed a sliding friction device specifically for steel frame structures, which generates sliding motion during earthquakes to mechanically dissipate most seismic energy through frictional resistance, thereby protecting the primary structure from yielding. In 1993, Grigorian et al. [5] developed slotted bolted connections that utilize friction between steel and copper plates to achieve energy dissipation during tension–compression deformation. Mualla and Belev [6,7] developed the RFD installed at the brace locations of steel frames. By increasing the number of steel plates and friction pads, the sliding force and energy dissipation capacity of the damper can be effectively enhanced. In energy dissipation devices, friction dampers are characterized by low maintenance costs, ease of installation and replacement, and a straightforward and well-defined energy dissipation mechanism. Using the RFD as an energy-dissipating component, our research team investigated its mechanical properties and evaluated its seismic mitigation performance in the bridge–track system of high-speed railways through quasi-dynamic hybrid simulation [8–10].

7.2.1 Fundamental Principle of RFDs

An RFD is assembled by connecting two fundamental rotational friction nodes through high-strength bolts. The fundamental energy dissipation principle of the RFD involves the combination of metal or other solid friction materials to enable relative motion between damper components to generate a friction force, thereby converting mechanical energy into heat through frictional work. As illustrated in Fig. 7.1, the RFD assembles multiple steel plates and friction plates into a rotational friction node using pretensioned bolts, and applies preload to the bolts to pressurize the friction interfaces between steel and friction plates, thereby generating rotational friction. The rotational friction node of RFD requires three machined long steel plates and two friction plates. A friction plate is inserted between every two steel plates, with bolt preload ensuring uniform pressure distribution across the friction interfaces. One end of the steel plates forms a friction node, while the other end is equipped with a hinge connection to the main structure. Relative displacement between the two hinge connections induces friction at the rotational node, enabling energy dissipation. The uniform pressure on friction surfaces is adjustable by bolt-tightening torque, and Belleville spring washers are installed at bolt connections to maintain a stable bolt preload.

First, the mechanical behavior of RFD sliding friction is mathematically derived. As the RFD follows Coulomb's friction law [11], the working

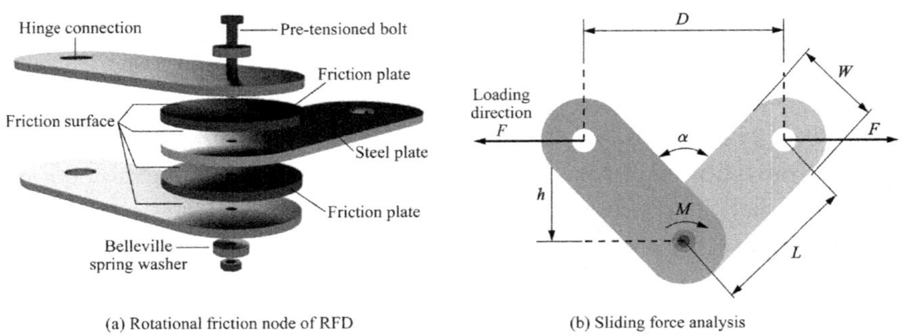

Figure 7.1 Construction and force analysis of RFD.

state of the RFD can be divided into two stages as described here. (1) Sticking stage. When the applied external force is below the sliding force (maximum static friction force) of the RFD, no relative motion occurs between friction interfaces, and the reaction force of the damper maintains equal magnitude and opposite direction to the applied external force. (2) Sliding stage. When the applied external force is above the sliding force of the RFD, relative motion occurs between friction interfaces. Therefore, the sliding force serves as the key metric of the RFD. The fundamental parameters of the RFD are defined as follows: the effective length of the steel plate L (distance between connection holes at both ends of the plate), the width of the steel plate W, the angle between adjacent plates α, the distance between hinge connections D, the frictional moment M, and the thickness of the steel plate and friction plate t.

The sliding force can be derived through structural force analysis. The force equilibrium of the RFD is expressed by the following equation, where n represents the number of friction interfaces at the rotational friction node.

$$n \cdot M = F \cdot h \tag{7.1}$$

where $h = \sqrt{L^2 - (D/2)^2}$; and F denotes the applied external force.

According to Coulomb's friction law, the frictional moment is proportional to the normal pressure acting on the friction surface, and the frictional moment on the friction interface can be derived through integration over the contact area. The sliding force F_{slide} of the RFD is ultimately expressed as:

$$F_{\text{slide}} = \frac{2}{3} \cdot \frac{n \cdot \mu \cdot P_{\text{bolt}}}{L \cdot \cos\left(\frac{\alpha}{2}\right)} \cdot \frac{R_2^3 - R_1^3}{R_2^2 - R_1^2} \tag{7.2}$$

where R_2 denotes the outer diameter of the friction surface; R_1 represents the inner diameter of the friction surface; μ indicates the friction coefficient of the friction surface; P_{bolt} corresponds to the vertical force of the pretensioned bolt; and α denotes the angle between adjacent steel plates.

When the RFD is in the sliding stage, the force–displacement relationship of the RFD can be derived. Fig. 7.2 illustrates the geometric relationship of RFD motion, which can be used to derive the geometric relationship between the rotation angle and horizontal displacement. According to the principle of virtual work, the work done by the horizontal

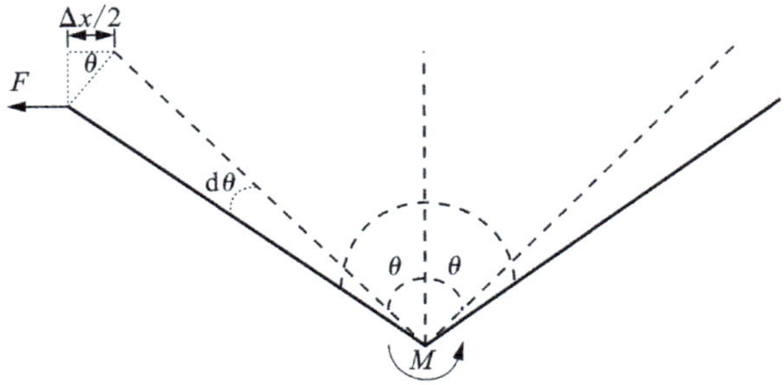

Figure 7.2 Geometric relationship analysis of RFD motion. θ, half of the rotation angle at the hinge connection; Δx, horizontal displacement.

force at the hinge connection equals the work done by the frictional moment at the rotational friction node, from which the force–displacement relationship of the RFD can be derived as follows.

7.2.2 Mechanical Properties of RFD

To determine the actual mechanical properties of RFD, mechanical property tests are conducted on RFD. The RFD specimen studied is shown in Fig. 7.3, with the steel plates in the RFD specimen machined from Q235 steel. Specific parameters of the RFD specimen are shown in Table 7.1. The friction plates in the RFD specimen are cut from copper plates into a circular shape and installed between two steel plates.

Research by Bijwe, Österle, and Bonfanti et al. indicates that in the RFD specimen, there exists a frictional contact interface between copper plates and steel plates. Due to the high thermal friction stability of the metallic friction materials used in the RFD and the enhanced thermal conductivity provided by the ductility of metallic materials, the mechanical stability of the friction material is ensured during prolonged operation [12–14].

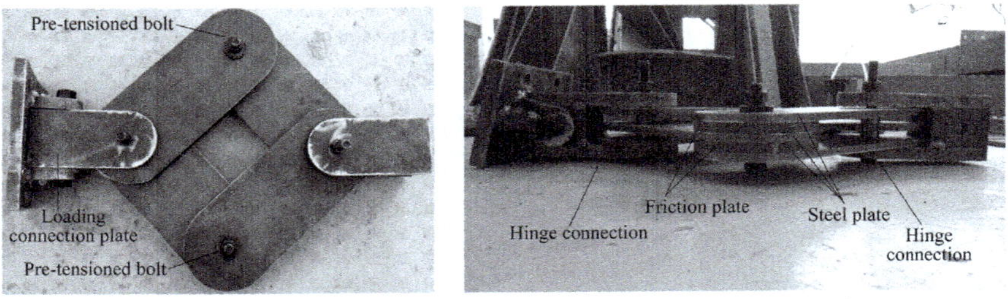

(a) Top view of a single-node rotational friction damper (b) Front view of a single-node rotational friction damper

Figure 7.3 RFD specimen.

Table 7.1 Parameters of the RFD specimen.

Distance Between Hinge Connections *D* (mm)	Width of the Steel Plate *W* (mm)	Effective Length of the Steel Plate *L* (mm)	Thickness of Steel Plate and Friction Plate *t* (mm)	Angle Between Adjacent Plates α (°)
353.5	150	250	12	90

The testing of the RFD specimen was conducted on a unidirectional shake table at Central South University. The unidirectional shake table is equipped with a hydraulic actuator produced by SERVOTEST, configured with a MOOG hydraulic servo valve. It can provide a maximum dynamic thrust of 72 kN, with a working stroke of 250 mm. The test setup is shown in Fig. 7.4. One end of the RFD specimen is hinged to the actuator via a connecting component, while the other end is hinged to the reaction base. A linear variable differential transformer (LVDT) is externally mounted between the actuator and the specimen to precisely measure the deformation of the damper. Additionally, a force sensor with a capacity of 50 kN is installed between the actuator and the specimen to measure the reaction force of the damper. Cyclic loading is applied to the RFD specimen to evaluate its hysteretic behavior and mechanical properties. The primary objective of the test is to characterize the hysteretic behavior of the RFD specimen under varying parameters (bolt tightening torque and loading frequency).

Three torque-specific bolts (50, 100, and 150 Nm) are used for testing. The hysteresis curves under different bolt preload torques are shown in Fig. 7.5. As shown in Fig. 7.5, the operation of the RFD can be divided into two stages: the sticking stage and the sliding stage. During the sticking stage, the RFD exhibits high initial stiffness, with its reaction force

Figure 7.4 Test setup.

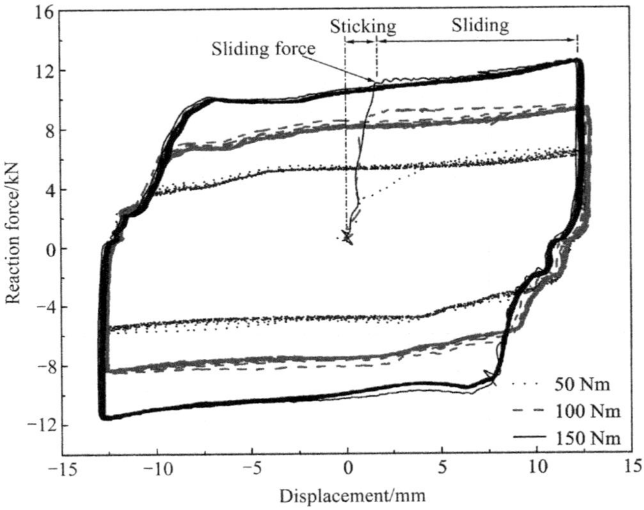

Figure 7.5 Hysteresis curves under different bolt preload torques.

increasing rapidly as deformation grows. This stage corresponds to static friction occurring at the rotational friction node. When the external force exceeds the sliding force, the sliding stage begins, where relative sliding occurs between the steel plate and the friction plate of the RFD, causing the reaction force to increase gradually with displacement. The RFD exhibits distinct two-stage stiffness characteristics: when deformation is small, the RFD maintains linear elastic behavior with high initial stiffness, which can enhance the local stiffness of the structure. When deformation is great, it transitions into the sliding stage, forming stable and plump hysteresis curves.

To validate the loading frequency dependence, another critical factor affecting the RFD, harmonic waves with primary frequencies of 2, 5, 10, and 15 Hz are applied to the RFD specimen according to the loading protocol. All RFD parameters remain consistent across different frequency tests. The test results indicate that loading frequency has a minimal impact on the hysteretic performance of the RFD specimen. As shown in Fig. 7.6, under different loading conditions, the hysteresis curves maintain consistent shapes, all exhibiting symmetrical bilinear characteristics. However, as the loading frequency increases, the hysteresis curves exhibit a slight lateral shift, which is particularly noticeable in Fig. 7.6D. This phenomenon is attributed to repeated impacts between the connecting components within

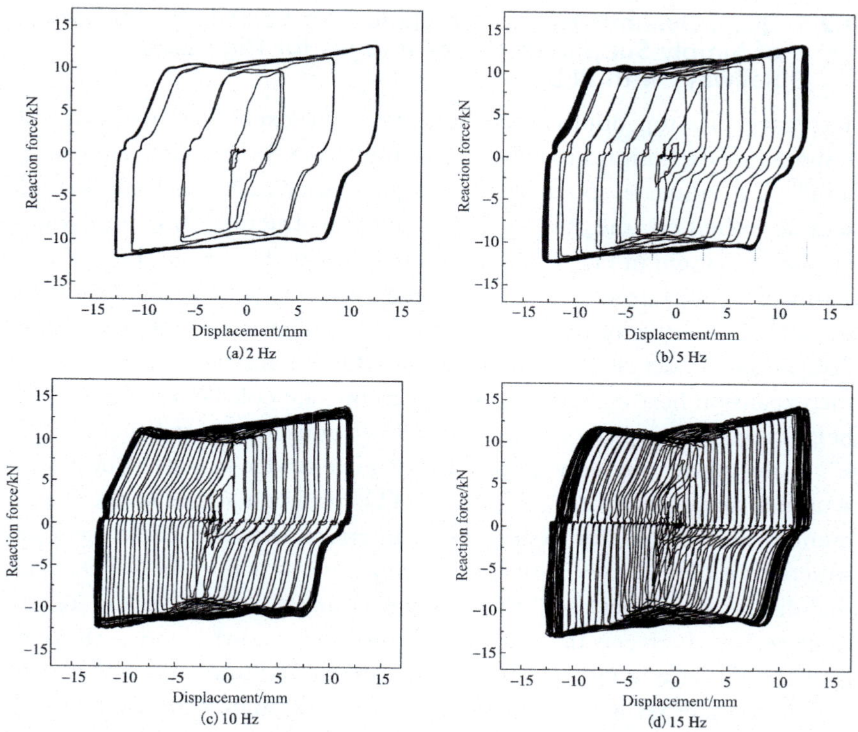

Figure 7.6 Hysteresis curves under varying loading frequencies.

the RFD specimen under high-frequency loading cycles. The inevitable gaps in these connecting components cause a certain degree of displacement during the transition between positive and negative loading stages.

The experimental results demonstrate that the RFD exhibits plump hysteresis curves with bilinear characteristics. By adjusting the bolt tightening torque at the rotational friction nodes, the sliding force, peak resistance, and energy dissipation capacity of the RFD can be effectively tuned. This indicates that the mechanical properties of the RFD can be designed by rapidly modifying the tightening torque of pretensioned bolts. The test results on the RFD loading frequency demonstrate that the RFD is a displacement-dependent damper, whose mechanical properties are almost unaffected by the loading frequency.

7.2.3 Quasi-Dynamic Hybrid Simulation for Seismic Performance of Simply Supported Girder Bridges for High-Speed Railways with RFD

Excessive pier-top displacement under seismic excitation may cause the response of expansion bearings at pier tops to exceed design deformation limits. To mitigate excessive pier-top displacement of bridges under seismic excitation, the RFD is employed as a structural vibration control device. As shown in Fig. 7.7A, the RFD is installed between the top of the 16-m pier and Girder #2 via a hinged connection. The relative deformation between them induces a frictional moment in the RFD for energy dissipation. As detailed in Fig. 7.7B, the RFD is installed in parallel with the expansion bearing at the 16-m pier to operate collaboratively, with the height adjustment plate serving solely as a connecting component.

In this study, the high-speed railway track–bridge system is modeled as a numerical substructure, while the RFD specimen serves as the experimental substructure for testing. The interface between these two substructures is defined by the two hinge connections within the RFD, where the displacement between these hinge connections equals the relative displacement between the top of the 16-m pier and the underside of the girder. The relative displacement at the partitioned interface of the bridge structure under seismic excitation is obtained through finite element (FE)

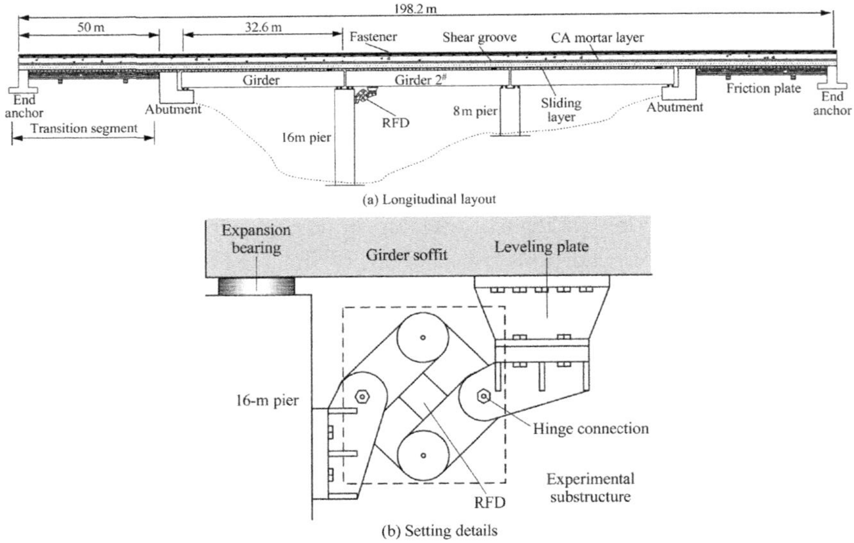

Figure 7.7 Layout of RFD in simply supported girder bridges for high-speed railways.

analysis, which is used as the command signal input to the actuator. The actuator drives the motion of the experimental substructure RFD, and the reaction force generated by the RFD specimen during its motion is fed back to the partitioned interface of the numerical substructure via a force sensor. This process iterates through each time step until the entire loading time history is completed. The interaction between the RFD specimen and the high-speed railway bridge model is achieved through data exchange between the two substructures.

To enhance the computational efficiency of the numerical substructure in hybrid simulation, it is necessary to simplify the original track–bridge interaction numerical model to improve the computational efficiency of the hybrid simulation. The substructure of the bridge is removed from the model, accounting solely for the effect of the track structure on the girder and applying cyclic reciprocating loading to the girder to obtain the force–displacement relationship of the track–bridge interaction. During loading, the bearings of the girder should be replaced with rolling bearings, which allow longitudinal displacement of the girder, while the remaining bearings are converted into hinge bearings, permitting only rotational movement at the girder ends. During loading, the external force–displacement relationship of the girder can be obtained. The force–displacement curve reflects the constraint effect of the track slab structure on the girder. Similarly, the constraint effect of the track slab structure for each span can be obtained by applying the same process to the girders of other bridge spans. The resulting force–displacement curve exhibits consistent patterns, as shown in Fig. 7.8. The curve exhibits four distinct performance points, each corresponding to different structural states of different components within the track slab system. Before the inflection point A, both the sliding layer and shear groove remain in a linear elastic state. After the inflection point A, the sliding layer yields and exhibits nonlinear behavior, while the shear groove remains linearly elastic. Upon reaching the inflection point C, the shear groove begins to yield, with internal reinforcement starting to slip. After the deformation at the inflection point D exceeds 12.6 mm, the reinforcement shears off, leading to the failure of the shear groove. After this, only the friction from the sliding layer within the bridge span contributes to resistance. Therefore, the constraint effect of the track slab structure on adjacent bridge members can be simplified as a nonlinear spring, which can be considered as a combined system composed of the sliding layer within a single bridge span and the shear groove. Through this approach, the complex track–bridge

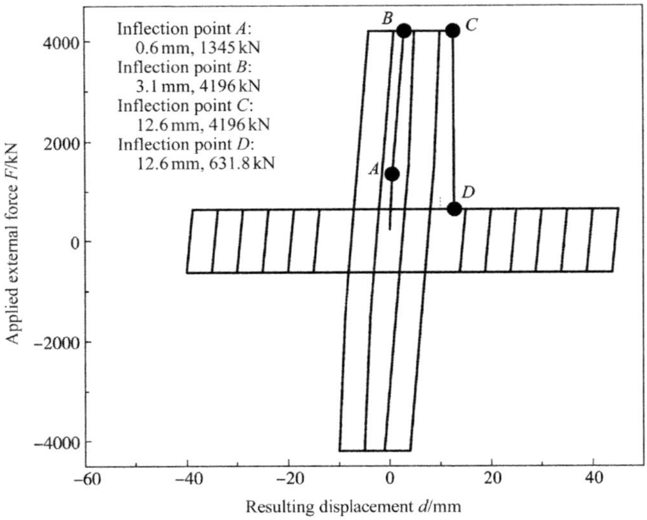

Figure 7.8 Force–displacement relationship in track–bridge interaction.

interaction can be simplified into a nonlinear track spring, and the accuracy of this simplified model can be validated. The simplification method is illustrated in Fig. 7.9. Finally, the seismic mitigation performance of the RFD in a three-span simply supported high-speed railway bridge is investigated through the quasi-dynamic hybrid simulation.

To carry out the quasi-dynamic hybrid simulation, three ground motion records are selected from the PEER database according to the criterion that a 2% probability of exceedance in 50 years corresponds to a rare ground motion. The three ground motions are scaled as required to achieve the peak acceleration corresponding to rare ground motions. They are referred to as GM1, GM2, and GM3, with their parameters shown in Fig. 7.10.

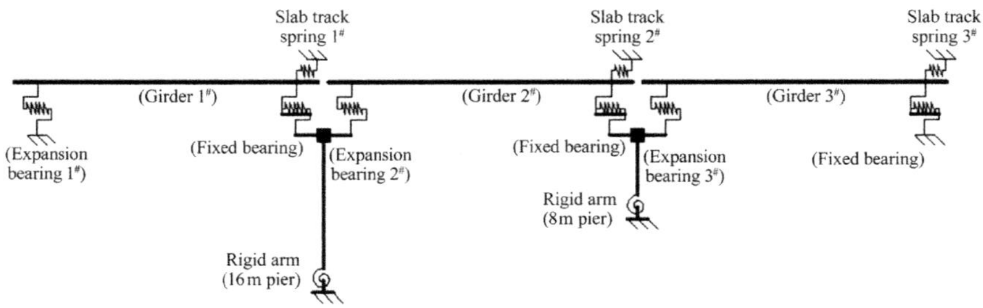

Figure 7.9 Simplified model for hybrid simulation.

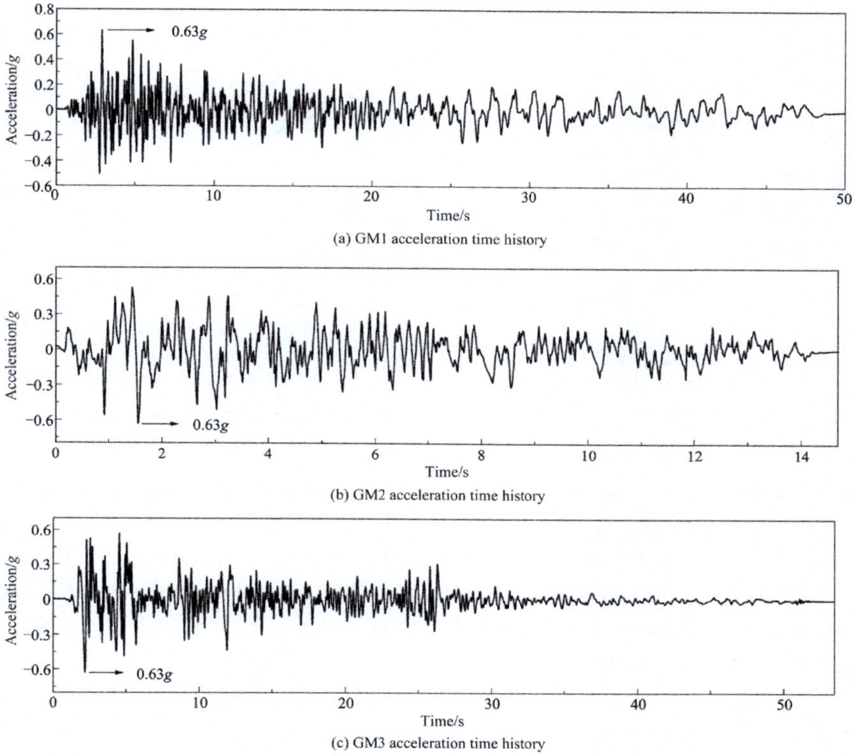

Figure 7.10 Selected ground motion parameters.

Due to the flexibility of the hybrid simulation method, the response of the RFD specimen can be multiplied by a corresponding positive integer coefficient before simulation. This represents the collaborative operation of multiple RFD specimens in a bridge structure, forming a multijoint rotational friction damper (MRFD). During the subsequent hybrid simulation under the three selected seismic motions, the effect of the sliding force of the MRFD on the vibration control effectiveness of simply supported girder bridges for high-speed railways is investigated. The sliding force values are set at 450, 900, 1350, 1800, and 2250 kN, respectively. The sliding force is determined based on its ratio to the gravity load borne at each bearing of the simply supported girder, with ratios set at 10%, 20%, 30%, 40%, and 50%, respectively. The experimental results are compared with the bridge response without MRFD (i.e., under zero sliding force).

The results of quasi-dynamic hybrid simulations with different sliding forces are presented in Fig. 7.11. As shown in Fig. 7.11A, the peak pier-top

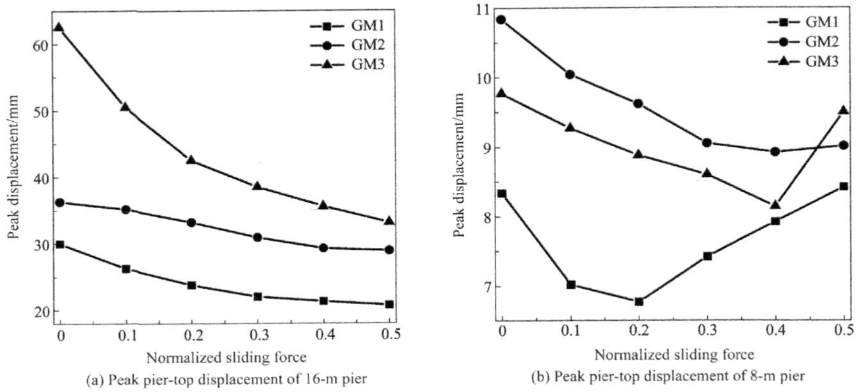

Figure 7.11 Pier-top displacement response under seismic excitation.

displacement response of the 16-m pier continuously decreases as the sliding force of the MRFD increases. Under GM3 ground motion, the peak pier-top displacement response of the 16-m pier decreases by 30 mm, while under ground motions GM1 and GM2, it only decreases by 5 and 10 mm, respectively. Fig. 7.11B presents the variation trend of the peak pier-top displacement response of the 8 m pier. As the experimental data show, the peak response undergoes a V-shaped trend as the sliding force increases. This indicates that, in addition to its excellent energy dissipation capability, the MRFD can also transfer a portion of the seismic force from the 16-m pier to the 8-m pier through the girder. As a result, the response of the 8-m pier increases under high sliding forces of the MRFD.

The peak responses of the expansion bearings and the simplified track slab springs under seismic action are shown in Fig. 7.12. Generally, the greater the sliding force of the MRFD, the smaller the peak responses of both the expansion bearings and the simplified track slab springs. However, at the 8-m pier, both Expansion Bearing $3^{\#}$ and Track Slab Spring $2^{\#}$ exhibit increased responses when the sliding force increases. The response variation patterns of Expansion Bearing $1^{\#}$ and Expansion Bearing $2^{\#}$ are consistent with those of the 16-m pier, with the peak reductions within the same range. The similarity in variation patterns can be attributed to the relationships between components. Expansion Bearing $1^{\#}$ is connected to the 16-m pier via Girder $1^{\#}$, while Expansion Bearing $2^{\#}$ is installed directly at the top of the 16-m pier. Consequently, the response of the pier inherently correlates with the responses of Expansion Bearing $1^{\#}$ and Expansion Bearing $2^{\#}$. For the same reason, the variation pattern of

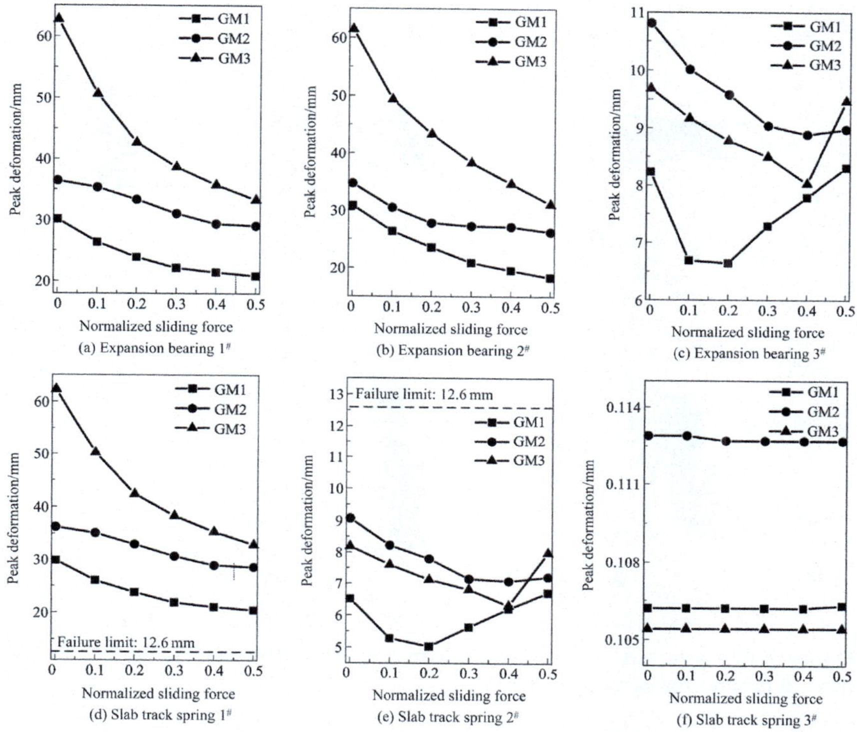

Figure 7.12 Peak responses of expansion bearings and simplified track slab springs under seismic action.

Expansion Bearing $3^{\#}$ aligns with that of the 8-m pier. This demonstrates that controlling the response of bridge piers is the key to governing the structural response of the bridge. To analyze the impact of different damper sliding forces on slab track springs, we compare the responses of three slab track spring locations, as shown in Fig. 7.12D and F. For Slab Track Spring $1^{\#}$, its peak deformation decreases as the MRFD sliding force increases. Although the application of MRFD reduced its deformation response to some extent, excessive deformation of Slab track spring $1^{\#}$ is observed in all tests, exceeding the failure limit of the shear groove. For Slab Track Spring $2^{\#}$, its deformation remains below the failure limit. However, an increase in deformation is observed. For Slab track spring $3^{\#}$, its deformation is minimal due to the fixed connection between Girder $3^{\#}$ and the abutment. It is evident that the responses of both the expansion bearing and slab track springs are directly correlated with the pier-top displacements of the pier.

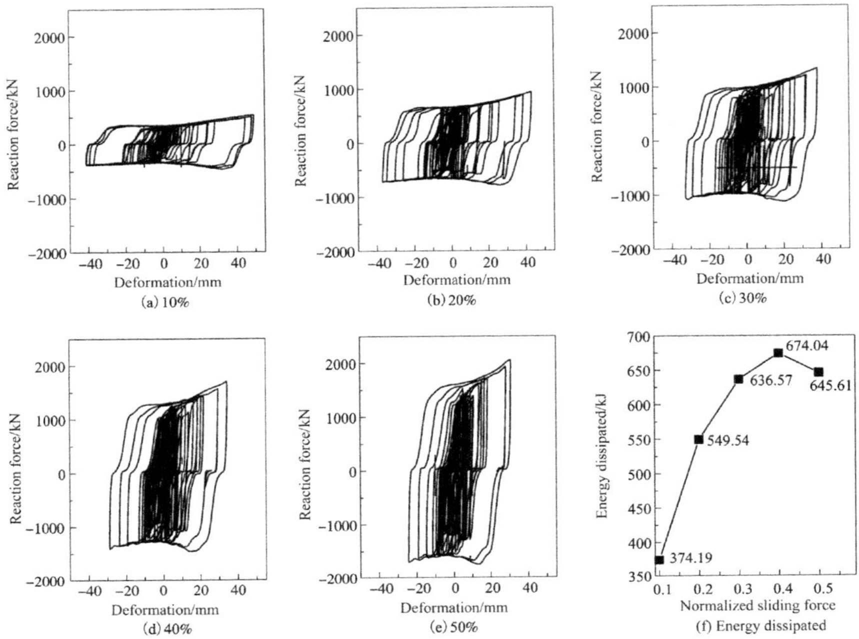

Figure 7.13 Hysteresis curves and dissipated energy of the MRFD under the GM3 ground motion for varying preload forces.

The results of the quasi-dynamic hybrid simulation under GM3 ground motion are specifically discussed to further analyze the seismic mitigation performance of MRFD in high-speed railway track–bridge systems. Fig. 7.13 presents the hysteresis curves of the MRFD under seismic excitation, along with the total energy dissipated during the ground motion. The deformation of the MRFD primarily falls within ± 20 mm. As shown in Fig. 7.13, as the sliding force increases, the hysteresis curves gradually elevate vertically and narrow horizontally. During the gradual increase of the sliding force, the pier-top displacement of the 16-m pier is progressively controlled. The total energy dissipated by the MRFD generally increases with higher sliding forces, but a decline in the total energy dissipated is observed when the normalized sliding force is 0.5. As a displacement-dependent damper, the MRFD features an energy dissipation capacity directly related to the relative deformation at its loading point. Under small displacements, the MRFD cannot achieve optimal energy dissipation performance. Therefore, the MRFD should be installed at locations with large relative deformations within the structure, and its sliding force should be adjusted based on specific structural configurations. In summary, the input energy dissipation is generally enhanced by increasing the sliding force of MRFD. However, an excessive sliding force reduces damper deformation, thereby decreasing energy dissipation.

An analysis is conducted for the MRFD with a normalized sliding force of 0.5. As shown in Fig. 7.14A, with the use of the MRFD, the pier-top displacement of the 16-m pier is effectively controlled, with significant

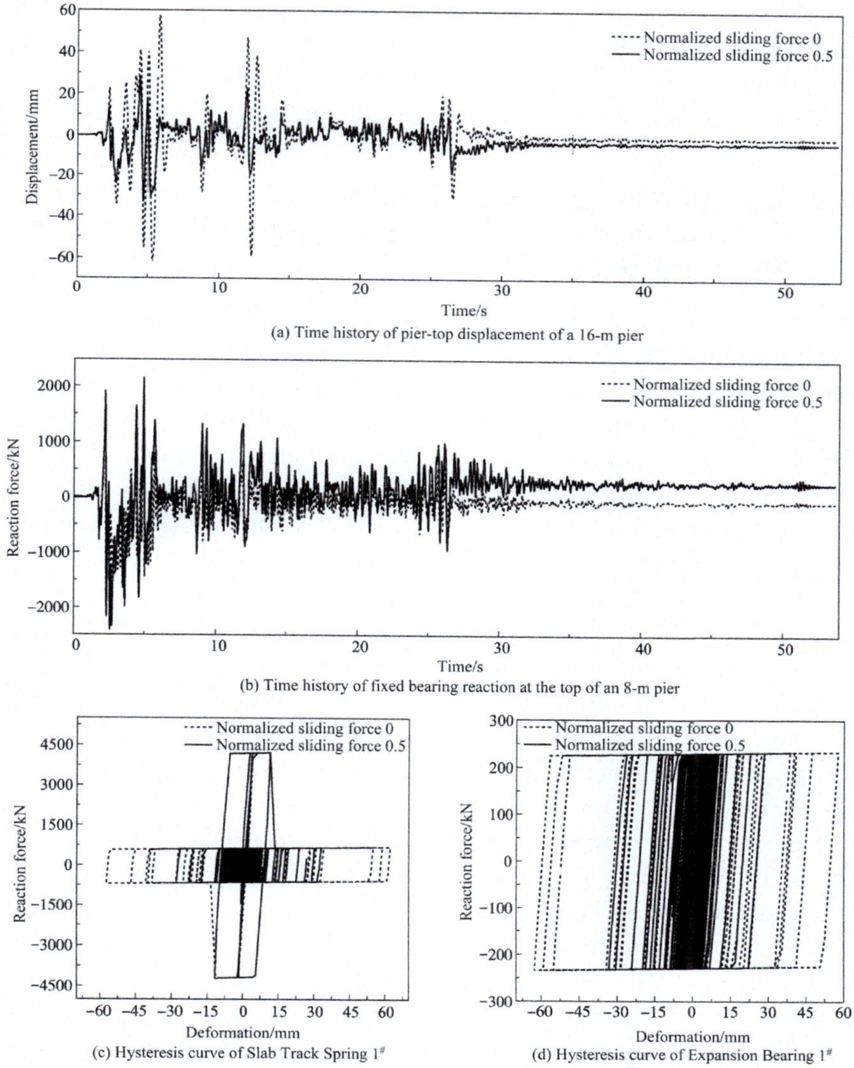

Figure 7.14 Response comparison of simply supported girder bridges in high-speed railways under GM3 ground motion.

reductions in several peak displacements. The peak displacement decreases from 62.57 to 33.15 mm; however, the residual deformation increases by 3.08 mm. As shown in Fig. 7.14B, the force transmitted to the expansion bearing at the 8-m pier increases with the rise in the sliding force of the MRFD installed at the 16-m pier, consequently leading to an amplified response of the 8-m pier. As shown in Figs. 7.14C and D, the responses of the slab track spring and expansion bearing in the first span also significantly decrease. In summary, the results of the quasi-dynamic hybrid simulation indicate that the RFD can effectively control pier responses under large deformations, but may amplify the responses of other bridge components. Detailed design of the sliding force of the MRFD should be carried out to better protect the bridge structure and track slab structure.

7.3 Seismic Mitigation Technologies for Track–Bridge Systems Based on Combined Energy-Dissipating Restraint Bearings

An analysis of the seismic responses of simply supported girder bridge–railway systems for high-speed railways reveals that under strong earthquakes, fixed bearings may fail. However, when fixed bearings transition into dry friction with piers, they can dissipate part of the seismic input energy, thereby reducing the seismic response of bridge piers. However, this also introduces some challenges, such as difficulties in repairing damaged bearings and time-consuming replacement, which may impede the rapid deployment of post-earthquake rescue efforts. Furthermore, research by Lin Qingli et al. indicates that the weakened constraint between the girder and piers leads to increased bearing displacement responses, causing girder misalignment and even girder falling [15].

In recent years, some scholars have proposed various improved designs for spherical bearings and pot bearings widely used in high-speed railway bridges to enhance their seismic mitigation performance and displacement-restraining capacity. Wang Hailong et al. [16] proposed a novel X-shaped steel damper pot bearing by integrating an X-shaped steel damper with unidirectional sliding pot bearings. Ouyang Liu [17] developed a novel translational-rotational composite friction bearing based on existing seismic spherical bearings, achieving both horizontal isolation and vertical vibration reduction. Guo Jin et al. [18] proposed a restrained-type seismic isolation bearing that integrates both displacement-limiting functionality and girder falling prevention measures into a bearing. This design also provides buffering against impact forces and self-centering capabilities. Nie Liying et al. [19] analyzed the seismic isolation and energy dissipation performance of elastoplastic spherical steel bearings with girder falling prevention through the engineering case study. Yuan Wancheng et al. developed a cable-reinforced seismic isolation and energy dissipation bearing based on pot bearings, conducted experimental studies and numerical simulations, and applied it to a four-span continuous girder bridge. The results show that this bearing can significantly reduce the seismic forces acting on the fixed pier during earthquakes, while the cables within the bearing effectively prevent the girder from falling. By studying a high-speed railway bridge with unequal pier heights and equipped with E-shaped steel damper bearings, Wang Tong et al. established an integrated nonlinear FE bridge–track analysis model. Using dual-failure criteria based on displacement and energy, along with probabilistic reliability analysis methods, they found that the fragility under seismic actions varies significantly across different bearing types at various damage states. Gao Kang et al. proposed a cable-friction pendulum bearing to mitigate seismic damage to curved girder bridges caused by bending–torsional coupling. They also analyzed its seismic mitigation performance in actual curved bridge projects, revealing that it can significantly enhance bending–torsional coupling effects and effectively decrease internal forces in both side piers and central piers of curved bridges.

To maximize the performance of the bearing, this section introduces a combined energy-dissipation and displacement-restriction bearing (hereafter termed "combined bearing") based on the research by Du Qiaodan and Wang Yang. Under seismic loads, this bearing exhibits stable energy-dissipation capacity and provides effective displacement restriction under large deformations, thereby effectively mitigating the risk of girder drop-off [20,21].

7.3.1 Composition and Working Principle

According to Gao Kang et al.'s research, the combined bearing integrates the friction energy-dissipation characteristics of spherical bearings with the plastic energy-dissipation performance of S-shaped steel plate dampers. It also utilizes the tensile behavior of S-shaped steel plate dampers under large deformations, enabling them to function as restraint devices similar to stay cables in bridge structures [22]. This design offers functional versatility. The three-dimensional structure of the combined bearing and its component breakdown are illustrated in Figs. 7.15 and 7.16, respectively.

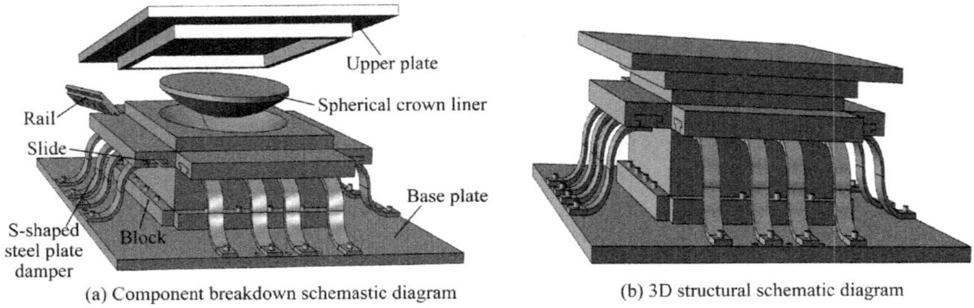

Figure 7.15 Schematic diagram of a combined bearing.

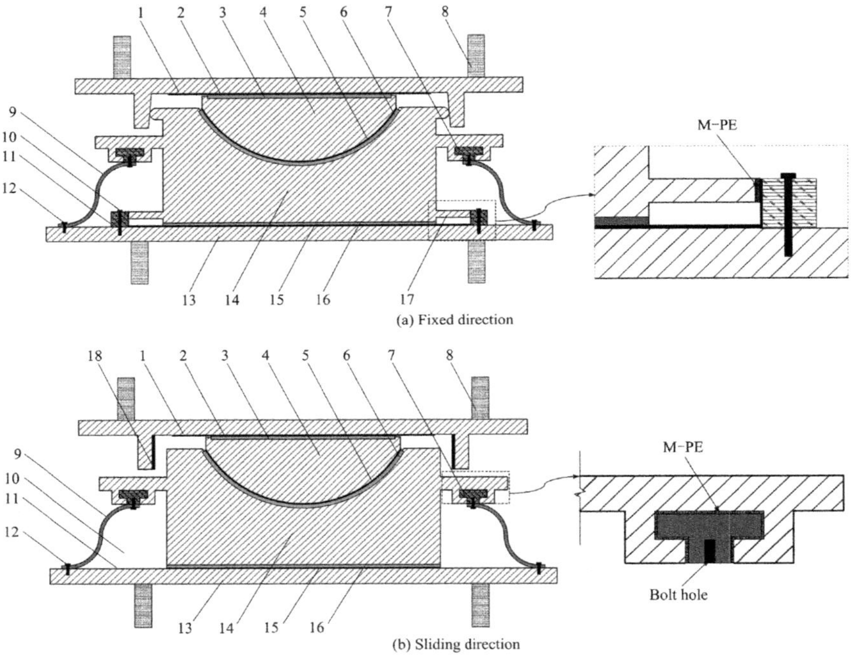

Figure 7.16 Schematic diagram of the structural components of a combined bearing. *1*, upper plate; *2*, flat stainless steel plate; *3*, flat wear-resistant sliding plate; *4*, spherical crown liner; *5*, spherical stainless steel plate; *6*, spherical wear-resistant sliding plate; *7*, sliding rail; *8*, anchor bolt; *9*, S-shaped steel plate damper; *10*, shear key; *11*, block; *12*, bolt; *13*, base plate; *14*, intermediate lining plate; *15*, additional stainless steel plate; *16*, additional friction surface; *17*, cantilevered top plate; *18*, buffer rubber.

While maintaining the fundamental structure of spherical bearings, this combined bearing exhibits the following features: (1) Four sliding tracks are integrated into the intermediate lining plate of the bearing, with their outer surfaces coated with ultra-high-molecular-weight polyethylene

(M-PE) wear-resistant sliding plates to ensure durability and reduce the friction coefficient during sliding within the guide channels. The lower end of the rail is equipped with multiple bolt holes for connecting to S-shaped steel plate dampers, allowing for replacement upon failure while enabling repeated reuse of the track. (2) The S-shaped steel plate damper consists of multiple S-shaped steel plates arranged in series, with the quantity determined by the design horizontal force. Its upper end is bolted to the rail, while the lower end is fixed to the base plate via bolts. The rail ensures that when the bearing moves in one direction, the S-shaped steel plate damper in the orthogonal direction remains free from out-of-plane deformation. (3) An M-PE friction surface is added between the intermediate lining plate and the base plate, and shear keys are installed in the fixed direction of the bearing. When the shear keys are sheared off and the block fails, sliding initiates in the fixed direction of the bearing. For the sliding direction of the bearing, a clearance is maintained between the upper plate and the intermediate lining plate as shown in Fig. 7.16B, accommodating the deformation of the girder caused by thermal loads. A layer of buffer rubber is bonded to the inner lower edge of the upper plate, providing cushioning when contact occurs between the upper plate and intermediate lining plate under seismic action.

Combined bearings can be classified into four types based on their functions: fixed bearings, transverse expansion bearings, longitudinal expansion bearings, and multidirectional expansion bearings. Figs. 7.16A and B illustrate the fixed and sliding directions of these bearings, respectively. By combining different configurations, the four types of bearings can be achieved. The horizontal force transmission path of the combined bearing is shown in Fig. 7.17. The incorporation of S-shaped steel plate dampers, additional friction surfaces, and shear keys alters the mechanical behavior of spherical bearings, effectively enhancing their seismic isolation and energy dissipation performance and restricting girder displacement. In this book , we use longitudinal bearings as an example to illustrate how the bearings work. For convenience, fixed bearings and transverse expansion bearings are collectively referred to as "fixed bearings," while longitudinal expansion bearings and multidirectional expansion bearings are termed "expansion bearings."

Its working principle is as follows. Under normal operating conditions, the shear keys introduced in the fixed bearings can provide sufficient horizontal stiffness to meet the requirements for ride smoothness and stability during train operation. Meanwhile, expansion bearings are designed

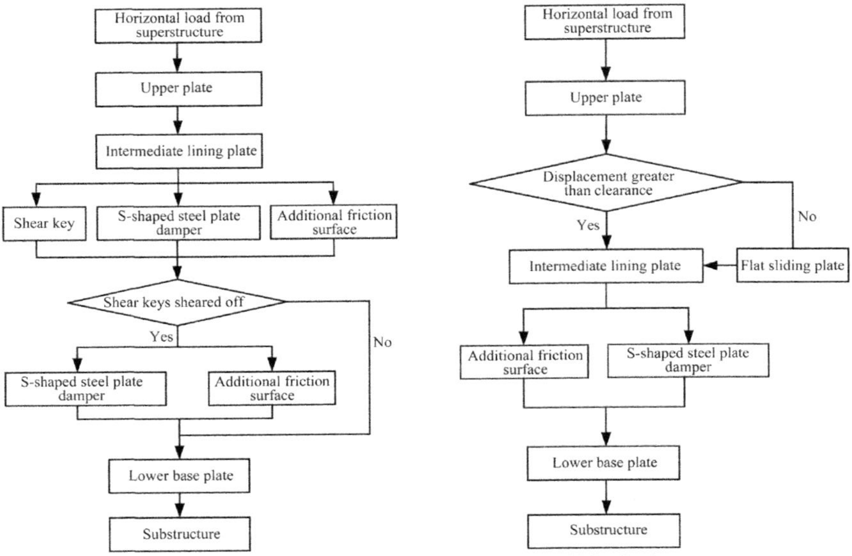

Figure 7.17 Horizontal force transmission path of the combined bearing.

with a gap between the upper plate and the intermediate lining plate to ensure free deformation under loads, temperature variations, concrete shrinkage, and creep. To guarantee the effective seismic isolation and energy dissipation performance of the combined bearing, the horizontal shear capacity of the connecting components between the bearing and the girder/pier must exceed the ultimate horizontal bearing capacity. Under seismic loading, when the relative horizontal displacement between the girder and pier top exceeds the sliding clearance of the expansion bearing, and the horizontal seismic force surpasses the sliding friction of the additional friction surface of the expansion bearing, the additional friction surface between the intermediate lining plate and lower base plate of the expansion bearing dissipates energy through friction. The S-shaped steel plate damper connecting the intermediate lining plate and lower base plate also dissipates energy through plastic deformation. The additional friction surface and the S-shaped steel plate damper form a dual-energy dissipation system that provides seismic damping for the structure. Under strong seismic excitation, when the horizontal seismic force exceeds the sliding yield force of the fixed bearing, the shear keys of the bearing are sheared off, thereby reducing the overall stiffness of the bridge structural system and providing seismic isolation for the bridge structure. For the entire bridge structural system, expansion bearings and fixed bearings can achieve staged energy dissipation under varying seismic intensities.

For the S-shaped steel plate damper in the combined bearing, on the one hand, its initial stiffness, which is relatively low, tends to cause plastic deformation due to bending, thus dissipating input seismic energy. On the other hand, as a flexural–tensional damper, the S-shaped steel plate damper transitions from bending deformation to tensile deformation when the damper is subjected to strong earthquakes with significant bearing displacement. Thus, it can provide higher stiffness and strength under strong earthquakes, functionally resembling stay cables used in bridge restraint systems [22,23]. The design achieves excellent displacement restraint, effectively preventing girder drop-off during intense seismic events. Conte proposed that for additional friction surfaces, different friction coefficients can be adjusted by modifying surface roughness and material composition ratios [24], thereby meeting requirements under different seismic fortification intensities.

7.3.2 Mechanical Model and Hysteresis Characteristics

Based on the structural composition and working principle of the combined bearing, this study establishes a simplified mechanical model of the bearing in the horizontal direction subjected to horizontal load, as shown in Fig. 7.18. The simplified mechanical model of the fixed bearing is a parallel system with a defined spatial arrangement composed of an S-shaped steel plate damper, an additional friction surface, and shear keys, as shown in Fig. 7.18A. The expansion bearing integrates a friction surface composed of planar sliding plates in series with the parallel system of the S-shaped steel plate damper and a additional friction surface. The sliding displacement of this planar plate is constrained within the clearance gap, as illustrated in Fig. 7.18B.

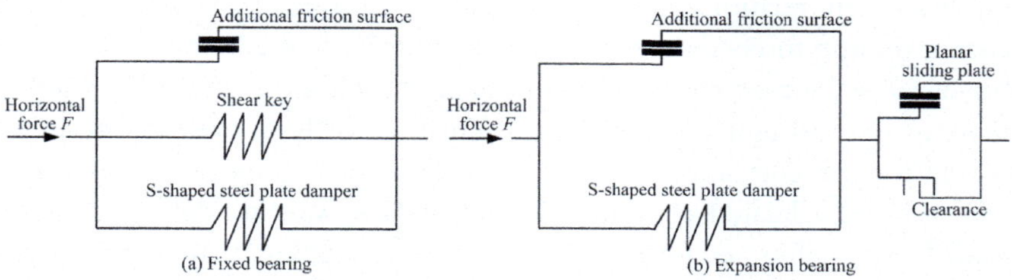

Figure 7.18 Simplified mechanical model for the horizontal direction of the bearing under horizontal load.

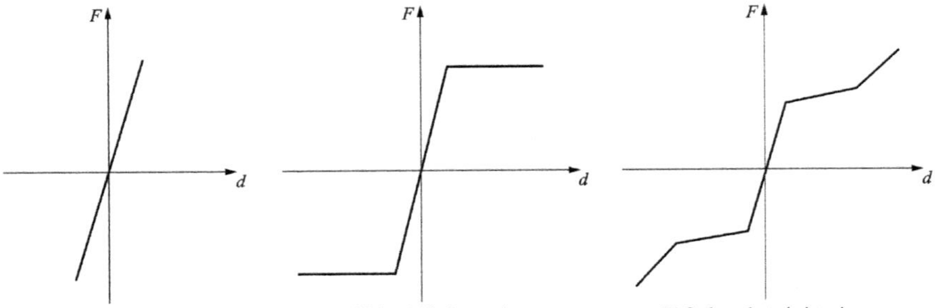

(a) Shear key (b) Additional friction surface (c) S-shaped steel plate damper

Figure 7.19 Skeleton curves of components in the combined bearing.

Fig. 7.19 illustrates the skeleton curves (force–displacement relationships) of components in the combined bearing. This book considers the shear keys in this bearing as brittle failure components. When the horizontal shear force reaches the ultimate capacity, the shear keys fail, and the bearing capacity drops abruptly to zero, as shown in Fig. 7.19A. The frictional behavior of the additional friction surface is modeled using the Coulomb friction model. In addition, to simplify computation, it is assumed that the static friction force equals the sliding friction force. An ideal elastoplastic bilinear model is used as the mechanical model, as shown in Fig. 7.19B. Similarly, the frictional behavior of the planar sliding plate in the expansion bearing is also simulated by the ideal elastoplastic bilinear model.

For the S-shaped steel plate damper, Zhai et al. [25] proposed a multilinear theoretical model through theoretical derivation and analysis, which was validated through FE models and relevant experiments. In this section, the multilinear theoretical model proposed by them is adopted, as illustrated in Fig. 7.19C.

Based on the mechanical calculation model of each component in the bearing and their interactions, the calculation equation for the force–displacement relationship under unidirectional horizontal loading and the corresponding force–displacement curve under unidirectional loading can be derived. Subsequently, the mechanical behavior of the bearing is investigated through FE numerical simulation analysis, and a physical FE model for the combined spherical bearing is established using ABAQUS software and validated against experimental data. After validation of the FE model parameters, unidirectional horizontal loading was applied in a displacement-controlled manner. The calculated parameters of the combined bearing model are substituted into the force–displacement

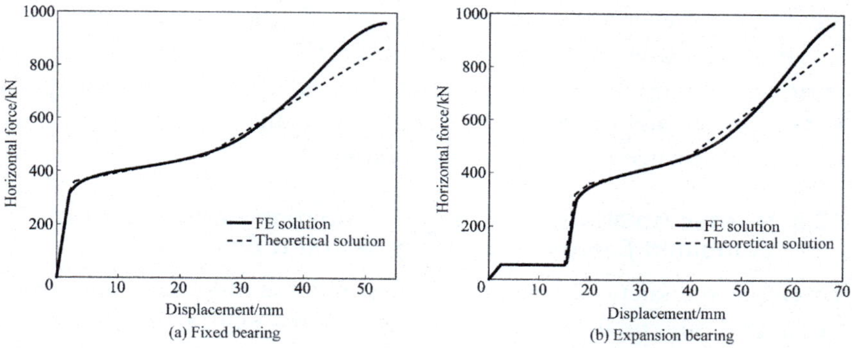

Figure 7.20 Force–displacement relationship curve of the combined bearing under unidirectional loading.

relationship formula, and the FE simulation results and theoretical calculations are plotted on the same graph, as shown in Fig. 7.20. It can be observed that the curves from both methods exhibit good agreement overall, with minimal discrepancies during the elastic working stage and bending energy dissipation stage. When the S-shaped steel plate damper within the combined bearing enters the tensile deformation stage, there is a slightly larger deviation between the theoretical solution and FE simulation results. This deviation arises from the errors in fitting the simplified mechanical model and the calculation formula for the force–displacement relationship of the S-shaped steel plate damper, but the deviation remains within an acceptable range. The FE numerical analysis results demonstrate that the derived theoretical formula effectively characterizes the force–displacement relationship of the combined bearing under unidirectional loading, thereby validating the rationality of its mechanical model.

Based on the theoretical analysis and FE modeling, we further study the hysteretic characteristics of the combined bearing under cyclic loading and analyze in detail the influence of multiple parameters of the bearing on its mechanical properties. Specifically, the quantity n, width b, and thickness t of the S-shaped steel plate damper are positively correlated with the initial stiffness, sliding yield force, post-yield stiffness, energy dissipation capacity, and ultimate bearing capacity of the bearing. The height (D) of the S-shaped steel plate damper primarily affects the deformation capacity of the bearing. The friction coefficient of the additional friction surface μ also significantly influences the hysteresis curve of the bearing. As the friction coefficient μ increases, the hysteresis loop becomes fuller, thereby enhancing the energy dissipation capacity of the bearing, though it does not affect its post-yield stiffness. Furthermore, the reserved clearance on the expansion bearing causes a time lag in the engagement of the additional friction surface and the S-shaped damper, thus affecting the energy dissipation performance of the expansion bearing.

7.3.3 Seismic Isolation and Energy Dissipation Performance of Combined Energy-Dissipating Restraint Bearings

To ensure the safety of bridge structures under earthquakes and prevent catastrophic structural damage, and in accordance with the relevant provisions on the operational states and damage levels of railway engineering under earthquakes outlined in the *Code for Seismic Design of Railway Engineering* (GB 50111—2006) [26], and based on the seismic response analyses of components in the simply supported girder bridge–track for high-speed railways, three performance objectives for seismic isolation and energy dissipation bridges under frequent earthquakes, design-level earthquakes, and rare earthquakes are proposed as follows. (1) Under frequent earthquakes, seismic isolation and energy dissipation bearings are designed to remain inactive, functioning identically to conventional bearings. Shear keys in fixed bearings for seismic isolation and energy dissipation remain intact, meeting Performance Level I requirements. (2) Under design-level earthquakes, the fixed bearings for seismic isolation and energy dissipation at the bridge abutment are designed to activate seismic isolation and energy dissipation functions before the earthquake intensity reaches the design level, reduce structural seismic response, and enable rapid resumption of operations after minor repairs, meeting Performance Level II requirements. (3) Under rare earthquakes, both the fixed bearings and expansion bearings for seismic isolation and energy dissipation at the upper part of bridge piers must enter the seismic isolation and energy dissipation state to reduce the overall structural response and prevent irreparable damage, meeting Performance Level III requirements.

7.3.3.1 Design Process

Based on research by Luo Hua et al., this section initially designs an isolation period and its corresponding equivalent damping ratio according to the types of seismic isolation and energy dissipation bearings, determines the bearing dimensions, and then conducts seismic response analysis to verify the validity of the results. If the outcomes do not meet expectations, the bearing parameters are adjusted, and the above process is repeated [27].

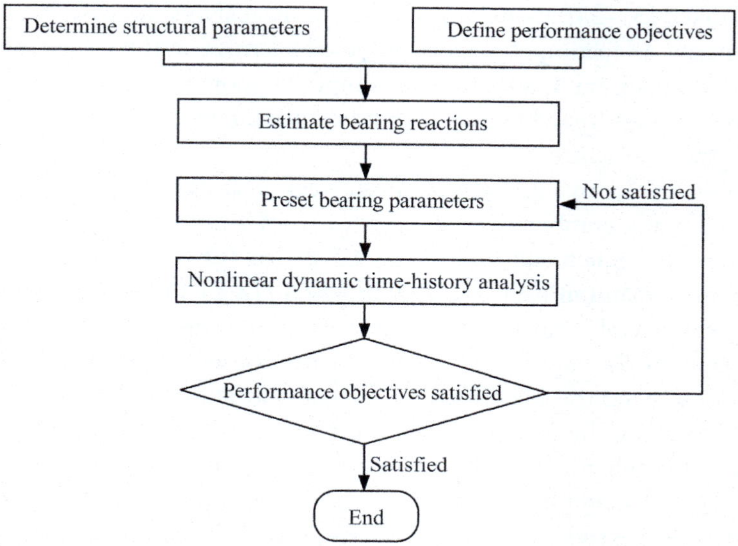

Figure 7.21 Computational process.

First, based on a single-pier model, the bearing reactions at each pier are approximately determined using the response spectrum method, thus preliminarily determining relevant bearing parameters. Subsequently, a nonlinear dynamic time-history analysis is performed on the seismic isolation and energy dissipation bridge overall to evaluate whether the results meet the performance objectives. If the objectives are not met, return to the previous steps for iterative calculations. The specific computational process is illustrated in Fig. 7.21.

7.3.3.2 Computational Model

Here, a five-span 32-meter simply supported girder bridge–track system for high-speed railway was taken as the subject of study. Research by Gao Jianqiang, Li Chen, Zhang Yongliang et al. indicates that as a critical component of the bridge–track system for a high-speed railway, the upper track structure has a strong constraining effect on the bridge structure [28–30]. Meanwhile, the post-seismic damage state of each track component directly impacts the resumption of train operations after earthquakes. However, in the study of seismic isolation and energy dissipation for high-speed railway bridges, the constraints of the upper track structure and the seismic responses of its individual components are often overlooked [31–34]. Consequently, this book analyzes the seismic isolation and energy dissipation effects of combined bearings on the bridge structure and the upper track structure in simply supported girder bridge–track systems for high-speed railways while accounting for the constraints of the upper track structure.

After the conventional spherical steel bearings in this system are replaced with combined bearings, the fixed bearings in the combined bearings are sequentially numbered GXZ1 to GXZ5 from left to right, while the expansion bearings are numbered HXZ1 to HXZ5 from left to right, respectively. An FE model of the simply supported girder bridge–-track system for high-speed railways with combined bearings has been developed using the FE software OpenSees, as illustrated in Fig. 7.22B.

To evaluate the seismic mitigation performance of combined bearings more comprehensively, the following three working conditions are analyzed. (1) Combined bearings are configured in the simply supported girder bridge–track system for high-speed railway. (2) Conventional TJQZ-5000 spherical steel bearings are installed in the simply supported girder bridge–track system for high-speed railway, while their potential failure under seismic loads is considered. (3) Conventional bearings with reinforced treatment are used in the simply supported girder bridge–track system for high-speed railway, with the assumption that no failure occurs under seismic conditions.

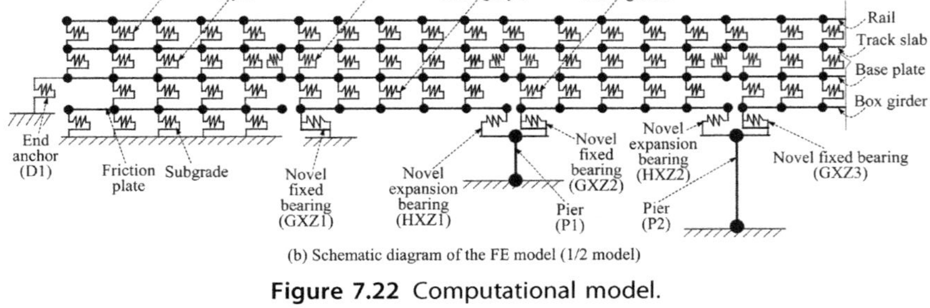

Figure 7.22 Computational model.

7.3.3.3 Analysis of Seismic Mitigation Performance

Based on the seismic fortification intensity, site type, and characteristic period of the study area, as well as the corresponding PGA for design-level earthquakes, five ground motions (G1–G5), as shown in Table 7.2, are selected for nonlinear dynamic time-history analysis. A comparative analysis of the longitudinal seismic responses of the bridge structure and upper track structure in the bridge–track system for a high-speed railway under three working conditions is conducted to verify the seismic mitigation performance of the bearings.

The results of the comparative analysis of nonlinear dynamic time history under three working conditions subjected to the five selected ground motions are presented. The seismic mitigation performance of combined bearings on the bridge structure within the track–bridge system for a high-speed railway is analyzed through time-history calculations of seismic responses, pier-top displacements, base shears, and girder displacements. The conclusions are as follows:

(1) Under rare earthquakes, both the fixed and expansion bearings in the combined bearings predominantly operate in the seismic isolation and energy dissipation phase, fulfilling their functions of isolation and energy dissipation. Moreover, the hysteresis curves form full loops, effectively dissipating the seismic energy input to the structural system, thereby largely achieving the predefined performance objectives in the design phase.

Table 7.2 Information on GM1–GM5.

S/N	Ground Motion Name	Year	Magnitude	Station Name
G1	Imperial Valley-02	1940	6.95	El Centro Array #9
G2	Northridge-01	1994	6.69	FAI095
G3	Borrego	1942	6.5	El Centro Array #9
G4	San Fernando	1971	6.61	LA–Hollywood Station FF
G5	Southern California	1952	6	San Luis Obispo

(2) When the combined bearings perform seismic isolation and energy dissipation, compared to the track–bridge system for a high-speed railway with conventional bearings, the combined bearings not only effectively reduce pier-top displacements and base shear force of the piers, but also minimize post-earthquake residual displacement, thereby enhancing the seismic resistance of piers under rare earthquakes.

(3) The combined expansion bearings enhance energy dissipation capacity while effectively constraining their own peak displacement responses. Compared to conventional expansion bearings, the combined ones demonstrate significantly reduced peak displacement responses overall. Meanwhile, they effectively limit the displacement of the girder and the peak response in expansion joint spacing, substantially decreasing the likelihood of girder collision.

(4) For the filling and connecting components in the upper track structure, the use of combined bearings results in varying seismic mitigation performance across different elements. Among them, the sliding layer and shear grooves, which exhibit significant responses under seismic loads, demonstrate good seismic mitigation performance. The peak displacement responses are significantly reduced, while the seismic mitigation performance on the CA mortar layers, shear reinforcement, and fasteners under seismic loads remains limited, with no notable amplification observed.

(5) For the longitudinal continuous components in the upper track structure, the overall seismic mitigation performance is not significant. It should be noted that under rare earthquakes, the peak response of the baseplate near the end anchors is significant and exceeds the average tensile yield stress of the section, and the longitudinal reinforcement in the baseplate may experience tensile yielding. It is recommended to extend the length of the road–bridge transition section to improve the mechanical performance of the baseplate in that section. The peak tensile stress of the track slab near the abutment slightly exceeds its cracking section stress, which may lead to cracking during seismic events.

7.4 TMD-Based Seismic Mitigation Technologies for Train–Track–Bridge Systems

Common structural vibration control methods in civil engineering can be achieved by supplemental damping, seismic isolation, or vibration absorption devices. In recent years, researchers have begun to focus on vibration control methods that account for operational safety. Chen et al. [35] investigated the seismic responses of high-speed railway train–bridge systems with and without seismic isolation bearings, considering structural nonlinearity. The results indicate that train operational safety is primarily influenced by seismic frequency components near the fundamental frequencies of the train and the bridge. However, seismic isolation devices increase the structural period and lead to excessive displacements, which may further elevate the risk of train derailment during earthquakes. TMDs offer advantages such as simple installation, cost-effectiveness, and reliable performance, making them a widely adopted seismic mitigation measure in bridge seismic resistance and train–bridge coupling vibration systems [36,37]. Chen Z et al. [36,38] designed the optimal stiffness and damping parameters of TMDs by utilizing bridge characteristics and employing random white noise as excitation instead of sinusoidal excitation. They also validated the control effectiveness of TMDs on train vibration under earthquakes through train–bridge coupling vibration analysis. Xu et al. [39] selected 20 groups of TMDs with varying mass ratios and damping ratios to conduct dynamic analysis of train–bridge systems under train loads, revealing the variation patterns of the peak bridge displacement with respect to the mass ratio and damping ratio of the devices. To enhance energy dissipation efficiency, Yin et al. [40] combined viscoelastic dampers with TMDs and validated their vibration reduction capability through a train–bridge system with multispan continuous rigid-frame bridges. While providing vibration damping, TMDs can also harness the kinetic energy generated during their operation for electricity generation. Hou et al. utilized TMDs [41], Multiple Tuned Mass Dampers (MTMDs) [42], and coil-based magnetic flux cutting to achieve vibration control and energy recovery in train–bridge systems. The authors' research group has designed seismic mitigation devices for high-speed railway bridges based on earthquake-induced operational safety. Parameter optimization design, layout analysis, and seismic mitigation performance calculation of TMDs were also conducted based on the natural period and modal characteristics of the bridge. The impact of TMDs on the operational safety of trains under minor earthquakes has been discussed [43].

7.4.1 TMD Model and Optimization Methods

The structure–TMD dynamic model is shown in Fig. 7.23. The main structure is connected to the ground via springs and dampers, while the additional TMD mass is also linked to the main structure through springs and dampers. The main structure may be subjected to loads from external force $F(t)$ excitation or base acceleration (U_g) excitation. Under the sinusoidal external force, the dynamic equation is given by Eq. (7.3). When the load vector is generated by inertial force due to base acceleration, the physical meanings of the variables are listed in Table 7.3.

$$\begin{bmatrix} m_1 & \\ & m_2 \end{bmatrix} \begin{pmatrix} \ddot{x}_1(t) \\ \ddot{x}_2(t) \end{pmatrix} + \begin{bmatrix} c_1 + c_2 & -c_2 \\ -c_2 & c_1 \end{bmatrix} \begin{pmatrix} \dot{x}_1(t) \\ \dot{x}_2(t) \end{pmatrix} + \begin{bmatrix} k_1 + k_2 & -k_2 \\ -k_2 & k_1 \end{bmatrix} \begin{pmatrix} x_1(t) \\ x_2(t) \end{pmatrix} = \begin{pmatrix} F_0 \cos\omega t \\ 0 \end{pmatrix} \tag{7.3}$$

The optimization design of TMD parameters under sinusoidal external force can be performed by referring to DenHartog's P–Q two fixed-points theory [44]. For the undamped structure ($c_1 = 0$) subjected to a sinusoidal external force with frequency ω, the dynamic response is typically evaluated using the maximum displacement ($y_{\max}$) or the displacement magnification factor R. The static displacement magnitude is defined as $y_{st} = P_0/k_1$, and the dynamic magnification factor R is expressed by Eq. (7.5):

$$R = \frac{y_{max}}{y_{st}} = \sqrt{\frac{\left(f^2 - \beta^2\right)^2 + \left(2\xi_2 f\beta\right)^2}{\left[\left(f^2 - \beta^2\right)\left(1 - \beta^2\right) - f^2\beta^2\mu\right]^2 + \left(2\xi_2 f\beta\right)^2\left(1 - \beta^2 - \beta^2\mu\right)^2}} \tag{7.4}$$

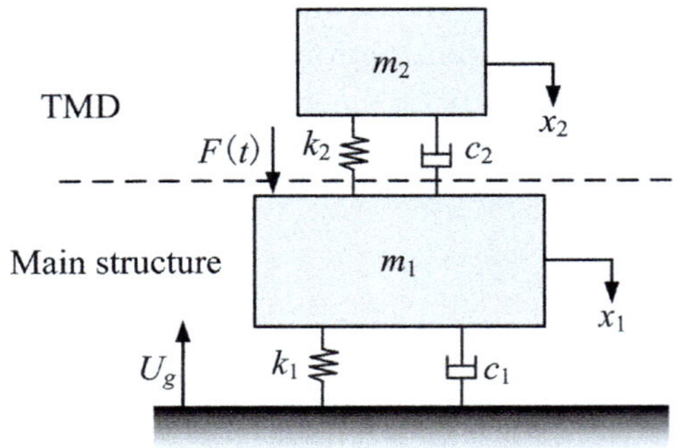

Figure 7.23 Structure–TMD dynamic model.

Table 7.3 Physical meaning of TMD expression.

Symbol	Physical Meaning	Expression
m_1	Structural mass	–
k_1	Structural stiffness	–
c_1	Structural damping	–
ζ_1	Structural damping ratio	$\frac{c_1}{2\sqrt{k_1/m_1}}$
ω_1	Structural fundamental frequency	$\sqrt{k_1/m_1}$
m_2	TMD mass	–
k_2	TMD stiffness	–
c_2	TMD damping	–
ω_2	TMD fundamental frequency	$\sqrt{k_2/m_2}$
ζ_2	TMD damping ratio	$\frac{c_2}{2\sqrt{k_2/m_2}}$
μ	Mass ratio	m_2/m_1
β	Frequency ratio	ω_2/ω_1
F(t)	External load on structure	–
ω	Sinusoidal frequency	–
F_0	Sinusoidal force amplitude	–

where $\beta = \omega/\omega_1$ denotes the ratio of excitation frequency to natural frequency of the main structure, and the displacement magnification factor R is a function of μ, ζ_2, f, and β.

The dynamic response of the two degrees of freedom (2-DOF) structure under external excitation is simulated using MATLAB, and the explicit Newmark method is employed for the solution. The calculation parameters are set as $\mu = 0.1$ and $f = 1.1$, with damping ratios selected at 0.5%, 10%, 30%, 50%, and ∞. Fig. 7.24 illustrates the variation pattern of the maximum displacement response of the structure with respect to the frequency of the sinusoidal load. The observation of the variation in the maximum displacement response of the structure with increasing damping ratios reveals that when the structural damping is zero, there are two resonance points at the fundamental frequency and the TMD frequency, respectively. This occurs because, without damping, the resonant response of the TMD amplifies the primary structure. As damping gradually increases, the structural response gradually diminishes. However, when damping reaches extremely high values, the structure and TMD become effectively coupled and behave as an single degree of freedom (SDOF) system with mass $(1+\mu)\times m_1$, which causes resonance at a new frequency point, and the response can tend to become infinite. The objective of

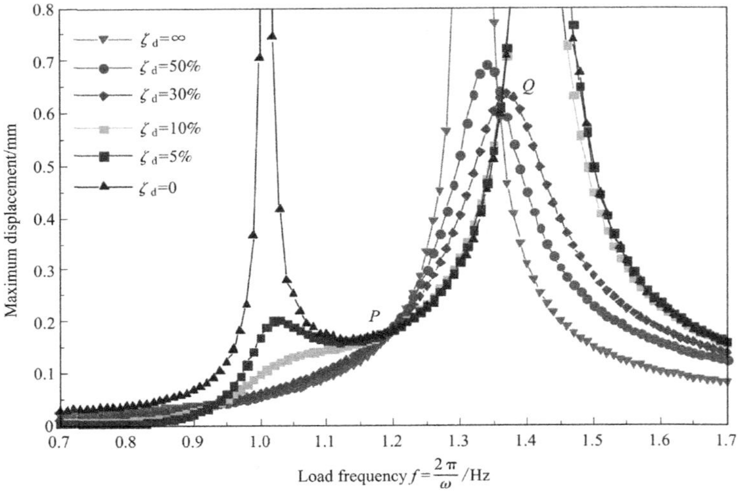

Figure 7.24 Variation of maximum displacement response of the structure with respect to the frequency of the sinusoidal load.

TMDs is to minimize the peak response at resonance frequency to maintain relatively low amplitudes across a broad frequency range. The figure reveals that regardless of the damping ratio variation, the curves consistently pass through two fixed points, denoted as Point P and Point Q. By adjusting the frequency ratio f, the heights of these two points can be equalized. Subsequently, the TMD damping ratio ζ_2 ensures the curve tangentially passes through both points. The optimal TMD frequency ratio f_{opt} and damping ratio ζ_{opt} are given by Eq. (7.6), while the optimal TMD stiffness and damping are defined by Eq. (7.7).

$$\begin{cases} f_{\text{opt}} = \dfrac{1}{1+\mu} \\ \xi_{\text{opt}} = \sqrt{\dfrac{3\mu}{8(1+\mu)}} \end{cases} \tag{7.5}$$

$$\begin{aligned} K_{\text{opt}} &= m_2 \frac{k_1}{m_1}\left(\frac{1}{1+\mu}\right)^2 \\ C_{\text{opt}} &= 2m_2\sqrt{\frac{k_1}{m_1}}\cdot\sqrt{\frac{3\mu}{8(1+\mu)^3}} \end{aligned} \tag{7.7}$$

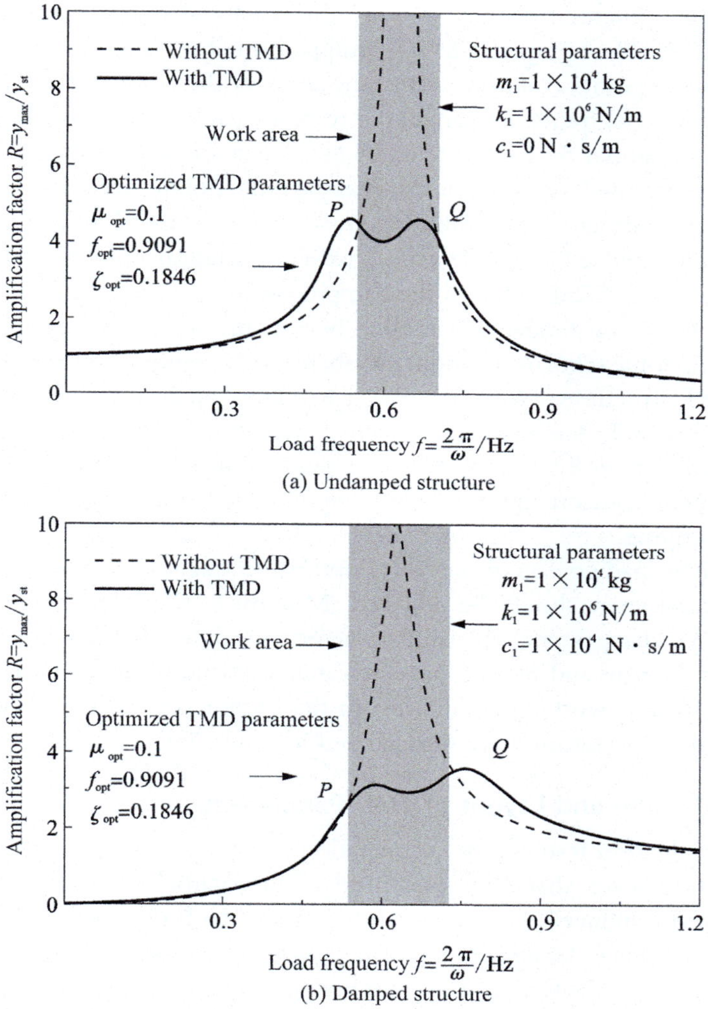

Figure 7.25 Seismic mitigation performance of optimized tuned mass damper design parameters.

Using the aforementioned optimization design method, the optimal parameters of TMDs for the structure are established. Fig. 7.25 compares the displacement magnification factor R of the structure with and without TMDs, while the structural parameters and optimized TMD parameters are listed in the figure. The intersection points between the curves with and without TMDs define the operational frequency range of the TMDs. Within this range, the structural response is significantly reduced.

When structural damping $c_1 = 0$, the structural response is shown in Fig. 7.25A. The optimized TMD-equipped structure reaches its maximum response at points *P* and *Q*, where the slope of the curve is zero. Outside the operational frequency range, the response slightly increases, but at the structural natural frequency of 0.628 Hz, the resonance points are significantly suppressed, resulting in a substantial reduction in structural response. When the structure has damping ($c_1 = 1 \times 10^4$), the TMD parameters designed based on points *P* and *Q* exhibit vibration reduction effects as shown in Fig. 7.25B. Due to the damping in the primary structure, points *P* and *Q* are no longer at the same height, meaning this solution is not optimal. Some scholars, including Mohebbi et al., optimized parameters for damped structures [45], but the parameters designed for undamped structures still demonstrated effective vibration reduction. Therefore, although originally proposed for undamped structures, the fixed-point *P*–*Q* theory remains applicable in practical engineering for structures with low damping.

When loading conditions vary, optimal frequency ratio and damping ratio parameters should be adjusted. Moreover, when the optimization objective changes, the optimal parameters also differ. Many scholars, such as Den Hartog and Soong, have researched selecting optimal parameters under various working conditions. In this section, the analytical solutions for TMD optimization are summarized in Table 7.4.

7.4.2 Design and Layout of TMD Parameters

The evaluation results of train operational safety on bridges under earthquakes indicate that high-pier bridges are particularly vulnerable to earthquake-induced threats. Specifically, trains on bridges with 24-m piers may experience derailment under design-level seismic loads. To enhance train operational safety on high-pier bridges under earthquakes, this section applies TMDs to high-pier bridges and verifies their impact on train operational safety on bridges. TMD parameters are designed following the workflow illustrated in Fig. 7.26, with the structural velocity *SI* response as the optimization objective. The determination of the *SI* metrics and SI_L is detailed in Section 6.4 of Chapter 6 (Safety Evaluation of Train Operation on Bridges under Earthquakes Based on Spectral Intensity Index). The design parameters to be determined for the device are the mass ratio μ, frequency ratio f, and damping ratio ζ_2. For economic considerations, μ is typically set within the range of 0–0.1, while f and ζ_2 are determined

Table 7.4 Analytical solutions for TMD optimization [44,46].

S/N	Excitation Type	Position of Application	Optimization Objective	Optimal Frequency Ratio f_{opt}	Optimal Damping Ratio ζ_{opt}	Optimal Amplification Factor R_{opt}
1	Sinusoidal force	Main structure	x1	$\frac{1}{1+\mu}$	$\sqrt{\frac{3\mu}{8(1+\mu)}}$	$\sqrt{1+\frac{2}{\mu}}$
2	Sinusoidal force		x1 $\ddot{x}_1$	$\sqrt{\frac{1}{1+\mu}}$	$\sqrt{\frac{3\mu}{8\left(1+\frac{\mu}{2}\right)}}$	$\sqrt{\frac{2}{\mu(1+\mu)}}$
3	Random force		X_1^2	$\frac{\sqrt{1+\mu/2}}{1+\mu}$	$\sqrt{\frac{\mu(1+3\mu/4)}{4(1+\mu)(1+\mu/2)}}$	$\sqrt{\frac{1+3\mu/4}{\mu(1+\mu)}}$
4	Sinusoidal acceleration	Base	x1	$\frac{\sqrt{1-\mu/2}}{1+\mu}$	$\sqrt{\frac{3\mu}{8(1+\mu)(1-\mu/2)}}$	$(1+\mu)\sqrt{\frac{2}{\mu}}$
5	Sinusoidal acceleration		x1 $\ddot{x}_1$	$\sqrt{\frac{1}{1+\mu}}$	$\sqrt{\frac{3\mu}{8(1+\mu)}}$	$\sqrt{1+\frac{2}{\mu}}$
6	Random acceleration		X_1^2	$\frac{\sqrt{1-\mu/2}}{1+\mu}$	$\sqrt{\frac{\mu(1-\mu/4)}{4(1+\mu)(1-\mu/2)}}$	$(1+\mu)^{3/2}\sqrt{\frac{1}{\mu}-\frac{1}{4}}$

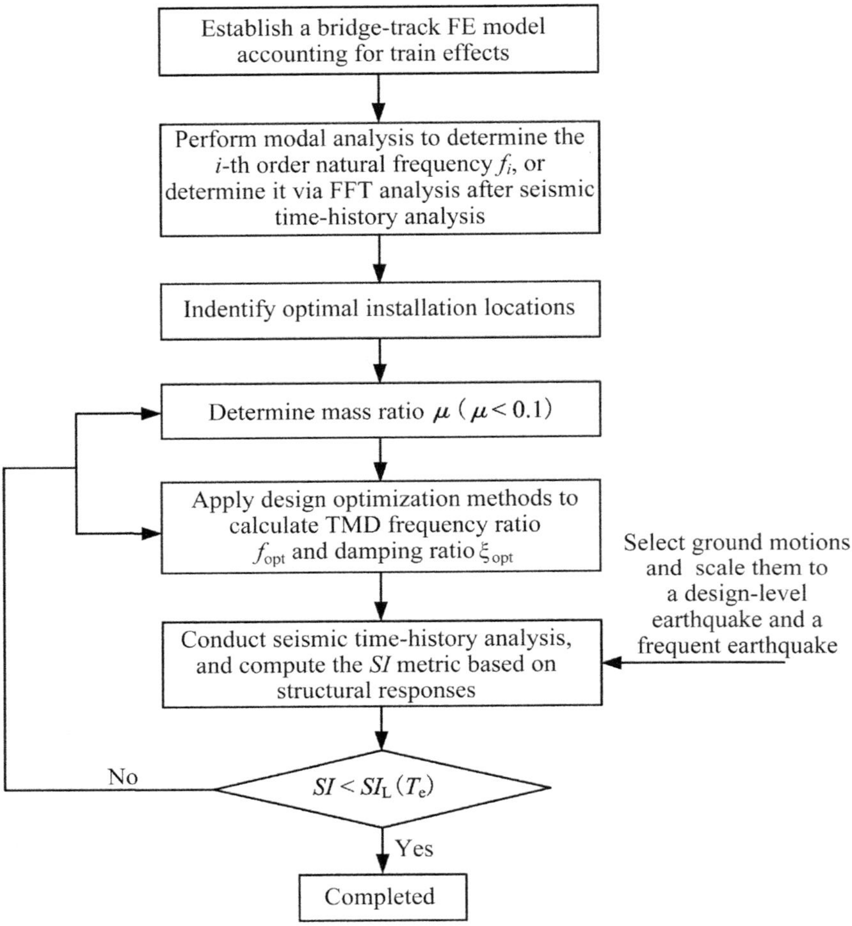

Figure 7.26 TMD design based on *SI* metrics.

through selected optimization design methods. After the optimal parameters of the TMD are determined, the dynamic time-history analysis of the structure–TMD model under seismic excitation is performed. *SI* metrics are calculated based on structural responses and compared with the SI_L. If the SI_L is not met, the mass ratio is adjusted or recalculated after the method is optimized until the SI_L is met.

Based on the modal parameters of the bridge, TMDs with different mass ratios are designed using the three methods listed in Table 7.4. The structural mass of the bridge is calculated as the sum of the box girder mass and half of the pier mass. The three optimization methods adopted are as follows: Method 1 is based on sinusoidal force excitation, with the optimization objective being structural acceleration; Method 2 is based on random force excitation, with the optimization objective being structural displacement; Method 3 is based on random base acceleration excitation, with the optimization objective being structural displacement.

To determine the reasonable installation location of TMDs, a modal analysis of the bridge model is first conducted to identify its fundamental frequency and vibration modes. The installation location is then determined based on the calculated structural modes. Research shows that placing TMDs at the points of maximum modal displacement of the *i*-th mode achieves optimal vibration control for the *i*-th mode, as these points typically exhibit the maximum vibration amplitude in the corresponding mode shape. The transverse vibration mode of the bridge with 24-m piers is shown in Fig. 7.27. As shown in the figure, the optimal installation position for vibration control of the *i*-th mode in tall-pier bridges is the pier top.

However, the limited space at the pier top precludes direct installation of the devices. As the girder is rigidly connected to the pier in the transverse direction through fixed bearings, which ensure synchronized

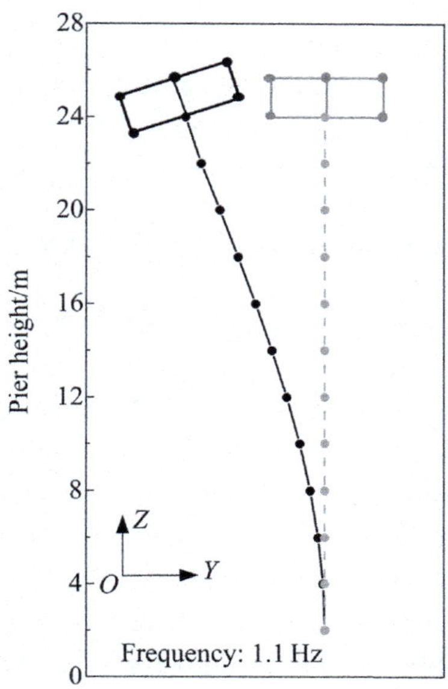

Figure 7.27 Transverse vibration mode of the bridge (with 24 m piers).

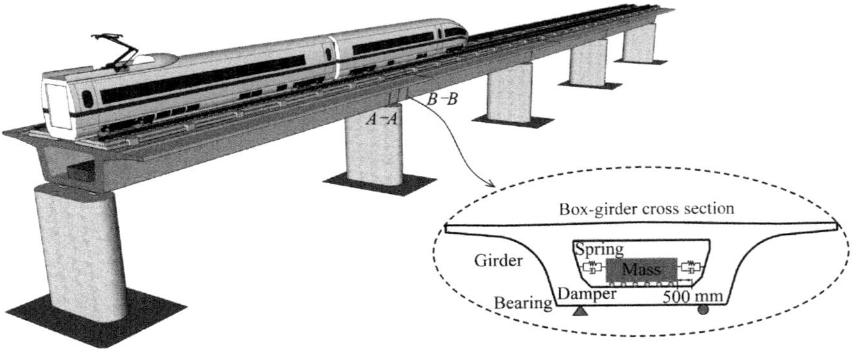

Figure 7.28 TMD installation layout.

movement before bearing failure, the available space inside the box girder on both sides of the pier top can be utilized for installation. The TMD installation layout [47] is illustrated in Fig. 7.28.

7.4.3 Seismic Mitigation Performance of TMDs

To evaluate the vibration mitigation performance of the devices, a damping performance parameter is defined: indicator reduction percentage (IRP), as expressed in Eq. (7.8).

$$IRP = \frac{x - x_r}{x} \times 100\% \qquad (7.7)$$

where x represents the structural response without TMDs, such as peak displacement, peak acceleration, and velocity spectrum intensity; x_r denotes the structural response with TMDs.

7.4.3.1 Seismic Mitigation Performance of Different Design Methods

First, the seismic mitigation performance of TMD parameters designed by the three selected methods is compared. The time-domain and frequency-domain responses of a 24-m pier bridge under seismic excitation are shown in Fig. 7.29, with all methods employing the same mass ratio ($\mu = 0.05$). The time-history comparison reveals that the structural displacement response is reduced under different TMD parameters, with a significant reduction observed at large displacements. The frequency-domain comparison shows a notable reduction near the fundamental frequency (approximately 1.1 Hz) in both ground motions, while the reduction of other frequency components is less pronounced. This is attributed to the

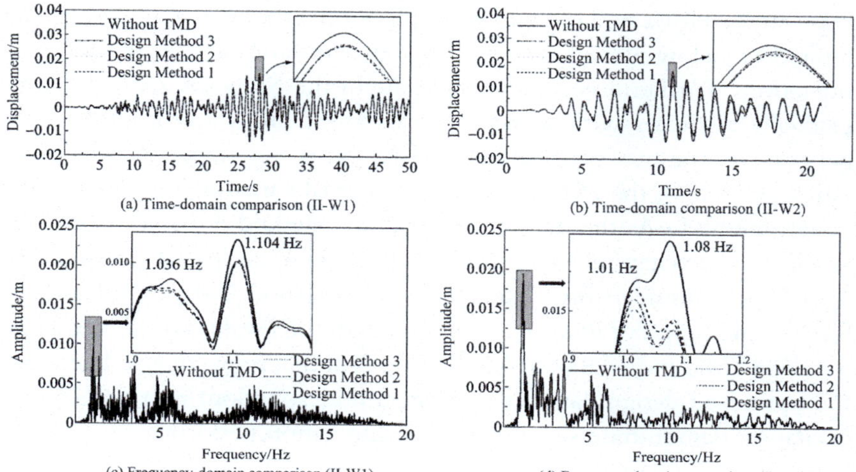

(a) Time-domain comparison (II-W1) (b) Time-domain comparison (II-W2) (c) Frequency-domain comparison (II-W1) (d) Frequency-domain comparison (II-W2)

Figure 7.29 Comparison of TMD seismic mitigation performance of different design methods ($\mu = 0.05$).

fact that a single-frequency TMD can only effectively operate within its designed frequency band. Under II-W1 ground motion, the three design methods demonstrate comparable seismic mitigation performance. However, under II-W2 ground motion, Method 3 exhibits superior seismic mitigation performance compared to Methods 1 and 2, as clearly evidenced in the frequency-domain plot (Fig. 7.29D).

To analyze the seismic mitigation performance of different design methods for the 24-m pier bridge under varying mass ratios μ, this book examines the relationship between the displacements and the *IRP* values of the *SI* metrics at mass ratios of $\mu = 0.1$, 0.08, 0.06, 0.05, 0.03, and 0.01. The variation of *IRP* with mass ratios under II-W2 ground motion is illustrated in Fig. 7.30. As shown in the figure, *IRP* increases with higher

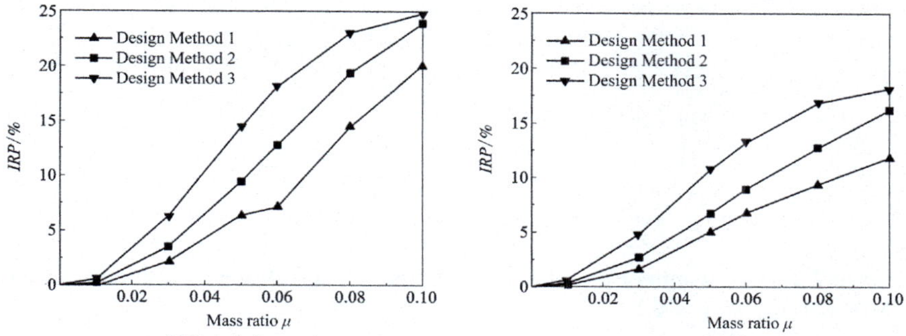

(a) Peak displacement (b) *SI* metric

Figure 7.30 Variation of indicator reduction percentage with mass ratio for different design methods (II-W2).

mass ratios. However, in practical applications, the mass ratio is typically limited to below 10% due to installation constraints and economic considerations. Furthermore, the growth rate of the *SI* metric reduction ratio gradually decreases with increasing mass ratio. From a cost-effectiveness perspective, the mass ratio should not be excessively large. At small mass ratios ($\mu = 0.01$), the *IRP* results show minimal variation among the three design methods. As the mass ratio increases, Method 3 (based on random acceleration excitation) yields higher *IRP* values compared to Method 1 (based on sinusoidal force excitation) and Method 2 (based on random force excitation). For instance, at $\mu = 0.05$, the *IRP* values of *SI* metrics are 5.09%, 6.75%, and 10.8% for Methods 1, 2, and 3, respectively. Therefore, the optimized design parameters derived under random base acceleration excitation demonstrate superior seismic mitigation performance.

7.4.3.2 Seismic Mitigation Performance of 24-m Piers

The seismic mitigation performance of 24-m piers is analyzed. The mid-span track displacement time-history responses of 24-m piers under various ground motions with and without TMDs are compared, as shown in Fig. 7.31 ($\mu = 0.1$). After TMDs with optimized parameters are installed, the bridge structure exhibits significant reductions in peak displacements under all seismic excitations. For instance, under II-W1 ground motion, the maximum displacement response of the structure with TMDs decreases from 0.0146 to 0.0115 m after TMD installation; under IV-W1 ground

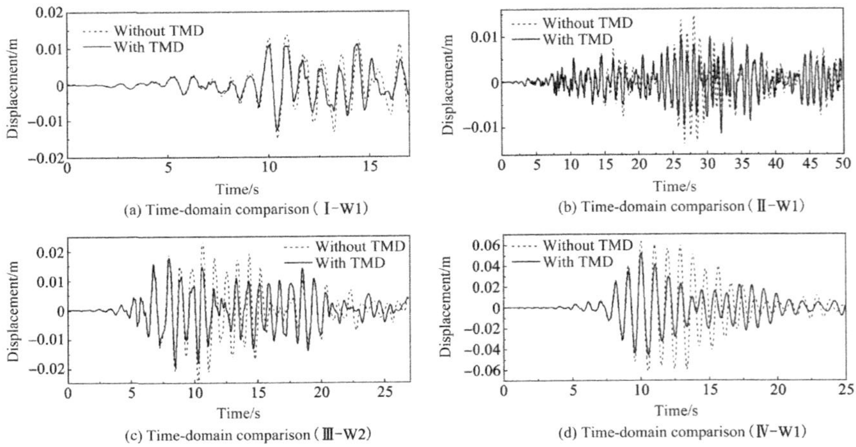

Figure 7.31 Displacement time-history responses of the structure with and without TMDs (24 m pier, $\mu = 0.1$).

motion, the peak displacement is reduced from 0.0635 to 0.0517 m. All peak displacements mentioned in this section are absolute structural displacements.

Fig. 7.32 illustrates the seismic mitigation performance of TMDs on mid-span track acceleration responses for the 24-m pier bridge under various ground motion excitations in the frequency domain. As is evident from the figure, the maximum structural response occurs near the fundamental frequency of 1.1 Hz, while the acceleration frequency components around this fundamental frequency show a significant reduction after TMD installation. Since Type IV sites consist of soft soil foundations where seismic excitations are predominantly low frequency and closely match the natural frequency of the bridge, this leads to amplified structural displacements and accelerations that are particularly detrimental to operational safety. Although ground motions rich in low-frequency components pose unfavorable excitations for train–bridge systems, their dominant frequencies typically concentrate in the low-frequency range and have a high probability of falling within the operational bandwidth of properly designed TMDs. Under such conditions, the TMDs demonstrate their most pronounced seismic mitigation performance. As evidenced in Fig. 7.32D by the seismic mitigation performance around the fundamental frequency, the design philosophy employing fixed P and Q control points has been effectively implemented [44].

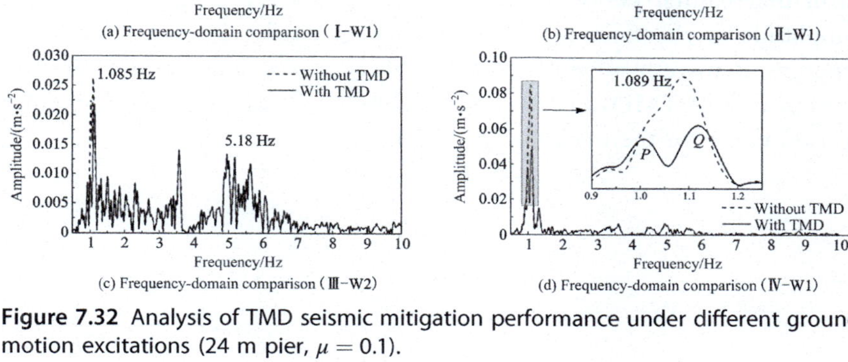

Figure 7.32 Analysis of TMD seismic mitigation performance under different ground motion excitations (24 m pier, $\mu = 0.1$).

7.4.4 Impact of Train Operational Safety on Bridges

The traditional evaluation criterion for TMD seismic mitigation performance based on bridge parameter design is the peak displacement of the bridge. However, bridge displacement responses cannot be directly translated into train operational safety metrics. To evaluate train operational safety, a coupling model for the train–track–bridge-vibration control system must be established, which involves substantial workload and is unsuitable for iterative optimization in TMD design. However, by conducting dynamic time-history analysis of bridge structures under seismic excitation, the seismic acceleration response time histories before and after TMD installation can be obtained. Subsequently, spectral analysis of the acceleration response time histories yields velocity response spectra. Through integral operations within the target period, the spectral intensity index—representing total energy within the target frequency range—can be derived for rapid vehicle operational safety evaluation. Fig. 7.33 presents the velocity response spectra derived from the acceleration time histories at the mid-span track of a 24-m pier bridge under different seismic excitations with TMDs of varying mass ratios. The spectral amplitude can reflect the energy magnitude of train-induced excitations. In the figure, the horizontal axis represents the period T_a of the velocity response spectrum, while the vertical axis denotes the mass ratio μ of the TMD (where $\mu = 0$ indicates the absence of damping installation). As shown in the figure, the S_v peaks are predominantly concentrated near the fundamental frequency of the structure. Under soft soil foundations motion (IV-W1), the energy distribution exhibits the highest concentration, with minimal energy components below or above the structural fundamental frequency. When ground motions contain certain high-frequency components (e.g., II-W1 and II-W2), elevated spectral peaks appear in the high-frequency region (left of the energy peak), while energy intensity beyond the structural natural period remains relatively low. As μ increases, the S_v amplitude at the fundamental frequency gradually decreases, while the energy-intensive bandwidth progressively narrows. This demonstrates the critical role of TMDs in reducing train-induced energy within the target period range for the track–bridge system.

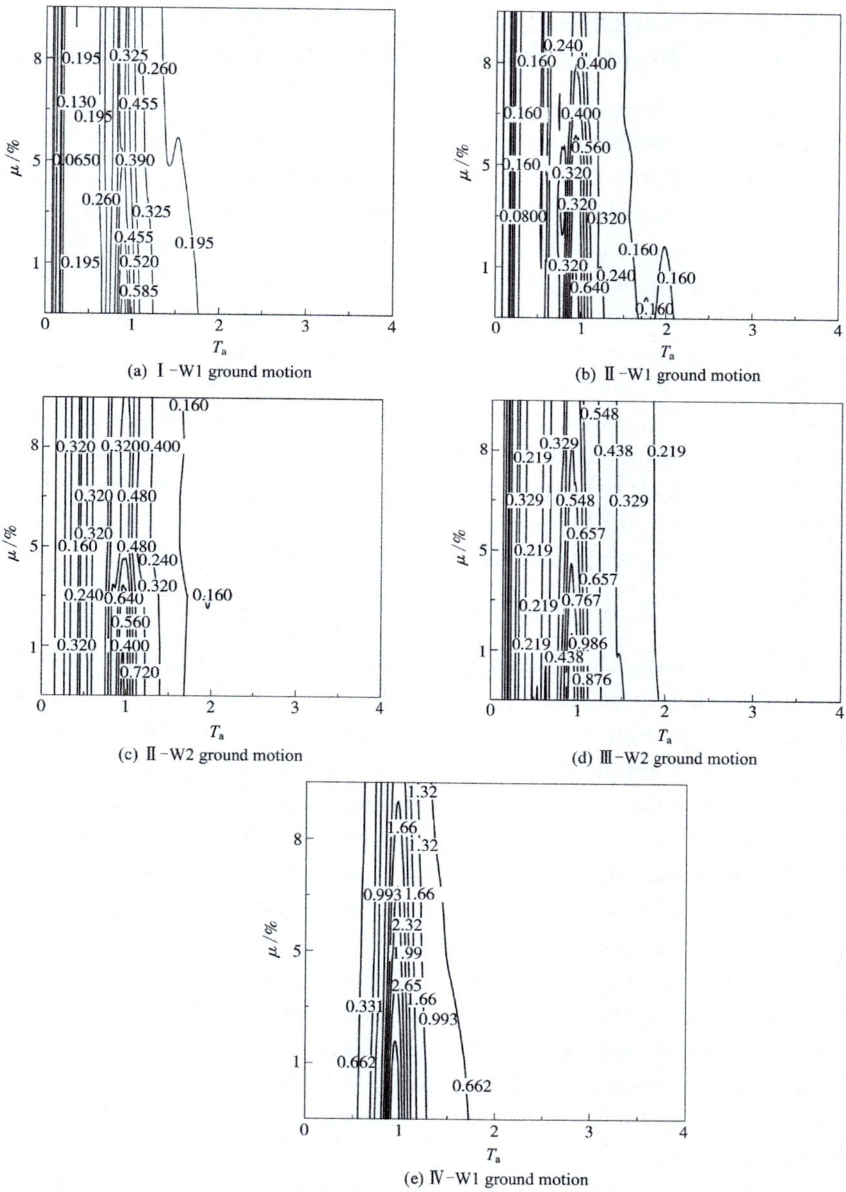

Figure 7.33 Velocity response spectra.

7.5 Performance-Based Design Method for High-Speed Railway Bridges Based on Energy Balance

Essentially, the process of ground motion involves the transfer, transformation, and dissipation of energy from ground movement to structures. At the first World Conference on Earthquake Engineering (WCEE), Housner proposed the energy-based seismic design philosophy [48]. The design of structural capacity curves and damping characteristics ensures that the energy dissipation capacity of the structure meets or exceeds seismic input energy, thereby achieving equilibrium between structural capacity and seismic demand. When the total seismic input energy to the structure is denoted as E_i, the energy dissipated by dampers is E_ξ, while the remaining energy is stored in the structure in the forms of kinetic energy E_k and strain energy E_a. If the structure remains elastic, E_a will be stored as elastic energy E_s. If the structure enters yielding, E_a will be partitioned into elastic energy E_s and hysteretic energy E_h. Uang and Bertero formulated the energy relationships [49] described above in Eq. (7.8).

$$E_i = E_k + E_\xi + E_a = E_k + E_\xi + E_s + E_h \tag{7.8}$$

Extensive research on energy-based performance design methods has been conducted by numerous scholars globally. Akiyama examined seismic input energy for SDOF and multiple degrees of freedom (MDOF) systems, established total input energy spectra applicable to these two system types, conducted the hysteresis and damping energy dissipation analysis of the structure, and proposed steps for energy-based design methods [50]. Xiao et al. conducted time-history analyses on various structural systems to obtain total input energy spectrum curves, while proposing simplified analysis methods for elastoplastic energy evaluation [51]. Jiang et al. analyzed near-fault ground motions on structures, deriving design spectra for seismic input energy in near-field earthquakes, and proposed an energy-based seismic evaluation and design method for bridge structures [52]. The aforementioned research collectively established input energy response spectra to determine the total energy dissipation of structures. Additionally, some studies have focused on establishing the proportion of hysteretic energy dissipation that causes structural deformation or failure relative to total energy dissipation. Fajfar et al. employed an instantaneous stiffness-proportional damping model, accounting for the influence of different hysteretic restoring force models. Statistical analyses using multiple seismic records as input indicate that for damping ratios of 5% and 2%, hysteretic energy dissipation constitutes up to 80% and 90% of total energy dissipation, respectively [53].

7.5.1 Principles of Energy-Based Design Methods

This section approaches seismic demand and seismic capacity from an energy perspective, which constitutes the foundation of energy-based performance design methods. Yang et al. further examined the energy relationship between linear SDOF systems and their equivalent nonlinear SDOF counterparts. For the linear SDOF structure, the structure remains elastic under seismic actions. The seismic input energy is stored as elastic energy, which can be directly calculated through simple computation. In contrast, for their equivalent nonlinear SDOF counterparts, after the structure enters plastic deformation, the input strain energy is ultimately dissipated as hysteretic energy, which is obtained through time-history analysis. For structures under design, such energy cannot be directly calculated but requires iterative computation procedures. However, these two energy components exhibit an equivalence relationship. Therefore, their seismic input energy of an equivalent nonlinear SDOF structure can be derived by calculating the seismic input energy of their linear SDOF counterparts. Meanwhile, when a monotonic pushover analysis on a nonlinear SDOF structure is performed until reaching seismic forces equivalent to those generated by an earthquake of the same magnitude, the input energy exhibits a proportional relationship with the hysteretic energy dissipation upon plastic deformation. Therefore, an energy proportionality coefficient can be established to correlate the elastic input energy of a linear SDOF structure under the same ground motion with the monotonic pushover energy of its equivalent nonlinear SDOF counterpart [54].

Based on the above energy balance concept, Yang et al. developed the EEDP, a novel performance-based design method for parallel seismic systems. This method selects design-level ground motions according to seismic demand, with strength and ductility of the SDOF structure as direct design parameters. By utilizing the energy proportionality coefficient of the structure, the relationship between elastic energy and monotonic pushover energy is established to derive the force–displacement curve of the SDOF structure. Subsequently, an ideal elastoplastic yield mechanism is determined for the primary structure and supplemental seismic system for detailed subsequent component design. This method eliminates the need to preselect initial structural periods or member dimensions and can directly obtain performance curves meeting target requirements without complex iterative calculations, thereby offering significant advantages for practical engineering applications. This book aims to achieve seismic mitigation and control objectives for train–track–bridge systems through effective seismic mitigation and control technologies and feasible design methods. The train–track–bridge system is a complex, large-scale system, but current research in this area remains insufficient. Therefore, in this book, an energy-based performance design method for high-speed railway bridges is proposed. This method applies to both track–bridge structure design and train–track–bridge design. Based on Huang et al.'s research, this book investigates relevant design methods by treating bridge piers as protective components [55,56].

7.5.1.1 Determination of Performance Objectives for Seismic Isolation and Energy Dissipation Design in High-Speed Railway Bridges

The seismic performance of high-speed railway bridge piers is evaluated by accounting for the macroscopic behavior (elastic, nonlinear, and elastoplastic) of the bridge structure under frequent earthquakes, design-level earthquakes, and rare earthquakes as specified in relevant standards. The performance states of bridge piers under three-level seismic fortification conditions are determined through experimental and numerical simulation results. The skeleton curve of bridge piers is simplified into an elastoplastic bilinear using experimental data, thereby deriving the yield shear force of the piers. By comparing it with the seismic force obtained from the response spectrum specified in design standards, it is concluded that high-speed railway bridge piers may undergo yielding under design-level earthquakes. Subsequently, by establishing an FE model of the bridge piers, the stress and strain data of reinforcing steel and concrete can be obtained, making it more reliable by determining the yield point from a mechanical perspective.

Based on the experimental data and numerical simulation of the railway bridge piers, it can be concluded that the bridge piers may yield under the designed seismic load. Given that pier designs typically prioritize stiffness control, without altering the original pier design, seismic isolation and energy dissipation bearings are used in place of fixed bearings for piers that would yield under seismic intensities below the design level. This approach achieves functional separation through the use of seismic isolation and energy dissipation bearings. When the seismic intensity is below frequent earthquakes, seismic isolation and energy dissipation bearings function equivalently to fixed bearings without activating their seismic isolation and energy dissipation functionality, thereby ensuring the operational performance of the bridge. When the seismic intensity reaches frequent earthquakes, the bearings begin to exert their seismic isolation and energy dissipation effects, enabling piers that would otherwise yield before the design-level earthquake to remain elastic before the design-level earthquakes. Under the performance-based objectives specified by the standards, the three-level seismic defense objectives for high-speed railway bridges with supplemental seismic isolation and energy dissipation bearings are further defined as follows. Under frequent earthquake conditions, the bridge structure achieves Performance Level I with both piers and bearings remaining elastic. When the seismic intensity reaches the frequent earthquake level, the bearings begin to slide. Under design-level earthquake conditions, the bridge structure achieves Performance Level II, with piers maintaining elasticity and bearings starting to slide. When the seismic intensity reaches the design-level earthquake level, the piers enter their target yielding state. Under rare earthquake conditions, the bridge structure achieves Performance Level III with both piers and bearings entering the yielding phase, where maximum displacements of piers and bearings are verified for calculation.

7.5.1.2 Establishment of an Equivalent SDOF System

The simply supported girder bridge for a high-speed railway can be simplified into a 2-DOF model, as shown in Fig. 7.34A, where m_1 and m_2 denote the masses of the bridge pier and superstructure, respectively. For performance-based design from an energy-based perspective, the calculation of the energy of an MDOF structure requires defining interstory shear forces and interstory displacements. For the structure under design, the use of the 2-DOF model or energy calculations inevitably involves iterative processes. Therefore, the 2-DOF model consisting of a pier–bearing–girder system in series is simplified into an SDOF model. Seismic isolation and energy dissipation bearings are incorporated into the bridge structure while the original design of the bridge piers is maintained. The constitutive model of the seismic isolation and energy dissipation bearings is simplified into a bilinear model as shown in Fig. 7.34B. By designing the sliding force F_{yb}, when the seismic force on the bearing reaches F_{yb} under frequent earthquake loads, the bearing begins to slide while both the pier and the bearing remain in the elastic stage, as illustrated in Fig. 7.34B. Through the design of the sliding stiffness K_{2p} of the seismic isolation and energy dissipation bearings, under design-level earthquake loads, the pier starts to yield when the corresponding top displacement of

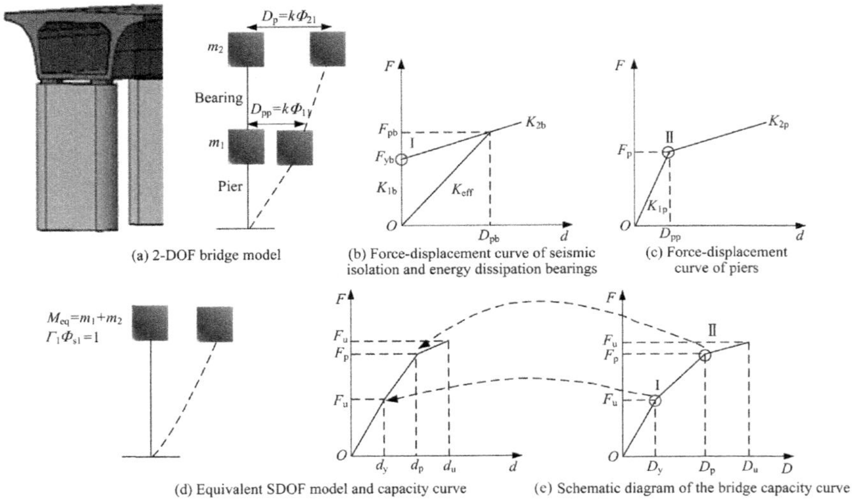

Figure 7.34 Simplified schematic diagram of a pier–bearing–girder model.

the pier reaches D_{pp}, the base shear force becomes F_P, the pier remains elastic, and the bearing begins to slide, as shown in Fig. 7.34C. This clearly defines the two performance points of the bridge structure under frequent and design-level earthquakes, with the structural capacity curve illustrated in Fig. 7.34E.

During the vibration process, it is assumed that the mode shape vector $\{\Phi\}$ of the MDOF system remains constant, and the lateral displacement pattern resembles the fundamental mode shape of the structure. The height of the equivalent SDOF system is determined by Eq. (7.9), as shown in Fig. 7.34D:

$$\Gamma_1 \times \Phi_{s_1} = 1 \tag{7.9}$$

And the first mode participation factor is denoted as Γ_1:

$$\Gamma_1 = \frac{\sum_{i=1}^{n} m_i \Phi_{1i}}{\sum_{i=1}^{n} m_i \Phi_{1i}^2} \tag{7.10}$$

where Φ_{s1} represents the first mode shape of the MDOF system at the equivalent SDOF structure height, and m_i denotes the mass and displacement at each node of the MDOF system. When both vibrate according to the first mode shape with maximum kinetic energy being equal, the following is obtained:

$$\frac{1}{2}M_{\text{eq}}(\omega_1 x_{\text{m}})^2 = \frac{1}{2}\sum_{i=1}^{n} m_i(\omega_1 x_i)^2 \tag{7.11}$$

where ω_1 denotes the natural circular frequency of the first mode shape; M_{eq} and x_{m} are the equivalent mass and displacement of the equivalent SDOF system, respectively; and x_i represents the displacement of the mass at each node of the MDOF system. From Eq. (7.11), we obtain:

$$M_{\text{eq}} = \frac{\sum_{i=1}^{n} m_i x_i^2}{x_m^2} \tag{7.12}$$

Let $x_i = k\Phi_{1i}$ and $x_{\text{m}} = k\Phi_{s1}$, and since $\Phi_{sl} = \frac{1}{\Gamma_1}$, it follows that

$$M_{\text{eq}} = \frac{\left(\sum_{i=1}^{n} m_i \Phi_{1i}\right)^2}{\sum_{i=1}^{n} m_i \Phi_{i1}^2} \tag{7.13}$$

The equivalent SDOF is $d = x_s = k\Phi_{ls}$, and the MDOF top displacement is $D_{\text{r}} = x_n = k\Phi_{1n}$. The conversion equation between the MDOF top displacement and equivalent monotonic top displacement can be derived as follows:

$$D = d\Gamma_1 \Phi_{1n} \tag{7.14}$$

When the seismic intensity is below the frequent earthquake level, the bearings remain fixed. In this case, the bearing stiffness $K_{1\text{b}}$ is much greater than the pier stiffness $K_{1\text{p}}$, with modal components $\Phi_{21} = \Phi_{11} = 1$, $M_{\text{eq}} = m_1 + m_2$, and $D = d$. Therefore, the 2-DOF model of the bearing–pier series system is simplified into an SDOF model with equivalent mass $M_{\text{eq}} = (m_1 + m_2)$ and initial stiffness equal to the pier stiffness $\Phi_{1\text{p}}$.

When the pier reaches the yielding state under the design-level earthquake action, the base shear force is F_{p} with corresponding pier-top displacement D_{pp}. When the bearing slips, the secant stiffness K_{eff} corresponding to the maximum displacement of the bearing D_{pb} is adopted. A modal analysis is then performed with K_{eff} and the pier stiffness $K_{1\text{p}}$. Assuming the first mode dominates the vibration, the modal shapes Φ_{11} and Φ_{21} are obtained. Then:

$$\frac{D_{pb}+D_{pp}}{D_{pp}}=\frac{1}{\Phi_{11}} \tag{7.15}$$

Meanwhile, the relationship between the top displacement of the MDOF system and its equivalent SDOF system is as follows:

$$D=d_{\mathrm{p}}\Gamma_1\Phi_{12} \tag{7.16}$$

$$\Gamma_1=\frac{m_1\Phi_{11}+m_2}{m_1\Phi_{11}^2+m_2} \tag{7.17}$$

This establishes the relationship between the displacement response and the equivalent secant stiffness of bearings for the bearing–pier series system and its equivalent SDOF system under design-level earthquake excitation, from which the capacity spectrum for the equivalent SDOF system is derived.

7.5.1.3 Design Under Different Seismic Fortification Levels

The seismic fortification levels of the bridge are established as frequent earthquake, design-level earthquake, and rare earthquake according to the standards. The corresponding peak ground accelerations are determined based on the seismic fortification intensity of the bridge, the appropriate response spectrum curves are selected according to the site type and characteristic period zone of the region, and the response spectrum curves are then transformed into the curves of bearing-top displacement D_y and pier base shear force F:

$$F=S_{\mathrm{a}}\times(m_1+m_2) \tag{7.18}$$

$$D=S_{\mathrm{a}}\frac{T^2}{4\pi^2} \tag{7.19}$$

where S_{a} represents the response spectrum acceleration corresponding to the simplified model, and T denotes the period of the equivalent SDOF structure.

(1) Frequent earthquakes

The top displacement d_{y} and base shear F_{y} of the structure under frequent earthquake action can be obtained from Eq. (7.17), as shown in Fig. 7.34. F_{y} is calculated using Eq. (7.20):

$$F_{\mathrm{y}}=S_{al}\times(m_1+m_2) \tag{7.20}$$

By distributing F_y to the piers and bearings according to their masses, the sliding force of bearings F_{yb} can be obtained:

$$F_{yb} = \frac{m_2}{m_1 + m_2} F_y \tag{7.21}$$

and

$$D_y = d_y = F_y / K_{p1} \tag{7.22}$$

where S_{a1} denotes the spectral acceleration corresponding to the frequent earthquake; F_y represents the base shear of the structure under frequent earthquake action; F_{yb} corresponds to the sliding force of the bearing, and D_y and d_y denote the top displacements of the 2-DOF and SDOF structures.

(2) Design-level earthquakes

Under the design-level earthquake action, the piers begin to yield where the corresponding base shear of the pier is the known value F_p, and the displacement of the equivalent SDOF system is denoted as d_p. Assuming the structure always remains elastic, when the seismic intensity increases from a frequent earthquake to a design-level earthquake, the incremental elastic energy input to the structure is ΔE_{E1}; whereas, in reality, when the seismic intensity increases from a frequent earthquake to a design-level earthquake, the incremental hysteretic energy dissipation after the structure yields is ΔE_{ND1}. When pushover analysis on the yielding structure is performed until reaching the predetermined performance objective, where the base shear reaches F_p, the incremental energy input to the structure is ΔE_{NM1}. The relationship among these three energy terms is as follows, where γ_a represents the ratio of the incremental elastic energy input to the structure under design-level earthquake action to the incremental energy input under monotonic shear load, as illustrated in Fig. 7.35.

$$\Delta E_{E1} = \Delta E_{ND1} = \gamma_a \Delta E_{NM1} \tag{7.23}$$

$$\Delta E_{E1} = \frac{1}{2} \left(F_y + F_{DE} \right) \cdot \left(d_{DE} - d_y \right) \tag{7.24}$$

$$\Delta E_{NM1} = \frac{1}{2} \left(F_p + F_y \right) \cdot \left(d_p - d_y \right) \tag{7.25}$$

By combining Eqs. (7.23)–(7.25), the following expression is obtained to determine d_p.

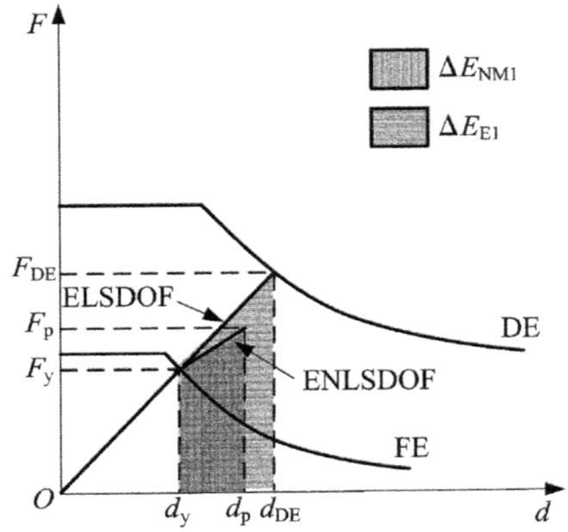

Figure 7.35 Schematic diagram of energy-based design under frequent earthquakes and design-level earthquakes. *d*, top displacement of the equivalent SDOF structure; *DE*, design-level earthquake; *ELSDOF*, elastic SDOF structure; *ENLSDOF*, elastoplastic SDOF structure; *F*, base shear force of the equivalent SDOF structure; *FE*, frequent earthquake.

$$d_p = \frac{2\Delta E_{E1}}{\gamma_a (F_p + F_y)} + d_y \tag{7.26}$$

where F_{FE} and d_{FE} represent the base shear force and top displacement of the elastic structure under the frequent earthquake, and F_p and d_p denote the base shear force and top displacement of the equivalent SDOF structure under the design-level earthquake. After the equivalent SDOF displacement is obtained, the secant stiffness corresponding to the bearing displacement can be determined using Eqs. (7.14)–(7.16), and its sliding stiffness can then be calculated. At this point, the two-parameter design of the bearing is completed, achieving Performance Objectives I and II.

(3) Rare earthquakes

In China's seismic design specifications, the seismic design of bridge piers under rare earthquakes is simplified to verify whether the ductility coefficient of the piers meets the requirements. The preceding section has completed the design of the bearings, and the mass and stiffness parameters of the bridge structural components have been determined. Consequently, the displacement of the bearings and the shear force at the pier base are determined using the energy-based design method.

When the seismic intensity acting on the bearing–pier model increases from the design-level earthquake to the rare earthquake, it is assumed that the seismic force increments at the pier base ΔF_1 and the bearing ΔF_2 are proportional to the mass, with the proportionality coefficient defined as a; Δd_u and ΔF_u represent the displacement increment at the top of the bearing and the seismic force increment at the pier base under the rare earthquake, respectively. Assuming the third-stage stiffness of the SDOF model used for energy calculation is k_u, the following relationship holds:

$$\frac{\Delta F_1}{\Delta F_2} = \frac{m_1}{m_2} = a \tag{7.27}$$

$$\Delta d_u = \frac{\Delta F_1 + \Delta F_2}{k_{b2}} + \frac{\Delta F_2}{k_{p2}} \tag{7.28}$$

$$k_u = \frac{\Delta F_u}{\Delta d_u} = \frac{\Delta F_1 + \Delta F_2}{\Delta d_u} = \frac{k_{p2} k_{b2}(a+1)}{k_{p2} + k_{b2}(a+1)} \tag{7.29}$$

If the structure remains entirely elastic, the increment in energy input to the structure when the seismic intensity increases from the design-level earthquake to the rare earthquake is ΔE_{E2}. However, in reality, both the bearing and the pier yield under a rare earthquake, resulting in an increment in hysteretic energy dissipation of ΔE_{ND2}. ΔE_{NM2} represents the increment in energy input when the seismic intensity increases from the design-level earthquake to the rare earthquake, assuming a monotonic seismic shear force is applied at the base of the structure after both the bearing and pier have yielded. γ_{b} represents the ratio of the energy input to the structure under rare earthquake excitation to the energy input under monotonic shear load, as illustrated in Fig. 7.36. The following relationship holds:

$$\Delta E_{\mathrm{E2}} = \Delta E_{\mathrm{ND2}} = \gamma_{\mathrm{b}} \Delta E_{\mathrm{NM2}} \tag{7.30}$$

$$\Delta E_{\mathrm{E2}} = \frac{1}{2}(F_{\mathrm{RE}} + F_{\mathrm{FE}})(d_{\mathrm{RE}} - d_{\mathrm{FE}}) \tag{7.31}$$

$$\Delta E_{\mathrm{NM2}} = \frac{1}{2}\left(F_{\mathrm{u}} + F_{\mathrm{p}}\right)\left(d_{\mathrm{u}} - d_{\mathrm{p}}\right) \tag{7.32}$$

$$F_{\mathrm{u}} = \frac{2\Delta E_{\mathrm{E2}}}{\gamma_{\mathrm{b}}\left(d_{\mathrm{u}} - d_{\mathrm{p}}\right)} - F_{\mathrm{p}} \tag{7.33}$$

$$F_{\mathrm{u}} = k_{\mathrm{u}} \Delta d_{\mathrm{u}} + F_{\mathrm{p}} \tag{7.34}$$

$$\Delta d_{\mathrm{u}} = d_{\mathrm{u}} - d_{\mathrm{p}} \tag{7.35}$$

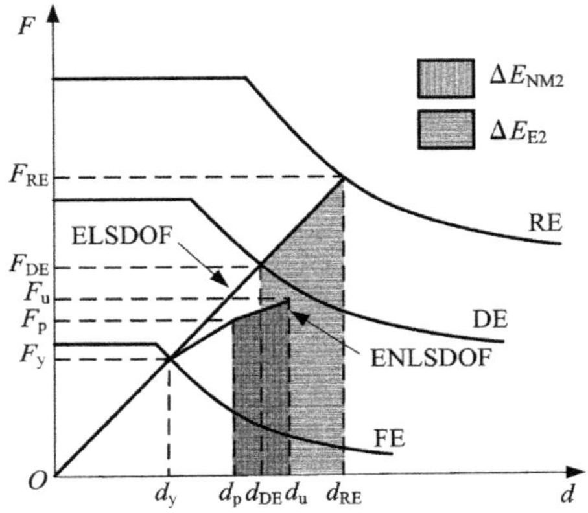

Figure 7.36 Schematic diagram of energy-based design method for rare earthquakes (RE).

In Fig. 7.36, F_{RE} and d_{RE} represent the base shear force and top displacement, respectively, F_u and d_u denote the base shear force and top displacement, respectively, of the elastic structure under rare earthquakes. Based on the previously established equivalence relationship, the base shear force F_u can be considered as the base shear force of the bearing–pier model. Meanwhile, the pier-top displacement is calculated to perform a ductility check, and the top displacement d_u is approximated as the bearing displacement, providing a reference for the maximum bearing displacement under a rare earthquake.

7.5.1.4 Energy Balance Coefficient

Housner suggested that the energy dissipated by the SDOF structure constitutes only a portion of the energy stored in its equivalent linear SDOF counterpart, but did not further investigate the proportional relationship between the two [57]. Goel and Chao adopted the inelastic response spectra derived by Newmark and Hall for idealized elastoplastic SDOF structures to quantify this relationship [58]. They expressed this relationship through a proportionality coefficient, deriving the following equation [59]:

$$\gamma = \frac{2\mu_s - 1}{R_\mu^2} \tag{7.36}$$

where μ_s denotes the structural ductility factor, and R_μ denotes the structural strength reduction factor. The energy balance coefficient proposed by Goel and Chao is based on the design method for elastoplastic structures.

The system must satisfy both the requirements of the maximum fortification earthquake and the basic design-level earthquake. However, simultaneously satisfying both seismic intensity demands is challenging, and the final design is typically governed by one of these strength requirements. The previous section designed the structural performance under three defined seismic fortification levels. Therefore, two energy balance coefficients are required to achieve the energy conversion relationship when the seismic intensity of the structure increases from the frequent earthquake level to the design-level earthquake level, and from the design-level earthquake level to the rare earthquake level. The core principle of the proposed design method in this section is to achieve distinct structural performance under different seismic fortification levels without requiring iterative calculations, where the most critical factor lies in the selection of energy balance coefficients. For predesigned bridge piers, the response spectrum can be determined based on their site type, characteristic period, and seismic fortification intensity, which in turn guides the selection of ground motion records. According to the site type, characteristic period, and seismic fortification intensity criteria provided in Code for Seismic Design of Railway Engineering (GB 50111–2006), the response spectrum is determined for a Type II site, Characteristic Period Zone 3 with a seismic fortification intensity VIII, corresponding to the PGA of $0.1g$ for frequent earthquakes, $0.3g$ for design-level earthquakes, and $0.57g$ for rare earthquakes. Based on the response spectrum specified in this standard, 16 ground motion records, as shown in Table 7.5, are selected, and the average response spectrum of these ground motions is used for subsequent calculations.

(1) Calculation of γ_a

Under design-level earthquake action, the performance objective (base shear) is known, and the yield ratio μ is defined as the ratio of target shear force F_p to the base shear force F_y of the pier under frequent earthquakes ($\mu = F_p/F_y$). The yield ratio must satisfy both energy balance and practical serviceability requirements. If the yield ratio is excessively small, the structure will yield under small seismic forces, and to dissipate the input seismic energy, the structure will undergo excessive displacements, whereas if the yield ratio is too large, the structure will develop excessive displacements. When the yield ratio is excessively large, the structure will only yield under relatively high seismic forces, thereby increasing the structural strength requirements. Consequently, for structures designed for a seismic fortification intensity of 8°, the yield ratio is selected within the range of 1.2–2.8. For a given yield ratio, multiple possible values of γ_a exist,

Table 7.5 Information on selected ground motion records.

Ground Motion Name	Year	Intensity	NGA#	Record
Chi-Chi, Taiwan, China China -05	1999	6.2	3160 3191	TCU014N TCU081N
Imperial Valley-06	1979	6.53	175 167	E12140 CMP015
Northridge-01	1994	6.69	970 1000	FAI095 PIC090
Kobe, Japan	1995	6.9	1100 1102	ABN000 CHY000
Landers	1992	7.28	850 3757	DSP000 NPF090
San Fernando	1971	6.61	57 83	ORR021 PUD055
Loma Prieta	1989	6.9	762 800	FRE000 SJW160
Superstition Hills-01	1987	6.2	718 726	IVW090 WLF225

as illustrated in Fig. 7.37. Within the range of possible γ_a values, an arbitrary value is selected, and the skeleton curve of the corresponding structure can be derived using Eq. (7.24). The average maximum top displacement under design-level earthquake is obtained by performing time-history analyses on this structure using the selected 16 ground

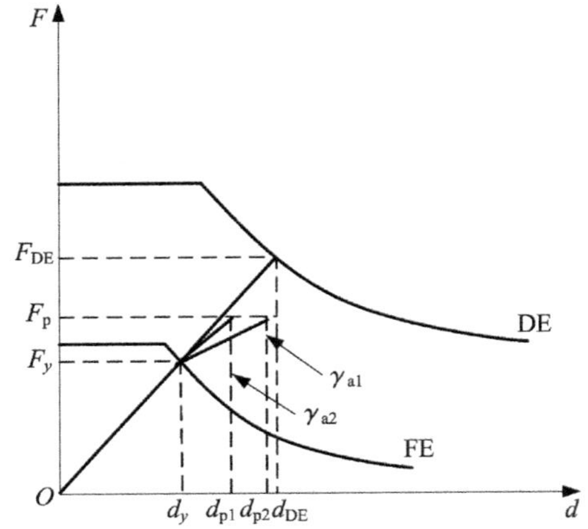

Figure 7.37 Schematic diagram of γ_a calculation.

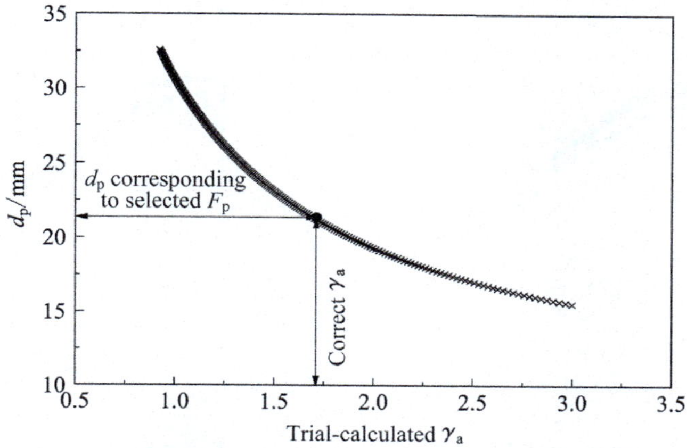

Figure 7.38 Influence of γ_a on top displacement.

motions. This average displacement value is then compared with the top displacement d_p calculated from Eq. (7.25), and the γ_a value corresponding to the closest matching case is selected as the optimal parameter for the initial period and yield ratio of the structure. Fig. 7.38 presents the relationship between γ_a and the top displacement under design-level earthquake conditions for $T = 0.39$ s and $\mu = 1.8$.

Fig. 7.39 presents the calculated results of γ_a. The relationship between γ_a and μ is fitted using a cubic curve based on the structural period, as expressed in Fig. 7.38. The correlation coefficients can be selected according to different periods.

$$\gamma_a = p_1\mu^3 + p_2\mu^2 + p_3\mu + p_4 \tag{7.38}$$

where p_1, p_2, p_3, and p_4 are fitting parameters, whose values can be obtained from the specified table.

(2) Calculation of γ_b

For a structure with an initial period and yield ratio, the post-yield stiffness ratio ν ($\nu = K_2/K_3$) must be introduced to determine its skeleton curve, which is defined as the ratio of the third-segment stiffness to the second-segment stiffness in the skeleton curve. As shown in Fig. 7.40, as with the calculation of γ_a, for a selected SDOF structure with an initial period, yield ratio, and post-yield stiffness ratio, its skeleton curve can be derived. Under the selected rare earthquake condition with a given base shear force F_u, the top displacement d_u can be calculated from the stiffness. The corresponding γ_b can be determined using Eqs. (7.30)–(7.35). The

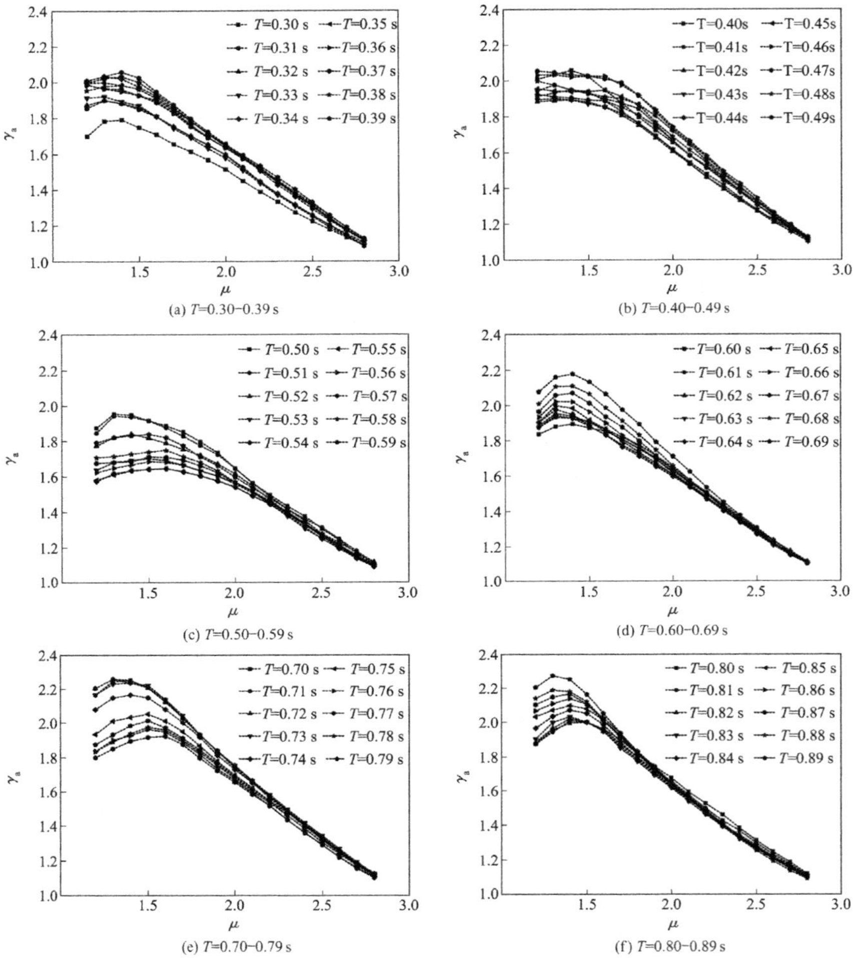

Figure 7.39 Calculation results of γ_a.

average maximum top displacement obtained through time-history analysis under rare earthquake conditions is compared with the calculated d_u. The γ_b value corresponding to the closest match is determined as the target value for the given initial period, yield ratio, and post-yield stiffness ratio. This section provides γ_b values for $T = 0.831$ s, $\mu = 1.9$, and $\nu = 0.1$–0.5, as shown in Fig. 7.41, along with the fitted empirical equation.

$$\gamma_b = -2.243\mu^2 + 1.775\mu + 1.755 \tag{7.39}$$

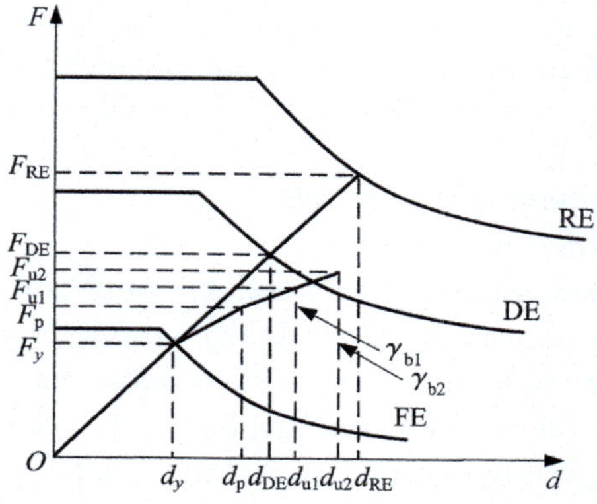

Figure 7.40 Schematic diagram of γ_b calculation.

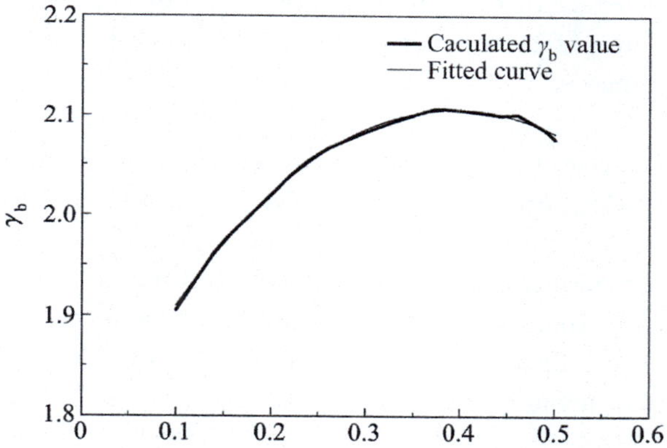

Figure 7.41 Value of γ_b when $T = 0.831$ s, $\mu = 1.9$.

7.5.2 Verification of the Performance-Based Design Method

To verify the accuracy of the EEDP method in high-speed railway bridges, this book designs seismic isolation and energy dissipation bearings for a 32 m simply supported prototype-based girder bridge pier. After the design parameters are obtained, a numerical model is established to perform nonlinear time-history analysis. The force–displacement results from the analysis are then compared with the design values to validate the accuracy of the method. The validation process is divided into two parts. In the first part, the response spectrum of a single seismic motion is used to design the model to avoid input errors introduced by the design spectrum. In the second part, to align with engineering practice, the design follows the principle of multiple scenarios. The average spectrum of the aforementioned 16 ground motions is adopted for design, thereby validating the effectiveness of the EEDP method in practical engineering applications.

7.5.2.1 Establishment of FE Models

(1) Track and girder structure

High-speed railway bridge girders exhibit significant lateral and vertical stiffness, generally remaining elastic under seismic conditions. Ballastless track is a form of track that replaces loose ballast beds with concrete or asphalt mixtures, offering superior stability, ride comfort, and uniform stiffness. The research by Yuan et al. and the Code for Seismic Design of Railway Engineering (GB 50111–2006) state that a single-pier model is generally used for seismic analysis of simply supported girder bridges with only the mass effect of the girder section considered [26]. Consequently, this section simplifies superstructures such as girders and tracks into concentrated masses.

(2) Piers

Based on a series of experimental data on high-speed railway piers and determined bridge performance objectives, the seismic force–displacement curve of the pier is simplified into a bilinear elastoplastic model, and the SDOF model is adopted to simulate the pier. In OpenSees FE software, the twoNodeLink element is used for pier simulation with the Steel01 material to represent the bilinear behavior of the pier, with the constitutive relationship shown in Fig. 7.42.

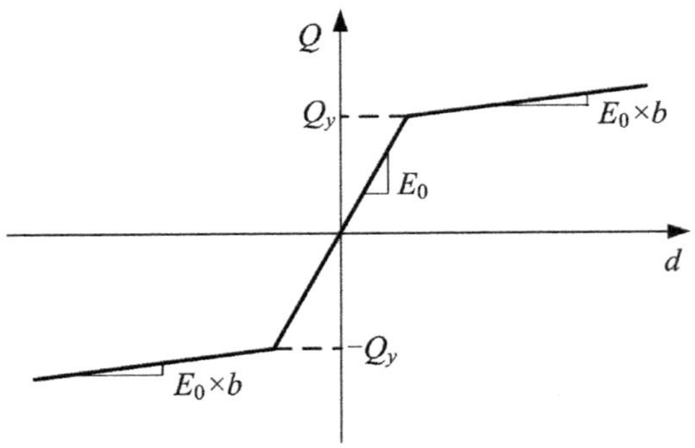

Figure 7.42 Constitutive relationship of Steel 01 material. *b*, hardening rate; *d*, deformation; E_0, initial stiffness; *Q*, internal force; Q_y, yield strength.

(3) Seismic isolation and energy dissipation bearings

Seismic isolation and energy dissipation bearings are also simulated using the twoNodeLink element with Steel01 material. Only horizontal seismic loads are considered in the analysis. Since bearings typically undergo shear failure under horizontal loads, the vertical load–displacement relationship is not investigated here. Additionally, a high vertical stiffness is assigned to the bearings for computational purposes.

7.5.2.2 Verification of the Single Ground Motion Response Spectrum

A simplified 2-DOF model for the "pier–bearing–girder" corresponding to a 32-m simply supported girder bridge is established. The bilinear skeleton curve model of an 8-m bridge pier is used, and Chi-Chi and San Fernando ground motions are selected, with their time-history responses shown in Fig. 7.43. The EEDP method is used to design the initial sliding force and sliding stiffness of seismic isolation and energy dissipation bearings. The model designed based on Chi-Chi ground motion is labeled as Model 1, while the model designed based on San Fernando ground motion is labeled as Model 2. Time-history analyses are performed using the corresponding ground motions, and the time-history responses of the piers and bearings under design-level earthquakes are shown in Fig. 7.44. The maximum values of pier base shear force and bearing-top displacement obtained from time-history analysis are compared with design values, as shown in Fig. 7.45. The hysteresis curves of the piers and bearings under various seismic fortification levels are illustrated in Figs. 7.46 and 7.47, respectively.

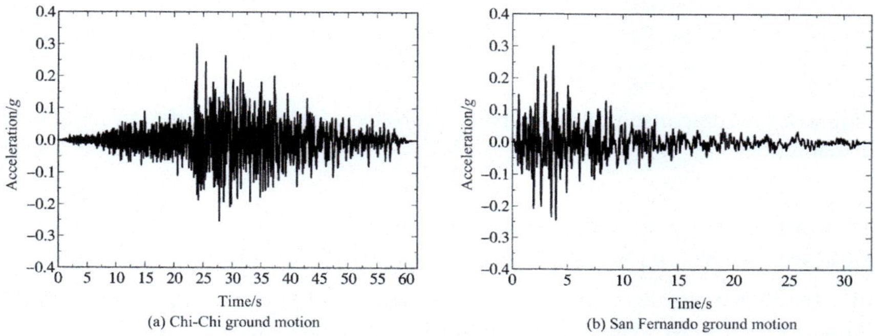

Figure 7.43 Time-history responses under ground motions.

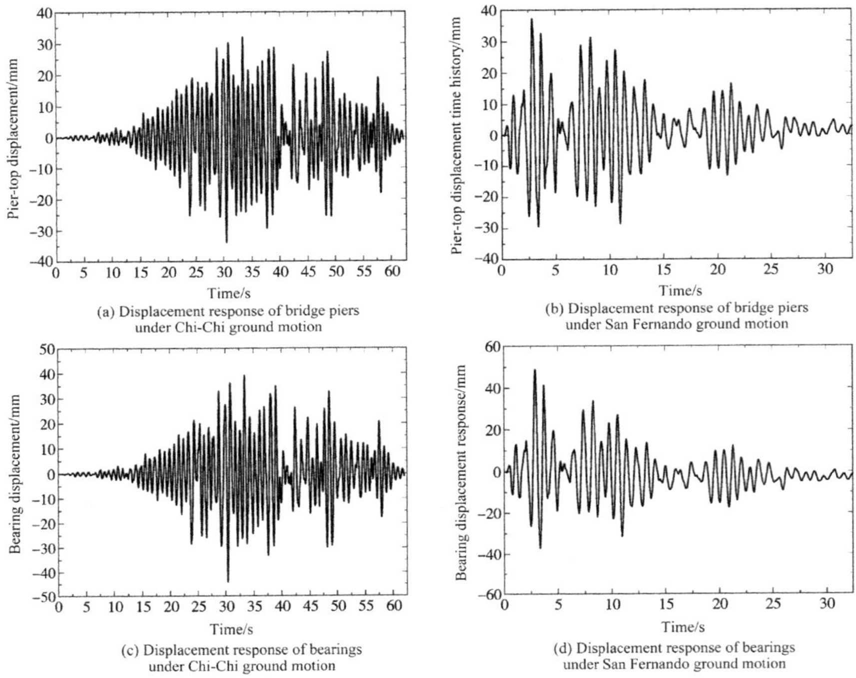

Figure 7.44 Seismic time-history responses of the bridge.

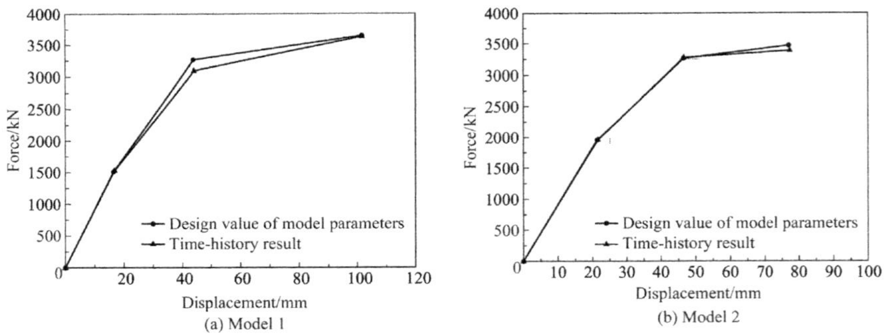

Figure 7.45 Comparison of performance objectives between the design model and actual model.

Under frequent earthquake excitation, the time-history values of Model 1 and Model 2 align with the design values, with negligible errors in pier base shear force, pier-top displacement, and bearing-top displacement. The piers remain elastic, while the bearings are on the verge of sliding,

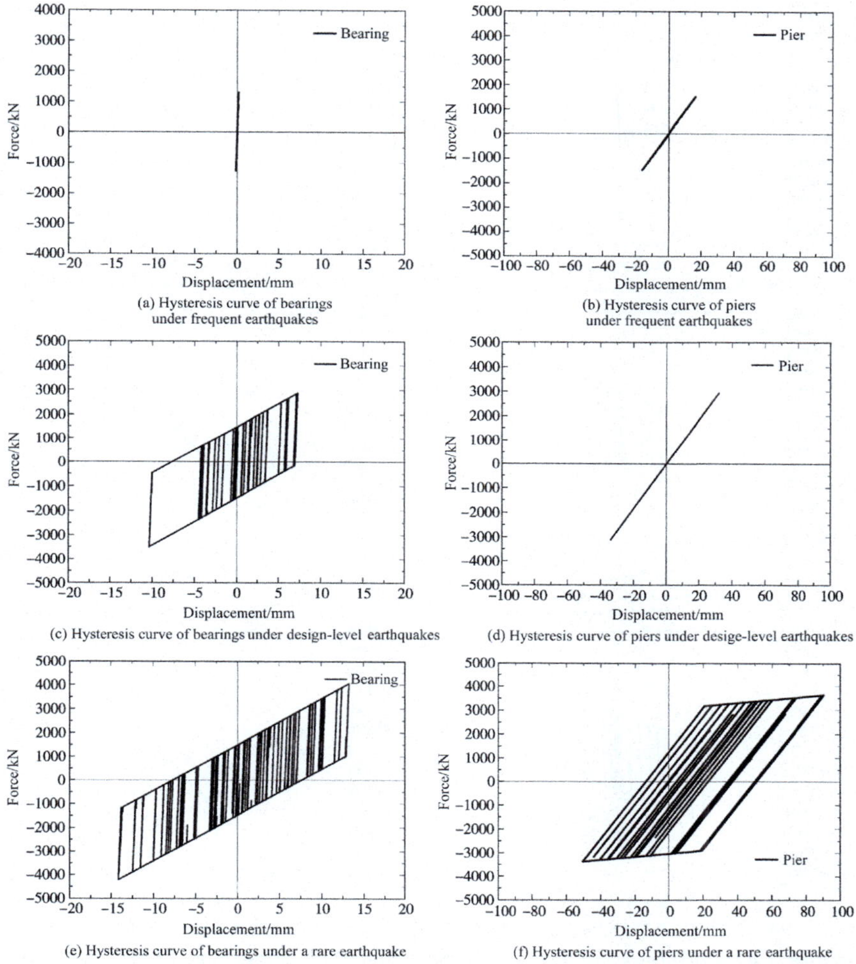

(a) Hysteresis curve of bearings under frequent earthquakes

(b) Hysteresis curve of piers under frequent earthquakes

(c) Hysteresis curve of bearings under design-level earthquakes

(d) Hysteresis curve of piers under desige-level earthquakes

(e) Hysteresis curve of bearings under a rare earthquake

(f) Hysteresis curve of piers under a rare earthquake

Figure 7.46 Hysteresis curves of structural members under Chi-Chi ground motion excitation.

because sliding initiates only when the seismic force on the bearing exceeds its friction force. Under this condition, it is appropriate to model the bridge as an equivalent SDOF and directly calculate displacement and shear force using response spectrum analysis. Under design-level earthquake excitation, for Model 1, the maximum pier base shear force obtained from time-history analysis exhibits a 5.71% error, below the expected yield shear force. As the pier remains elastic under design-level earthquakes, the corresponding pier displacement and bearing displacement are slightly

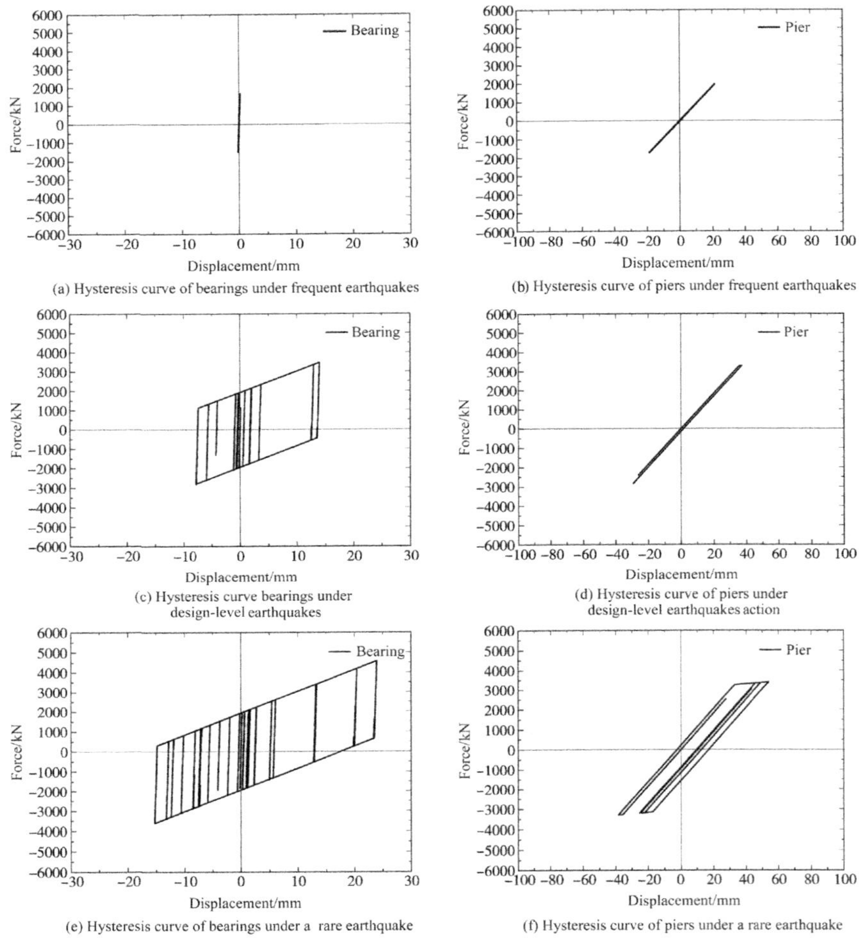

Figure 7.47 Hysteresis curves of structural members under San Fernando ground motion excitation.

smaller than the design values, with the bearing in a sliding state. For Model 2, the pier base shear force shows a 0.34% error, slightly above the expected yield shear force. The pier-top displacement and bearing displacement have errors of 4.67% and 4.76%, respectively, both slightly above the design values. In addition, as the pier yielded, the bearing also remained in a sliding condition. Under design-level earthquake excitation, the errors arise from simplified calculations during the design process, such as equivalent linearization, which introduce certain errors. In practice, it is challenging to precisely control the pier to reach the yield state point, its intended performance target. Once the pier begins to exhibit nonlinear behavior under design-level earthquakes, significant deviations will arise between time-history pier-top displacements and design pier-top displacements. Under rare earthquake excitation, the calculated pier base shear forces for both Model 1 and Model 2 are slightly lower than the design values, while the bearing-top displacements align well with the design values. Piers remain in a yielded state, and the bearings remain in a sliding state. As demonstrated by the analysis, when the same ground motion is used for model design and time-history analysis and input errors are avoided, under frequent earthquake excitation, the models achieve the intended performance targets with high precision by comparing the time-history results with the design values. Under design-level and rare earthquake excitations, the maximum time-history values of pier base shear force and displacement deviate within 10% of the design values, confirming the effectiveness of the method in accurately realizing the expected design objectives.

7.5.2.3 Design and Analysis of the Average Spectrum Under Multiple Ground Motions

Conditions under multiple ground motions should be considered for engineering design. To verify the feasibility of the method in practical engineering design, the average spectrum of the 16 ground motions mentioned earlier is selected to design an 8-meter-high bridge pier in this section. The bridge design parameters obtained using the EEDP method are presented in Table 7.6, and the EEDP design of the 32-m simply supported girder bridge is illustrated in Fig. 7.48.

After the initial sliding force and sliding stiffness of seismic isolation and energy dissipation bearings are obtained, a 2-DOF model of the pier–bearing–girder system is established. The pier model adopts a bilinear skeleton curve model for the prototype pier. Sixteen ground motions are applied to this model for time-history analysis. The average of the maximum values of pier base shear force, pier-top displacement, and bearing-top displacement from the time-history analysis are compared with their design values, as shown in Figs. 7.49–7.51.

Under frequent earthquake excitation, the average of the maximum pier base shear force, pier-top displacement, and the time history of bearing displacement is close to the design value, showing good agreement with the results. Under design-level earthquake excitation, the pier base shear force exhibits a 4.88% error below the design value, and the pier remains elastic. Consequently, the bearing-top displacement is less than the design value, with errors within 2%. Under rare earthquake excitation, the

Table 7.6 Design parameters of the 32-m simply supported girder bridge.

Parameters	Values	Remarks
m_1 (kg)	246,694	Pier mass, known
m_2 (kg)	1,360,200	Girder mass, known
m_{eq} (kg)	1,606,894	Equivalent mass, Eq. (7.12)
F_y (kN)	1721.41	Pier base shear force under frequent earthquakes, Eq. (7.19)
d_y (mm)	18.76	Equivalent SDOF top displacement under frequent earthquakes, Eq. (7.21)
D_y (mm)	18.76	Bearing-top displacement under frequent earthquakes, Eq. (7.21)
F_{yb} (kN)	1454.44	Initial sliding force of the bearing, Eq. (7.20)
γ_a	1.729	Energy balance coefficient, Eq. (7.36)
F_p (kN)	3266.47	Pier yield shear force (performance objective under design-level earthquakes)
d_p (mm)	48.72	Equivalent SDOF top displacement under design-level earthquakes, Eq. (7.25)
D_p (mm)	50.40	Bearing-top displacement under design-level earthquakes, Eqs. (7.13)–(7.15)
K_{2b} (kN/mm)	91.75	Bearing sliding stiffness, Eq. (7.15)
γ_b	1.937	Energy balance coefficient, Eq. (7.37)
F_u (kN)	3620.81	Pier base shear force under rare earthquakes, Eqs. (7.32)–(7.34)
D_u (mm)	105.42	Bearing-top displacement under rare earthquakes, Eqs. (7.32)–(7.34)

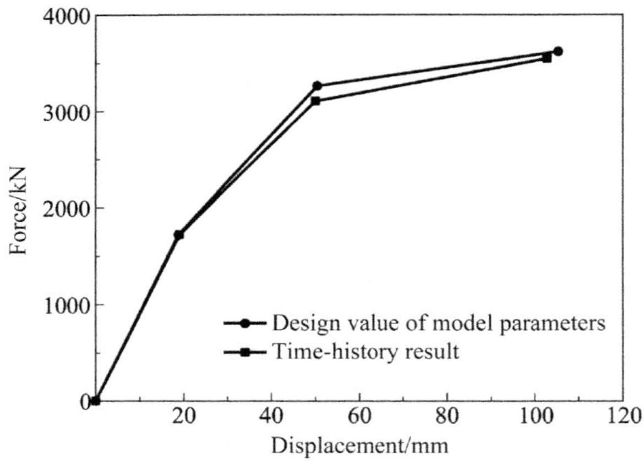

Figure 7.48 Design schematic of a 32-m simply supported girder bridge with EEDP.

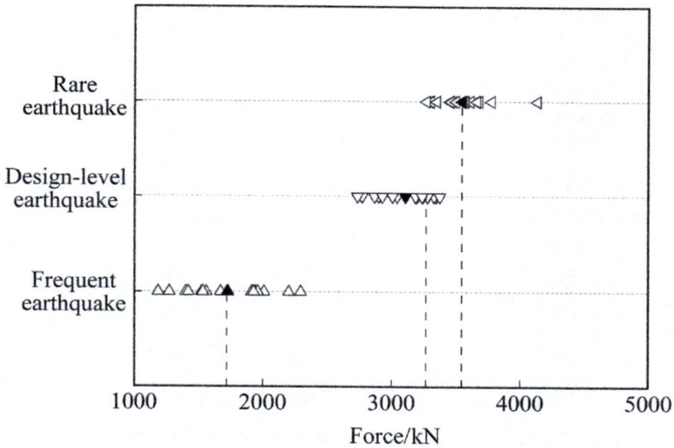

Figure 7.49 Comparison between design values and time-history values of pier base shear. △, maximum response under frequent earthquakes in time-history analysis; ▽, maximum response under design-level earthquakes in time-history analysis; ◁, maximum response under rare earthquakes in time-history analysis; ▲, average maximum response under frequent earthquakes in time-history analysis; ▼, average maximum response under design-level earthquakes in time-history analysis; ◀, average maximum response under rare earthquakes in time-history analysis.

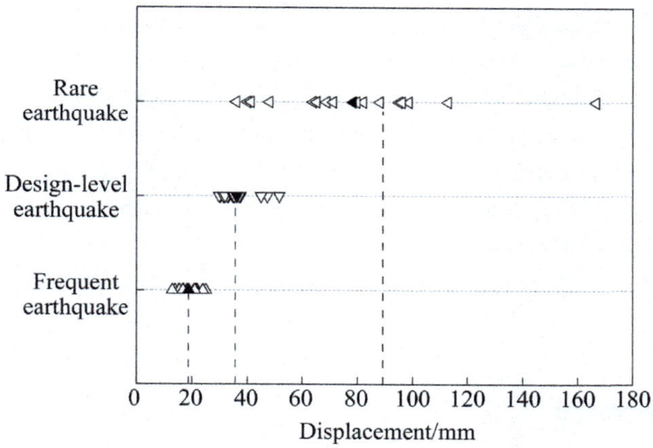

Figure 7.50 Comparison between design values and time-history values of pier-top displacement. △, maximum response under frequent earthquakes in time-history analysis; ▽, maximum response under design-level earthquakes in time-history analysis; ◁, maximum response under rare earthquakes in time-history analysis; ▲, average maximum response under frequent earthquakes in time-history analysis; ▼, average maximum response under design-level earthquakes in time-history analysis; ◀, average maximum response under rare earthquakes in time-history analysis.

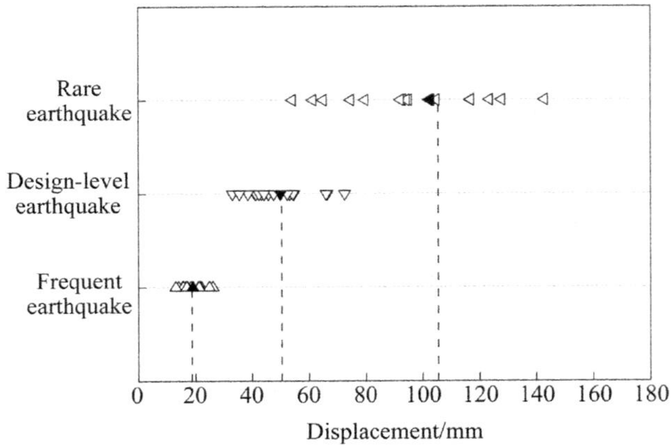

Figure 7.51 Comparison between design values and time-history values of bearing displacement. △, maximum response under frequent earthquakes in time-history analysis; ▽, maximum response under design-level earthquakes in time-history analysis; ◁, maximum response under rare earthquakes in time-history analysis; ▲, average maximum response under frequent earthquakes in time-history analysis; ▼, average maximum response under design-level earthquakes in time-history analysis; ◀, average maximum response under rare earthquakes in time-history analysis.

bearing-top displacement closely matches the design expectation, while the pier base shear force is slightly lower than the design value. The pier-top displacement exhibits a 12.12% error below the design value. As demonstrated by the above results, the performance-based design values of simply supported girder bridges for high-speed railways based on average response spectra of multiple ground motions show errors within 15% compared to time-history analysis values, confirming that the energy-based design method effectively achieves the intended performance objectives for such bridges.

7.6 Summary

As earthquake-prone zones cover vast areas of China, high-speed railway lines are inevitably exposed to seismic risks, posing significant risks to their operational safety. In this chapter, the work of our research group on seismic mitigation and control design for train–track–bridge systems under earthquakes was presented.

(1) Under minor earthquakes, high-speed railway bridges typically remain intact, but trains running on them may still derail. Based on our study of train operational safety on simply supported girder bridges under far-field seismic conditions, this book presents an optimized parameter design for TMDs, with experimental validation of their seismic mitigation performance. The results demonstrate that properly designed TMDs can effectively reduce vibrations near the structural fundamental frequency under minor seismic events and enhance operational safety for high-pier bridges.

(2) Our research group employs nonlinear time-history analysis to investigate the seismic response of bridge structures under track system constraints, and proposes a method to use the novel RFD as a vibration control device for high-speed railway bridges. Simplified modeling and quasi-dynamic hybrid simulation show that RFDs effectively suppress pier responses but may amplify structural responses in other components. Therefore, meticulous design of the initial sliding force of the RFD is required.

(3) To mitigate seismic responses of simply supported girder bridges for high-speed railways under strong earthquakes, a novel combined energy-dissipation and displacement-restriction bearing is proposed. The bearing is applied to the simply supported girder track–bridge system for high-speed railways under three working conditions, and a comparative analysis of the seismic mitigation performance of the combined bearing is conducted. The comparative analysis reveals that the implementation of this bearing not only effectively reduces peak responses of pier-top displacement and base shear force but also significantly enhances the seismic capacity of piers under rare earthquakes and reduces the post-seismic residual displacement of piers, demonstrating pronounced seismic mitigation performance.

(4) Our research group proposes the EEDP method for seismic isolation and damping design of high-speed railway bridges. The most notable feature of this method lies in its ability to achieve the desired multilevel seismic performance objectives without iteration. Based on energy balance principles, the method defines energy balance coefficients to quantify the energy relationship between nonlinear and equivalent linear SDOF structures when seismic intensity increases from frequent to design-level earthquakes, and further to rare earthquakes. Through extensive time-history analyses and data fitting, recommended formulas for these coefficients are derived. Our research team applied the EEDP method to design a 32-meter simply supported girder bridge. An FE model was developed for analysis, and the time-history analysis results were compared with the design values, thereby validating the accuracy of the proposed method.

References

[1] Jiang LZ, Zhou WB, Wei B, et al. Research progress of train–track–bridge system safety of high-speed railway under earthquake action. China Civ Eng J 2020;53(9):1–13 [In Chinese].
[2] Yeh K. Adaptive fuzzy control for earthquake-excited buildings with lead rubber bearing isolation. Int J Phys Sci 2011;6(6):1283–92.
[3] Ou JP. Structural vibration control-active, semi-active, and intelligent control. Beijing: science press; 2003 [In Chinese].
[4] Pall AS, Marsh C. Response of friction damped braced frames. J Struct Eng 1982;108(9):1313–23.
[5] Grigorian CE, Yang TS, Popov EP. Slotted bolted connection energy dissipators. Earthq Spectra 1993;9(3):491–504.
[6] Mualla IH, Belev B. Performance of steel frames with a new friction damper device under earthquake excitation. Eng Struct 2002;24(3):365–71.
[7] Mualla I, Belev B. Analysis, design and applications of rotational friction dampers for seismic protection. J Civil Eng Environ Archit 2016;XXXII:335–46.
[8] Zeng C. Hybrid experimental study on seismic mitigation control and post-earthquake operational safety of simply-supported girder bridges in high-speed railways. Changsha: Central South University; 2020 [In Chinese].
[9] Guo W, et al. Pseudodynamic hybrid simulation of high-speed railway bridge-track system with rotational friction damper. Int J Struct Stabil Dynam 2020;20(6):28.
[10] Guo W, Zeng C, Gou HY, Hu Y, Xu HC, Guo LL. Rotational friction damper's performance for controlling seismic response of high speed railway bridge-track system. Comput Model Eng Sci 2019;120(3):491–515.
[11] Mirzabagheri S, Sanati M, Aghakouchak AA, et al. Experimental and numerical investigation of rotational friction dampers with multi units in steel frames subjected to lateral excitation. Arch Civ Mech Eng 2015;15(2):479–91.
[12] Bijwe J, Kumar M, Gurunath PV, et al. Optimization of brass contents for best combination of tribo-performance and thermal conductivity of nonasbestos organic (NAO) friction composites. Wear 2008;265(5):699–712.
[13] Österle W, Prietzel C, Kloss H, et al. On the role of copper in brake friction materials. Tribol Int 2010;43(12):2317–26.
[14] Bonfanti A. Low-impact friction materials for brake pads. University of Trento; 2016.
[15] Lin QL, Lin JQ, Liu JL. Analysis of seismic damage to the railway system. In: Proceedings of the 9th National Conference on Earthquake Engineering; 2014. p. 566–74 [In Chinese].
[16] Wang HL, Cao WJ, Tang SC. Design and analysis of pot bearing with X-shaped steel damper. Railw Eng 2018;58(9):23–6 [In Chinese].
[17] Ouyang L. Dynamic performance study of translational-rotational composite friction isolation bearings. Sichuan: Southwest Jiaotong University; 2017 [In Chinese].
[18] Guo J, Wang G, Du YL, et al. Test and parameter sensitivity analysis on seismic isolation bearing with the function of displacement limitation used in bridges. J Harbin Eng Univ 2017;38(7):1114–20 [In Chinese].
[19] Nie LY, Zhu GF. Application of elastic-plastic falling-off prevention ball steel bearings in seismic design of bridge. J Jiang Univ (Nat Sci Ed) 2013;12(2):200–4 [In Chinese].
[20] Du QD. Seismic performance study of novel combined energy-dissipation and displacement-restriction bearing in high-speed railway simply-supported girder bridges. Changsha: Central South University; 2021. p. 4 [In Chinese].
[21] W. Guo , Y. Wang , Q. Du. Seismic control of high-speed railway bridge using S-shaped steel damper bearing Smart Struct Syst 2022;30(5):479-500.

[22] Gao K, Yuan WC. Seismic response analysis of cable-sliding friction pendulum bearings in curved girder bridges. Earthq Eng Eng Vib 2014;34(3):41–6 [In Chinese].
[23] Yuan WC, Wang BB. Numerical model and seismic performance of cable-sliding friction a seismic bearing. J Tongji Univ Nat Sci 2011;39(8):1126–31 [In Chinese].
[24] Conte M, Igartua A. Study of PTFE composites tribological behavior. Wear 2012;296 (1/2):5.
[25] Zhai Z, Guo W, Yu Z, et al. Experimental and numerical study of S-shaped steel plate damper for seismic resilient application. Eng Struct 2020;221:111006.
[26] Code for seismic design of railway engineering of the ministry of construction of the People's Republic of China (GB 50111—2006). Beijing: China Planning Press; 2009 [In Chinese].
[27] Luo H, Li ZH, Wang ZY, et al. Research progress on seismic isolation devices of high-speed railway bridges. J Hunan Instit Scie Technol 2020;33(2):37–41 [In Chinese].
[28] Gao JQ. Research on seismic analysis methods for ballastless track bridges. Lanzhou: Lanzhou Jiaotong University; 2015 [In Chinese].
[29] Li C. Study on the influence of track constraints on the seismic isolation performance of high-speed railway continuous beams. Beijing: Beijing Jiaotong University; 2018 [In Chinese].
[30] Zhang YL, Yang SJ, Chen XC. Analysis of seismic responses of high-speed railway bridges based on an integrated track and bridge model. Bridge Constr 2016;46(4):23–8 [In Chinese].
[31] Zhang L. Study on seismic response of friction pendulum-supported isolation bridges on high-speed railways. Chengdu: Southwest Jiaotong University; 2018 [In Chinese].
[32] Xia XS, Zhao HD, Ouyang HL. Study on friction pendulum bearing for seismic isolation of high-speed railway bridges. Earthq Resist Eng Retrofit 2014;36(3):21–6 [In Chinese].
[33] Bai QA. Application research on the new elastic-plastic limit seismic isolation device in high-speed railway simple supported girder bridges. J Railw Eng Soc 2019;36 (10):66–71 [In Chinese].
[34] Zhou Y Q, Wen Q, Kang W, et al. Research on the effect of steel bar damper and steel bar restrainer applied to a railway simply supported bridge. Railway Standard Design 2020; 64(S1):79–84.
[35] Chen L, Jiang L, Qin H, et al. Nonlinear seismic assessment of isolated high-speed railway bridge subjected to near-fault earthquake scenarios. Struct Infrastruct Eng 2019;15(11):1529–47.
[36] Chen Z, Fang H, Han Z, et al. Influence of bridge-based designed TMD on running trains. J Vib Control 2019;25(1):182–93.
[37] Leandro LFF, Lopez RH, Torii AJ, et al. Robust design optimization of TMDs in vehicle–bridge coupled vibration problems. Eng Struct 2016;126:703–11.
[38] Chen Z, Han Z, Zhai W, et al. TMD design for seismic vibration control of high-pier bridges in Sichuan-Tibet Railway and its influence on running trains. Veh Syst Dyn 2019;57(2):207–25.
[39] Xu JY, Hu JJ, Zhang C, et al. Vibration load identification and vibration control of bridge under heavy-haul train. J Huazhong Univ Sci Technol Nat Sci Ed 2018;46 (2):46–50 [In Chinese].
[40] Yin X, Song G, Liu Y. Vibration suppression of wind/traffic/bridge coupled system using multiple pounding tuned mass dampers (MPTMD). Sensors 2019;19(5):1133.
[41] Hou W, Li Y, Guo W, et al. Railway vehicle induced vibration energy harvesting and saving of rail transit segmental prefabricated and assembling bridges. J Clean Prod 2018;182:946–59.

[42] Hou W, Li Y, Zheng Y, et al. Multi-frequency energy harvesting method for vehicle induced vibration of rail transit continuous rigid bridges. J Clean Prod 2020;254:119981.
[43] Wang Y. Operational safety evaluation and control of simply-supported girder bridges in high-speed railways under far-field earthquakes. Changsha: Central South University; 2021 [In Chinese].
[44] Hartog D. Mechanical vibration. 4-th Edition. McGraw-Hill; 1956.
[45] Mohebbi M, Joghataie A. Designing optimal tuned mass dampers for nonlinear frames by distributed genetic algorithms. Struct Des Tall Special Build 2012;21:57–76.
[46] Soong TT, Dargush GF. Passive energy dissipation systems in structural engineering. John Wiley and Sons; 1997.
[47] Liang LT. Research on wind-induced vibration control of high-voltage transmission towers based on TMD. Changsha: Changsha University of Science and Technology; 2016 [In Chinese].
[48] Housner GW. Limit design of structures to resist earthquakes. In: Proceedings of the 1st World Conference on Earthquake Engineering; 1956. 5.1-5.13.
[49] Uang C, Bertero VV. Use of energy as a design criterion in earthquake-resistant design. Report No. UCB/EERC-88/18. Berkeley, California: College of Engineering; 1988.
[50] Akiyama H. Earthquake-resistant building structure design based on energy balance. Beijing: Tsinghua University Press; 2010 [In Chinese].
[51] Xiao MK, Liu B, Bai SL. Analysis of the total energy and its influencing factors for seismic structures. J Civil Architect Environ Eng 1996;18(2):20–33 [In Chinese].
[52] Jiang H, Zhu X. Energy input design spectra for near-fault regions and application in energy-based seismic design. Earthq Eng Eng Dynam 2006;26(5):102–8 [In Chinese].
[53] Fajfar P, Vidic T, Fishinger F. Seismic demand in medium and long period structures. Earthq Eng Struct Dynam 2010;18(8):1133–44.
[54] Yang TY, Tung DP, Li Y. Equivalent energy design procedure for earthquake resilient fused structures. Earthq Spectra 2018;34(2):795–815.
[55] Huang Z. Equivalent energy method for seismic performance design of seismic isolation and damping design in high-speed railway bridges. Changsha: Central South University; 2018. p. 5 [In Chinese].
[56] Guo W, et al. An improved equivalent energy-based design procedure for seismic isolation system of simply supported bridge in China's high-speed railway. Soil Dynam Earthq Eng 2020:134.
[57] Housner GW. The plastic failure of frames during earthquakes. In: Proceedings of the 2nd World Conference on Earthquake Engineering; 1960. p. 997–1012.
[58] Newmark NM, Hall WJ. Earthquake spectra and design. California: Earthquake Engineering Research Institute, El Cerrito; 1982.
[59] Goel SC, Chao S. Performance-based plastic design earthquake-resistant steel structures. Washington, DC: International Code Council; 2008.

CHAPTER 8

Research Conclusions and Outlook

Contents

8.1 Research Conclusions

The high-speed railway train-track-bridge system is a complex integrated system. Under earthquakes, its behavior involves intricate coupling of excitations between the earthquake and train vibrations. Beyond structural integrity concerns in the seismic design of conventional building structures, more attention should be paid to train operational safety and performance, such as derailment risks under earthquakes, and functional failure of railway lines due to post-earthquake impassability of trains. While earthquakes predominantly induce 10-20 Hz low-frequency structural vibrations, train-track interactions generate high-frequency coupling vibrations. Although trains in motion appear smooth, they actually exhibit "hunting" movement patterns. The low-to-medium-frequency structural vibrations induced by earthquakes, and the high-frequency vibrations and the "hunting" motion of trains can impact the safety and performance of trains running at high speeds on the bridge. Therefore, seismic disasters in high-speed railway train–track–bridge systems consist of structural failures and operational failures of high-speed trains. This is analogous to primary (building structure)-secondary (equipment facilities) systems in buildings formed by large auxiliary equipment; however, in this case, the "equipment" travels at speeds exceeding 300 km/h and is highly sensitive to millimeter-level track irregularities and structural dynamic responses. Therefore, in the design and disaster prevention of the train-track-bridge system, train factors should be considered.

High-Speed Railway Train–Track–Bridge Systems
ISBN 978-0-443-49310-2
https://doi.org/10.1016/B978-0-443-49310-2.00003-X

The technical approach and highlights of this book lie in the sharing of open-source programs and integrated utilization of established simulation platforms. Through co-simulation (CS) technology, platforms such as OpenSees and Simpack are effectively integrated. OpenSees excels in bridge structural simulation with its rich element library and mature algorithms, but it cannot simulate high-speed trains or wheel–rail coupling interactions. On the other hand, Simpack features extensive expertise in the train multibody system dynamics and wheel–rail coupling calculations, but it lacks the capability to simulate elastoplastic bridge structures. The method presented in this book integrates the advantages of OpenSees and Simpack, enhancing work efficiency and avoiding low-level repetitive tasks in areas such as Finite Element (FE) simulation and train dynamics. Furthermore, the OpenSees moving element developed in this book leverages its open-source and shared advantages, with its wheel–rail coupling simulation program directly accessible for use and modification on the OpenSees official website and repository.

Only a limited number of high-speed train derailment cases on bridges during earthquakes have been reported. Research on train-bridge operation under seismic conditions is largely confined to numerical simulations and theoretical studies. From the perspective of scientific research, in which theory, experiments, and field measurements are mutually verified in a closed loop, experimental methods serve as important validation tools. To address this limitation, this book proposes a physical simulation technology for train-bridge operation under seismic conditions. The proposed physical simulation technology for train operation on bridges under earthquakes represents a specialized shake table test methodology, typically used for testing at a reduced-scale model. The developed system employs a 1:10 scale ratio while addressing challenges associated with scale effects, including gravity distortion, fabrication errors, and similitude design issues. The theoretical analysis and experimental validation of the physical testing system show that this physical simulation technology can capture key characteristics of train operation on bridges under earthquakes, making it suitable for theoretical verification and simulation validation.

The Real-Time Hybrid Simulation (RTHS) technology for high-speed train operation on bridges under earthquakes represents an effective approach to achieving full-scale train testing. Unlike physical simulation technology, this hybrid simulation testing treats the train as the experimental substructure and the bridge as the numerical substructure. By adopting the logical framework of hybrid simulation and incorporating the characteristics of train-bridge relative motion, a laboratory-based hybrid testing platform for high-speed train operation on bridges under earthquakes has been established. Since the bridge and track are calculated in real time throughout numerical simulation, while only the train serves as a physical test component, full-scale testing for train operations on bridges under earthquakes can be achieved. The hybrid simulation technology reproduces the deformation response of the bridge under the earthquake at each moment through an actuator, while simulating wheel-rail interactions via variable-sized rollers. The core technical challenges primarily involve two aspects: First, in the real-time computation of complex large-scale numerical models, bridge response output must be completed within physical time, with computational time per time step typically falling within a millisecond order of magnitude. For complex numerical structures with tens of thousands of degrees of freedom (DOFs), real-time processing requires not only hardware enhancement but also algorithmic optimization. Second, the actuator typically employs large-scale hydraulic servo systems. Mitigating the time-delay effects in controlling hydraulic servo systems can effectively ensure both stability and accuracy in hybrid simulation. Therefore, higher-order nonlinear control algorithms are often adopted to enhance the reproduction precision of actuation commands. Meanwhile, the complexity and multi-DOF characteristics of the actuator often pose control challenges.

As previously mentioned, in the seismic failure analysis of high-speed railway train-track-bridge systems, both structural behavior and train safety should be considered. Current high-speed railway bridge designs generally prioritize the stiffness of core components, such as piers and girders, and control static and dynamic deformations to meet the demand for high-speed train operations. Consequently, high-speed railway bridge structures typically exhibit significant stiffness, with seismic performance resembling that of gravity piers and demonstrating flag-shaped self-centering characteristics under gravity but limited ductility under seismic loads. The existing requirement for a ductility coefficient of 4.8 in railway design standards appears incongruent with this structural behavior. The failure characteristics of high-speed railway piers under flexural-shear loads differ from those of highway piers due to low axial compression ratios and reinforcement ratios, as effectively evidenced by the flexural-shear strength formulas for piers provided in this book. In addition, the upper track structure often introduces constraint effects and inertial effects between adjacent girders. The influence of constraint effects becomes more pronounced when adjacent piers are of unequal height, which to some extent alters the simply supported girder characteristics of the bridge, resulting in seismic responses that fall between those of simply supported girders and continuous girders. The damage to the track structure itself can further impact train operational safety and performance during and after earthquakes. This book presents a simplified design model for the track structure precisely to account for this aspect in engineering design and analysis. This book establishes a corresponding operational performance evaluation framework, evaluates safe operating speeds under varying seismic intensities, and provides recommended operating speeds, thereby offering valuable guidance for design and maintenance.

How to consider operational safety in design is a critical issue that must be addressed in high-speed railway bridge engineering. The high-speed train-track-bridge system constitutes a complex, large-scale system, making it challenging to integrate train dynamics into design operations under simplified modeling approaches. Given the maturity of numerical simulation techniques for such systems, adopting analysis-driven design methodologies has become a rational paradigm. However, the book establishes a mapping relationship between train derailment states and bridge structural responses by correlating derailment conditions with structural seismic performance. Specifically, it defines safety thresholds where, within certain response ranges, train operational safety is maintained. The core metric is the response of structural velocity, for which the *SI* velocity spectrum intensity index for the structural response is proposed. The adoption of the *SI* velocity spectrum index means that, in bridge structural design, in addition to considering displacement deformation, it is also necessary to control the maximum velocity response. Therefore, this book incorporates train operational safety into high-speed railway bridge design through the established dynamic mapping relationship.

The primary requirement for high-speed railway bridge design is to ensure train operational safety and ride comfort while meeting functional performance needs. As such, bridge piers, girders, and track structures typically adopt standardized configurations, making it challenging to strengthen or weaken them to meet seismic disaster prevention needs. For seismic mitigation and control, the only adjustable component in high-speed railway bridges is the bridge bearing. It employs functional separation through shear pin mechanisms to ensure normal operational requirements, while seismic isolation and energy dissipation devices are activated during earthquakes. The Rotational Friction Dampe (RFD) proposed in this book is integrated into expansion bearings, which primarily serve to adjust the ratio of inertial force distribution between fixed piers and movable piers, while dissipating seismic energy, thereby effectively controlling pier deformation and damage levels. The integrated seismic bearings ensure postfailure behavior of fixed bearings while preventing extreme damage, such as girder drop-off, caused by massive inertial forces. The inertial mass damper is introduced to control girder vibration velocity through dynamic energy absorption, thus ensuring train operational safety under earthquakes. Here, *SI* metrics are used for the design of the inertial mass damper.

Performance-based seismic design is the dominant philosophy in modern earthquake-resistant engineering, where the rational design of components and structures, as well as dampers, enables the achievement of predefined multilevel performance objectives. However, current railway design standards remain conservative in this aspect, because they are still at the allowable stress design stage. The energy-based design method proposed in this book achieves multiobjective performance-based design for high-speed railway bridges, fulfilling seismic defense objectives under minor, moderate, and major earthquakes. Through rational parameter design of bearings and RFDs, it reduces damage to bridge piers. The energy-based design method offers advantages by eliminating the need for iterative optimization while maintaining unidirectional design, thereby enhancing engineering practicality. Although this book does not demonstrate how to apply this method to designs considering operational safety, i.e., performance-based design without explicitly incorporating operational safety as an objective, the established framework and method for pier-centric performance-based design can facilitate its application to broader seismic defense objectives.

8.2 Research Outlook

China's high-speed railway development has now passed its peak construction phase. At present and for a long time to come, the development will progressively shift from construction to operation and maintenance, with a focus on intelligent high-speed railways. This transition aims to address the preservation and enhancement of existing line functionalities while exploring the intelligent development of high-speed railway bridge engineering, including construction, inspection, and monitoring. Additionally, from a national strategic perspective, high-speed railway development in the coming years will focus on two major engineering initiatives: (1) The Sichuan–Tibet Railway, plagued by multi-hazard risks such as strong earthquakes, is under construction. In its construction, safety risks of the train-track-bridge system under more extreme seismic conditions should be addressed; (2) the CR450 EMU with a design speed of 400 km/h has been designated as a national key and technology application program. The construction of the Chengdu–Chongqing Central Line High-Speed Railway has also been approved by the National Development and Reform Commission. The higher design speed imposes stricter technical standards and requirements on bridge engineering, while the dynamic coupling of the train-track-bridge system becomes more pronounced. Against this backdrop of national strategy and development, technological research will focus more on core issues, emphasize intelligence, informatization, and visualization, and actively integrate rapidly evolving robotics technology. Here, we propose the following key areas for future research directions and priorities.

In the operation and maintenance of high-speed railway bridges, the way to conduct performance evaluation of existing lines and implement performance enhancement strategies will be a key focus of future research. This includes the incorporation of current deep-learning-based AI technologies. Under the constraints of traditional expertise frameworks and driven by big data, novel methods for evaluating the operational safety performance of trains on bridges will be proposed. By leveraging robotics and cybernetics, novel technologies for enhancing high-speed railway bridge performance will be developed.

The Sichuan–Tibet Railway will pose new challenges to the high-speed railway train-track-bridge system. The bridges along the Sichuan–Tibet Railway will face threats from near-fault and cross-fault earthquakes. Without seismic mitigation and control measures, extreme disasters such as girder drop-off and train derailment may occur. The development of new seismic isolation and damping devices along with safety control measures capable of resisting near-fault and cross-fault seismic impacts will be the cornerstone of disaster prevention and mitigation. Meanwhile, the derailment mechanism of trains under near-fault and cross-fault earthquakes remains an unresolved critical issue, requiring systematic theoretical, numerical, and experimental research. Additionally, although the Sichuan–Tibet Railway faces significant seismic hazards, its historical earthquake records are limited. Due to its rugged terrain, the impact of valley earthquakes must be considered. Furthermore, laboratory simulations of earthquake initiation and occurrence mechanisms may be necessary to develop physics-based methods for simulating near-fault and cross-fault ground motions.

The CR450 EMU also faces some outstanding technical challenges, such as train-track-bridge coupling vibrations induced by 400 km/h high-speed railway operations, the optimization of bridge structural design due to higher speeds, and the performance upgrades required for existing lines to achieve higher speeds. Resolving these challenges will drive updates to design standards for lower bridge structures to accommodate even higher train speeds.

Intelligentization has become a new driving force in many sectors, and its application in high-speed railway bridges is manifested in many areas such as construction, inspection, and monitoring. The fundamental implementation logic lies in introducing emerging algorithms and technologies such as deep neural networks (DNNs), graph neural networks (GNNs), generative adversarial networks (GANs), and recurrent neural networks (RNNs). By leveraging big data advantages during the high-speed bridge construction and operation process, these innovations enable rapid enhancement of traditional capabilities, a field poised for significant future development.

Informatization technologies and cybernetics can provide new impetus for high-speed railway construction and operation, enabling process automation and digitalization. Among them, the adoption of robotics stands as the most iconic application achievement. How to promote industrial development by aligning technologies with actual engineering scenarios is a question that both engineers and researchers should actively address. Meanwhile, in laboratory-based testing environments, leveraging intelligentization, informatization, and cybernetics to enhance the testing capabilities of large-scale actuators and conducting laboratory-based testing that more closely replicates field engineering conditions will significantly advance parametric research and bridge technology iteration. Among these approaches, hybrid testing combining numerical and physical methods, also known as hardware-in-the-loop simulation, stands as a representative and highly promising laboratory technology.

Machine vision utilizes image-processing techniques to extract valuable information, such as combining intelligent algorithms for object recognition, three-dimensional reconstruction, and deformation measurement. It can replace human labor or address challenges that are beyond the capability of traditional methods in certain aspects. Compared to conventional measurement techniques, machine vision offers greater precision and operates through noncontact methods. The in-depth application of machine vision will drive technological innovation in high-speed railway bridge inspection, detection, and safety monitoring. It synergizes with the emerging digital twin technology, providing robust technical support for digital construction, operation, and maintenance. Moreover, in noncontact laboratory measurement scenarios, machine vision has demonstrated clear advantages over traditional techniques through continuous technological iteration.

In summary, in an era where technological advancements are accelerating at an unprecedented pace, the interdisciplinary integration of intelligentization, informatization, visualization, and cybernetics will provide new impetus for high-speed railway development. Meanwhile, the construction and operational demand of high-speed railways will offer massive application scenarios for emerging technologies. Future research should not only address engineering challenges but also proactively explore the application and underlying principles of emerging technologies.

Index

Note: 'Page numbers followed by *f* indicate figures and *t* indicate tables.'

C

D

E

F

G

H

W

Printed and bound by CPI Group (UK) Ltd, Croydon, CR0 4YY
14/09/2026
02216409-0004